Gas Well Deliquification

Gas Well Deliquification

Third Edition

James F. Lea Jr
**Consultant, PLTech LLC,
Lubbock, TX, United States**

Lynn Rowlan
**Engineer, Echometer,
Wichita Falls, TX, United States**

Gulf Professional Publishing
An imprint of Elsevier

Gulf Professional Publishing is an imprint of Elsevier
50 Hampshire Street, 5th Floor, Cambridge, MA 02139, United States
The Boulevard, Langford Lane, Kidlington, Oxford, OX5 1GB, United Kingdom

Notices
Knowledge and best practice in this field are constantly changing. As new research and experience broaden
our understanding, changes in research methods, professional practices, or medical treatment may become
necessary.

Practitioners and researchers must always rely on their own experience and knowledge in evaluating and
using any information, methods, compounds, or experiments described herein. In using such information or
methods they should be mindful of their own safety and the safety of others, including parties for whom
they have a professional responsibility.

To the fullest extent of the law, neither the Publisher nor the authors, contributors, or editors, assume any
liability for any injury and/or damage to persons or property as a matter of products liability, negligence or
otherwise, or from any use or operation of any methods, products, instructions, or ideas contained in the
material herein.

British Library Cataloguing-in-Publication Data
A catalogue record for this book is available from the British Library

Library of Congress Cataloging-in-Publication Data
A catalog record for this book is available from the Library of Congress

ISBN: 978-0-12-815897-5

For Information on all Gulf Professional Publishing publications
visit our website at https://www.elsevier.com/books-and-journals

Publisher: Brian Romer
Senior Acquisition Editor: Katie Hammon
Editorial Project Manager: Mariana L. Kuhl
Production Project Manager: Bharatwaj Varatharajan
Cover Designer: Christian J. Bilbow

Typeset by MPS Limited, Chennai, India

Contents

Introduction

1

James F. Lea's experience includes about 20 years with Amoco Production Research, Tulsa, OK; 7 years as Head PE at Texas Tech; and the last 10 years or so teaching at Petroskills and working for PLTech LLC consulting company. Lea helped to start the ALRDC Gas Dewatering Forum, is the coauthor of two previous editions of this book, author of several technical papers, and recipient of the SPE Production Award, the SWPSC Slonneger Award, and the SPE Legends of Artificial Lift Award.

1.1 Introduction

Liquid loading in a gas well is the inability of the produced gas to lift the produced liquids from the wellbore. Under this condition, produced liquids will accumulate in the wellbore leading to reduced production and shortening of the time till the well no longer produces.

According to EIA, there are about 600,000 gas wells in the United States (see Fig. 1.1).

By some estimates, 70%−80% of gas wells are low rate and below about 300 Mscf/D. Therefore perhaps 400,000−500,000 gas wells are at risk of lower or no production from liquid loading unless artificial lift (AL) is properly applied.

Methods of diagnosing the occurrence of liquid loading will be presented here for both near vertical conventional wells and horizontal rapidly declining unconventional wells. Methods of solution will be presented and discussed in detail to help optimize the solution of liquid loading using various forms of AL including:

1. Newer techniques of rod design and rod protection in deviated wells using sucker rod systems
2. New methods for SRP (sucker rod pump) systems to allow deeper intake for the systems in horizontal wells
3. Design of gas lift systems for conventional and also declining unconventional wells using conventional gas lift with bracketed valves for anticipated changing rates
4. Use of high-pressure gas lift to allow more drawdown initially and to eliminate some downhole equipment
5. New techniques of tracking plungers, various forms of plunger lift, new plunger optimization techniques, new equipment, and plungers in horizontal wells

Gas Well Deliquification. DOI: https://doi.org/10.1016/B978-0-12-815897-5.00001-9

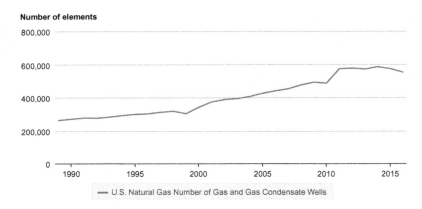

Figure 1.1 Number of gas wells.
Source: US Energy Information Administration.

6. Use of electric submersible pumps (ESPs) to dewater including design for lower rate wells requiring needed cautions
7. Optimization of progressing cavity pumpings (PCPs) that usually operate in shallower wells. Rod protection in deviated and horizontal wells
8. The latest in application of foamer chemicals and methods of application
9. Details and methods of application for gas separation for all the pumping systems
10. New advances in automation are presented in a separate chapter. Automation is a necessity if optimum conditions are to be achieved

1.2   Multiphase flow in a gas well

To understand the effects of liquids in a gas well, we must understand how the liquid and gas phases interact under flowing conditions.

Multiphase flow in a vertical conduit can be described by a number of available flow regime maps. These can be used to decide whether or not a well is predicted to be in a loaded condition. However, the well would have to be evaluated at both the surface and depth for a complete analysis. The flow regime of annular mist would be where one would like to flow a gas well and if it drops out of the flow regime, AL (artificial lift) would be required to remove liquid and lighten the gradient in the tubing. In the mist flow the effects of liquid production are felt the least by the well (Fig. 1.2).

Coincidentally the rate of 320 Mscf/D at 100 psi is the critical for 2 3/8 tubing. When pressure (200 psi), the point drops below critical and the line between annular and slug/churn for both values of bpd/Mscf. When the pressure is less than 100 psi, the velocity is more than critical for both 88 and 200 bbls/Mscf liquid/gas fractions. In Chapter 3 and Appendix B the expression derived and used for critical velocity and rate is independent of the liquid/gas fraction and

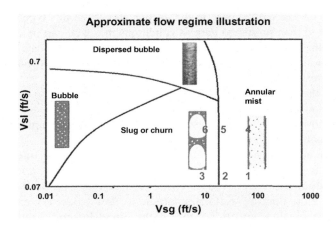

Figure 1.2 Illustration of possible flow regime map for vertical flow.
For above what numbers in bold indicate: (all for 2 3/8's tubing)
1: 88 bbls/Mscf, 50 psi, velocity for 320 Mscf/D, 120°F;
2: 88 bbls/Mscf/100 psi, velocity for 320 Mscf/D, 120°F;
3: 88 bbls/Mscf, 200 psi, velocity for 320 Mscf/D, 120°F;
4: 200 bbls/Mscf, 100 psi, velocity for 320 Mscf/D, 120°F;
5: 200 bbls/Mscf, 50 psi, velocity for 320 Mscf/D, 120°F;
6: 200 bbls/Mscf/50 psi, velocity for 320 Mscf/D, 120°F.

this shows why the critical, without this dependency, is still shown to work in this example.

This example uses an approximate flow regime map and if one is to use the flow regime chart to suggest if an operational point is above/below critical (in Annular Mist or not), then one should find a flow regime chart that is tested to agree with well data.

More details will be shown on the critical velocity and critical rate. Also it will be shown that Nodal Analysis (Chapter 4) can infer above/below critical or not. However, this example ties the critical to the multiphase aspects of the calculated critical rate. It is shown that the flow regime map, the calculated critical rate/ velocity model, and Nodal Analysis will be predictive techniques for critical rate and under what conditions liquid loading can occur.

A well may initially have a high gas rate so that the flow regime is in mist flow in the tubing near the surface, but is more liquid rich flow regimes in the tubing at depth. As time increases and production declines, the flow regimes from perforations to surface will change as the gas velocity decreases. Liquid production may also increase as the gas production declines.

Flow at surface will remain in mist flow until the conditions change sufficiently at the surface so that the flow exhibits a more liquid rich regime such as slug flow. At this point, the well production will be observed to become somewhat erratic, progressing to slug flow as gas rate continues to decline. This will often be accompanied by a marked increase in the decline rate. Note this type of analysis is more complicated than presented here as conditions in a well can be different from the surface to the bottom hole of the well. For instance liquids may be starting to accumulate in a more liquid-rich flow pattern downhole and the conditions uphole can still be in mist flow.

Eventually, the unstable slug flow at surface will transition to a stable, fairly steady production rate again as the gas rate declines further. This occurs when the gas rate is too low to carry liquids to surface and simply bubbles up through a stagnant liquid column at the bottm of the well.

If corrective action is not taken, the well will continue to decline and eventually log off. It is also possible that the well continues to flow for a long period in a loaded condition with gas produces up through liquids with no liquids coming to the surface. Note that the well can continue to flow below critical, sometimes for a long time, but it would flow more if the liquid loading problem could be solved.

1.3   Liquid loading

Nearly all gas wells produce some liquids even if the rate of liquid production is small; if the gas velocity is below the critical (to be defined in more detail in Chapter 3), then the well will experience liquid loading. In other words, liquids will accumulate in the wellbore and reduce production. This is shown by the fact that there are many gas wells on plunger lift that produce 5 or less bpd of liquids. If not on plunger, they will produce less or no gas. Critical velocity correlations (Chapter 3) do not require the liquid rate as an input. If more liquid is being produced then once below the critical the well can load faster. If little liquids are still being produced below critical, the well will eventually liquid load. Liquid loading modeling is sensitive to the liquid rate when using Nodal Analysis (Chapter 4).

1.4   Deliquification techniques

The below list[1] (modified) introduces some of the possible methods to deliquefy gas wells that will be discussed in this book. These methods may be used singly or in combination in some cases.

- *Initial high rates (for unconventional well on sharp decline)*
     Unconventional wells may come in high rates initially which are well above critical rate. For maximum PVP (present value profit) use Nodal to look at flow up casing, flow up casing/tubing annulus and to look at tubing size effect on flow. Some operators are considering annular gas lift and high-pressure gas to boost the high rates. Most of the profits from unconventional wells are in the first year or two, and then later with low rates and required AL, the returns are much lower.

- *Lower rates but still above critical (perhaps 300—1000/2000 bpd)*

Being above critical, the choice is available to flow the wells. However, the use of AL can exceed the rates above even the most optimum flow situation (which can be determined by Nodal). So higher rate ALs such as ESP, gas lift, or other pumping systems may be used to accelerate the rate above flowing conditions. Lower surface pressure helps all ALs and all above-critical flowing situations. Once AL is selected for the mid-range or even higher rates, then the process of AL staging begins (from higher rate systems to lower rate systems).

Conditions may determine the method/s of lift. For instance, if the well is very gassy or has solids then one may choose gas lift for long-term lift as excessive gas makes use of pumping systems more difficult. Also excesssive solids production and excesssive deviation or dog-leg severity can make one consider something other than pumping systems.

On the other hand, SRP systems are one of the most profitable methods of lift and with not too much gas, solids, and well-deviation SRPs can be used for a wide range of production rates.

Other systems have particular advantages that may help with selection.

- *Below critical production*

Here AL is required to prevent loading.

Plunger, surfactants, compression, pumping systems, and gas lift can be considered.

Pumps and then plunger could be sequentially applied. Or gas lift and then plunger could be a strategy. Discussion of the lift possibilities in the following chapters will help with the selection or staging of lift possibilities.

- *Low to very low production rates*

Beam pumps, hydraulics, and ESPs can be used to low rates. However, low rates for ESPs can be damaging and a lower limit on production or special precautions should be considered for EPSs at low rates. PCPs can be used for moderate to low rates but are more depth-limited. Ineffective gas separation can lead to poor drawdown with pumping systems.

Gas lift can be used to low rates but a smaller tubing allows for more efficient lift.

Plungers are designed for low rates and modified plunger lift sysstem such as GAPL (gas-asssisted plunger lift) and progressive plunger lift (use of two or more plungers running in one well, i.e. one plunger moving up and down in the top part of the well and another plunger moving up and down in the lower part of the well) can be efficient for lower rates and lower available well pressures. Sand limits or prevents the use of plunger lift. Again sand/deviation/gas production tends to lead to gas lift for longer life and fewer problems but most likely not as good of drawdown under certain circumstances.

1.5   Most used systems for deliquification

The top most used four AL methods of dewatering are as follows:

- Plunger
- SRPs
- Surfactants
- Gas lift

Augmenting the choices includes intermitting the well with no AL system, special pumping systems, heating systems, velocity strings, compression (which should be used in combination with all systems for maximum performance), combination systems, and more.

The best economical system or the most profitable system should be used. One such economics decision might be to use plunger lift because it is cheaper to install and operate but it may not give the best drawdown, especially with an on/off system. You might then choose a pumping system but it would cost more initially, also most likely cost more to operate, but might get better drawdown depending on conditions and time spent operationally. It seems that the industry chooses a lot of plunger because of low initial/operating costs, and to save the time spent to judge if another system may produce more or slightly more. Plunger does pretty well for a long time for producing rates for the lower range of liquids production. Some conditions such as sand/solids, not enough gas/liquid ratio, or low-pressure build up can rule out plunger.

All of these factors are covered in more detail in the following chapters.

Reference

1. Coleman SB, et al. A new look at predicting gas well liquid load-up. *J Petrol Technol* 1991;329−32.

Further reading

Lea JF, Tighe RE. Gas well operation with liquid production. In: *SPE 11583, presented at the 1983 production operation symposium*, Oklahoma City, OK, February 27−March 1; 1983.
Libson TN, Henry JT. Case histories: identification of and remedial action for liquid loading in gas wells-intermediate shelf gas play. *J Petrol Technol* 1980;685−93.

Recognizing symptoms of liquid loading in gas wells

2

James F. Lea's experience includes about 20 years with Amoco Production Research, Tulsa, OK; 7 years as Head of PE Department at Texas Tech; and the last 10 years or so teaching at Petroskills and working for PLTech LLC consulting company. Lea helped to start the ALRDC Gas Dewatering Forum, is the coauthor of two previous editions of this book, author of several technical papers, and recipient of the SPE Production Award, the SWPSC Slonneger Award, and the SPE Legends of Artificial Lift Award.

Lynn Rowlan, BSCE, 1975, Oklahoma State University, was the recipient of the 2000 J.C. Slonneger Award bestowed by the Southwestern Petroleum Short Course Association, Inc. He has authored numerous papers for the Southwestern Petroleum Short Course, Canadian Petroleum Society, and Society of Petroleum Engineers. Rowlan works as an Engineer for Echometer Company in Wichita Falls, Texas. His primary interest is to advance the technology used in the Echometer Portable Well Analyzer to analyze and optimize the real-time operation of all artificial lift production systems. He also provides training and consultation for performing well analysis to increase oil and gas production, reduce failures, and reduce power consumption. He presents many seminars and gives numerous talks on the efficient operation of oil and gas wells.

2.1 Introduction

As gas rate declines in a gas well (conventional or unconventional), a point will be reached where liquids will begin to accumulate in the tubing and either further diminish the production or possibly stop the production altogether.

If the liquid loading in wellbore goes unnoticed, the liquids can accumulate in the wellbore and the adjoining reservoir, possibly causing temporary or even permanent damage. It is important that the effects caused by liquid loading are detected early to prevent costly losses during production and possible reservoir damage.

This chapter discusses methods to recognize the occurrence of liquid loading. Methods can be predictive or can be observations of field symptoms. Actual field symptoms are more preferable as predictive methods may have some deviation from reality. However, field symptoms can also have their problems as drop in production could be due to formation damage, sand accumulations, or scaling which could be erroneously attributed to liquid loading.

Gas Well Deliquification. DOI: https://doi.org/10.1016/B978-0-12-815897-5.00002-0

Indications of loading can be from:
Predictive methods

- Use of critical velocity/rate correlations
- Use of Nodal Analysis
- Use of multiphase flow regime maps

Field symptoms

- Slugs of liquid begin to be produced
- Erratic flow and production dropping below the target decline curve
- Difference between the surface measured tubing/casing pressures begins to increase
- Pressure survey or acoustic analysis indicates presence of a liquid level in the well
- Shoot fluid level or do a wireline survey down the tubing and diagnose for posssible liquid loading

2.2   Predictive indications of liquid loading

Predictive indications of loading can be quick and easy. However, there can be a difference in what actually goes on and the predictions made. Instead of relying only on the predictive indications, it is a good idea to compare field symptoms with the results of predictive methods before making a final conclusion. Another use of predictive methods is to see what happened in the past occurrences of liquid loading in wells and see which predictive or adjusted predictive method fits best in that scenario and then such standalone predictions can be viewed with more confidence in future.

2.2.1   Predict or verify liquid loading using critical velocity correlations, Nodal Analysis, and multiphase flow regimes

See Chapter 4 where critical velocity relationships are reviewed and the details of development are presented. Chapter 4 and Appendix review Nodal Analysis and Nodal Analysis concepts. Chapter 5 reviews compresssion and discusses about how to analyze it using critical velocity and Nodal Analysis. Chapter 1 briefly reviews multiphase flow regimes which can be used to predict or verify liqiuid loading.

Critical velocity

Critical velocity correlations predict at what rate liquid loading will occur as the well rates decline. It is not a function of liquid production or bbl/mmscf. It is (for some widely used correlations) based on what rate or velocity will carry the liquid droplets up and when they can no longer be foreseen to travel up, then liquid loading is predicted. Turner and Coleman are two widely used methods but there are many other models.

The critical flow chart, shown in Fig. 2.1, shows that the critical rate for 2 3/8's tubing at 100 psi is about 320 Mscf/D with water production. If liquid is only condensate, the critical rate (and velocity) would be less. If any water is produced with condensate, use the water chart. Therefore, critical velocity correlations is a quick method to assess if flow is critical or not or is approaching critical. Note X-axis should be in psi.

Use of Nodal Analysis to predict if flow is above/below critical

Nodal Analysis is a model of the well. It usually has a reservoir inflow relationship and an outflow curve plotted. The outflow curve shows what pressure is needed at the bottom of the tubing to overcome the friction in the tubing (or other flow path), weight of gas/liquid in the tubing (gravity effects), and WHP. Some tubing correlations also account for fluid acceleration which is important only at high flow rates. Fig. 2.2 shows some of the possibilities for the relationship of the tubing performance curve relative to the inflow (reservoir) curve.

So a nodal tubing performance is stable toward the right of the minimum in the tubing curve. If the tubing curve intersects the inflow performance relationship (IPR) curve at the right of the minimum then a stable rate is predicted at the intersection. Even with no IPR if the tubing curve is slanting up and toward the right, the tubing is stable for that range of flows.

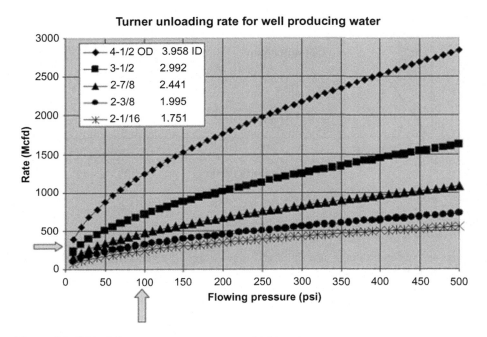

Figure 2.1 Critical flow rate versus pressure and tubing size.

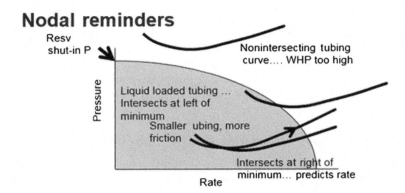

Figure 2.2 Some Nodal concepts.

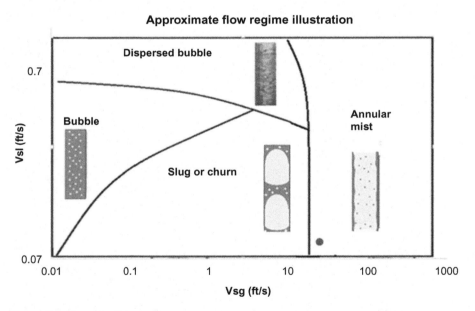

Figure 2.3 Example flow regime map.

Multiphase flow regimes

Based on various authors and multiphase pressure drop prediction models, there are a number of flow regime maps available in the literature. One must check the accuracy of the flow regime map with the performance of the well before selecting the map. One such example of the flow regime map is shown in Fig. 2.3 as discussed in the Introduction, Chapter 1. This map has entries regarding superficial velocity of gas and liquid. The superficial velocities are calculated as if only liquid and only gas are flowing in the conduit. Fig. 2.3 is followed by the calculations used to enter the example flow regime map and generate the round dot flow condition.

Input data and calculations for flow regime map to generate the round dot in Fig. 2.3.

Input API	30
Input Tbg ID	1.995 in.
Input WG	1.00
Input bbls/mmscf	88.00
input WC	0.95
Input Mscf/D	320.00
Calculated lip density	0.99
Calculated BPD	28.16
Calbulated Tbg area	0.0217 ft^2
Input pressure	100.00 psi
Input temperature	120.00°F
Input Z factor	0.90
Calculated Scf/D	320,000.00
Calculate Vsl	0.08 ft/s
	0.03 m/s
Calculate Vsg	170.70 ft/s 25.18998 in situ ft/s
	52.03 m/s 7.677905 in situ m/s

where Vsl and Vsg are calculated using the following formulas:

$$\text{Vsl} = \text{BPD} \times 5.615/(86,400 \times \text{area}_{\text{tbg}})\text{Vsg} = \text{Scf}/D/(86,400 \times \text{tubing area})$$
$$\text{Vsg} = \text{Scf}/D/(86,400 \times \text{tubing area})$$
Multiply by: $14/7/P'(T + 460) \times Z/520$ to convert to in situ gas velocity
WG = Water gravity; Vsl = superficial liquid velocity; and
Vsg = superficial gas velocity

From the above example the calculated round dot in the flow regime map corresponding to the above input data is inside the annular-mist flow regime. Thus, according to the example flow regime map, the well is not liquid loaded in this condition. Actually, this condition is for the critical velocity as predicted by Turner's model for critical velociy and critical rate. However, critical velocity, flow regime maps, and Nodal Analysis are all predictive techniques and it is wise to verify these methods with field symptoms which are discussed next.

2.3   Field symptoms of liquid loading

The shape of a well's decline curve can be an important indication of downhole liquid loading problems. Decline curves should be analyzed for long periods looking for changes in the general trend. Fig. 2.4 shows a smooth target "goal" that is fit to earlier time data. The round dots are from the production data. One can see that the production data near the bottom of the plot are falling below the "goal" decline curve. Since this is a well that has been flowing above critical before the data falls below the goal, the production falling off can be an indication of declining below the critical flow and the well is now flowing below critical. If so, then it is time to consider some artificial lift to try to return production to the goal. Actually, most operators, once they get the idea at what rate critical occurs for wells in the field, will install artificial lift before critical so they do not have to experience the drop in production shown later. Also one can do some work when wells seem to drop off and see what critical correlation, if any, best predicts when the well may drop below the goal decline curve. Possibilities on the graph, shown in Fig. 2.4, are Turner/Coleman critical models taken at the surface/bottom hole.

One aspect of comparing data to the goal decline curve to detect liquid loading is that there could be something else dropping the flow rates. For instance, the well may be sanding up and needs to be cleaned out. A hole in the tubing is another posssibility. The operator has to always consider that initial drops in production may be due to well damage or solids, etc. before concluding that it is due to the rates dropping below critical.

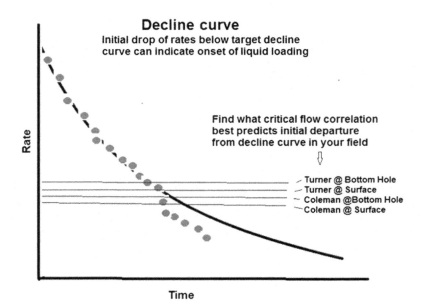

Figure 2.4 Decline curve analysis.

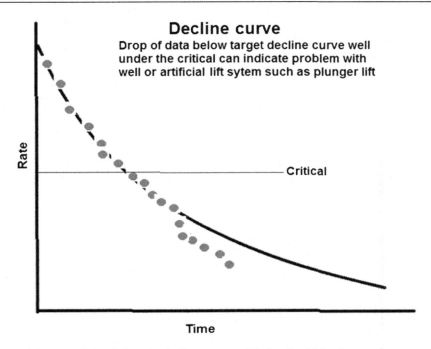

Figure 2.5 Drop of data below the decline curve well below liquid loading.

Fig. 2.5 shows a drop below the goal decline curve well below the critical. This is a well where the rate declined below critical and an artificial lift is installed to keep the well producing along the target decline curve. The drop below the decline curve is an indication that the artificial lift (perhaps plunger) is experiencing some problems which need to be diagnosed in attempt to bring production back to the target or decline curve. If plunger lift is being used, the problems could be a worn or sticking plunger or the cycle for the plunger needs adjustment back to more optimum control. If other methods of AL are being used then the AL system must be diagnosed with troubleshooting techniques particular to that system.

2.3.1 Increase in difference between surface values of casing and tubing pressures

If liquids begin to accumulate at the bottom of the wellbore, the casing pressure will rise to support the additional liquids in the tubing.

In packerless completions, where this phenomenon can be observed, the presence of liquids in the tubing is shown as an increase in the surface casing pressure as the fluids bring the reservoir to a lower flow, higher pressure production point. The gas produced from the reservoir percolates into the tubing casing annulus. This gas is exposed to the higher formation pressure, causing an increase in the surface casing pressure. Therefore an incease in the difference between tubing and casing pressures is an indcator of liquid loading. These effects are illustrated in Fig. 2.5 but the changes may/may not be linear with time as shown.

Finally, estimates of the tubing pressure gradient can be made in a flowing well without a packer by measuring the difference in the tubing and casing pressures. In a packerless production well the free gas will separate from the liquids in the wellbore and rise into the annulus. The fluid level in a flowing well will remain depressed at the tubing intake depth, except when "heading" occurs or a tubing leak is present.

During "heading" the liquid level in the annulus periodically rises and then falls back to the tubing intake. In a flowing well, however, the difference in the surface casing and tubing pressures is an indication of the pressure loss in the production tubing. The weight of the gas column in the casing can be computed easily. Comparing the difference between casing and tubing pressures with a dry gas gradient for the casing can give an estimate of the higher tubing gradient due to liquids accumulating or loading in the tubing. This will also allow the comparison to multiphase flow pressure drop correlations to check for accuracy for different correlations (Fig. 2.6).

2.3.2 Pressure survey showing liquid level

Flowing or static well pressure surveys are available to determine the liquid level in a gas well and thereby whether the well is loading with liquids. Pressure surveys measure the pressure with depth of the well either during shut-in or flowing. The measured pressure gradient is a direct function of the density of the medium and the depth, and the pressure with depth should be nearly linear for a single static fluid.

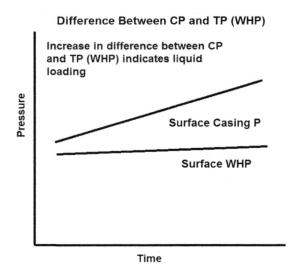

Figure 2.6 Casing and tubing pressure indications.

Pressure survey to determine liquid loading

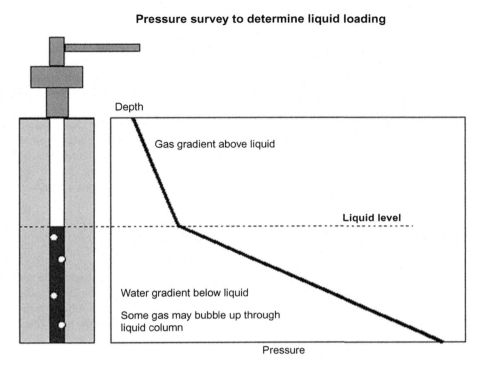

Figure 2.7 Pressure survey schematic.

Since the density of the gas is significantly lower than that of water or conden-sate, the measured gradient curve will exhibit a sharp change of slope when the tool encounters standing liquid in the tubing. Thus the pressure survey provides an accu-rate means of determining the liquid level in the wellbore. If the liquid level is higher than the perforations, liquid loading problems are indicated.

Fig. 2.7 illustrates the basic principle associated with the pressure survey. Note that the gas and liquid production rates can change the slopes measured by the survey, giving a higher gas gradient because of some liquids dispersed and a lower liquid gra-dient due to the presence of gas in the liquid. Also note that the liquid level in a shut-in gas well can be measured acoustically by shooting a liquid level down the tubing. Although it was previously done with a wireline pressure survey, a fluid level can be shot down the tubing with special cautions (echometer technique) to detect a fluid level with no wireline pressure survey (see Section 2.3.4).

2.3.3 Appearance of slug flow at surface of well

Fig. 2.3 shows that if you are in the annular-mist regime at first, you can move to the slug flow regime as you move to less gas on the X-axis. In practice, if you are operating a gas well when it is strong, you can see mist flow. But if it liquid loads then you move into the slug flow regime. One indication of liquid loading is that you see slugs of liquid being produced (you can hear them at the well) where there were no slugs of liquid before. Actually by the time you start seeing the slug flow at the surface, a good portion of the well downhole is most likely already liquid loaded, so this indicator is sort of an after-fact indicator. However, it is still an indicator and if you see slug flow at the surface, the well is liquid loaded (unless well damage dropped the gas flow and put you into the slug flow regime).

In the past, this has also caught some operators by surprise as they had smooth data from the flow measurement orifice while in mist flow, but suddenly got spikes in the readings when slugs start coming through. This may not be an issue now as most will have liquid KO's in front of the gas measuring orifice to handle the liquids before measurement. However, you can still hear slugs if they are present in the wellhead before the liquids are knocked out.

2.3.4 Acoustic fluid level measurements in gas wells (Echometer)

Determination of liquid loading can be made by other methods mentioned earlier including critical velocity, field symptoms, and wireline pressure survey to look for a liquid level in the well. However, a nonintrusive fluid shot down the tubing, right after the well is shut-in, can also be used to look for liquids that might have been accumulating in the well if it is in the process of being liquid loaded.

An acoustic fluid level in a gas well starts when the microphone attached at the surface records the sudden change in the pressure at the surface of the wellbore when the fluid-level shot is initiated. The sudden change in pressure creates a pressure wave that travels away from the surface at the speed of sound, through the gas composition in the wellbore. Echoes of pressure waves back to the microphone are produced as slices of the traveling pressure waves are reflected by changes in diameter in the path of the traveling wave inside the tubing or annulus of the wellbore. The greater the change in diameter of the wellbore, larger are the echoes at the surface, because more energy from the traveling wave is reflected back to the surface. The echoes displayed on an acoustic trace have traveled from the surface to the wellbore diameter change and back to the surface. When the recorded acoustic trace is displayed, a reduction in the wellbore cross-section area is seen as a down-kick, an increase as an up-kick, and typically the fluid level is the largest kick because almost all of the traveling wave energy is reflected back to the surface.

Acoustic surveys on gas wells (shooting fluid levels down the tubing) answer the following questions:

1. Is gas flowing? At what rate?
2. What is the depth to the top of the liquid?
3. Does liquid exist above the formation? In tubing?
4. What is the percentage of liquid in the fluid column?
5. Does the liquid above the formation restrict production?
6. Does surface pressure restrict production? How much is the flow rate restricted due to backpressure from liquid loading?

7. What are the producing and static bottom-hole pressures (BHPs)?
8. Does tubing gas/liquid pressure push liquid out of tubing?
9. What is the maximum rate available from the well?

Acoustic surveys on wells:

1. Require stabilized conditions
2. Determination of liquid level
3. Measurement of surface pressure
4. Measurement of surface pressure buildup rate
5. Wellbore description
6. Oil, water, and gas densities
7. Oil, water, and gas production rates
8. Identify if shot acquired down tubing or casing annulus
9. Type of shot—implosion or explosion

Fig. 2.8 shows the stages of conditions that exist in gas wells with liquids. Type 1 is when the velocity is above critical and liquids are in the mist flow regime creating little pressure drop. Type 2 is when gas flow is below critical and liquids are accumulating in the well but some liquids are still coming to the surface. A liquid level is in the well at the bottom with gas coming through it. A Type 3 well is where liquids are accumulating in the well and gas is flowing to the surface but no liquids are transported to the surface. In the condition, a quick glance at this lads to the conclusion that the well does not/never did produce liquid. This, however, would be false conclusion.

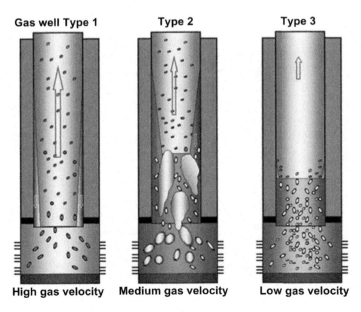

Figure 2.8 Types of flow that occur in the tubing when shooting acoustic fluid levels.
Source: Echometer.

A Type 1 well

1. *Liquid* being produced with the gas or condensing due to temperature and pressure changes is *uniformly distributed in the wellbore.*
2. Gas velocity is sufficient to continuously carry *liquid as a fine mist* or small droplets to the surface (above critical).
3. Gas velocity is sufficient to establish a relatively low and *fairly uniform flowing pressure gradient.*

If you shut the well in and quickly shoot a fluid level down the tubing the apparent fluid level (top of the mist in the tubing) will appear to be at the surface. The longer you keep the well shut-in, the more the top of the mist flow will move toward the bottom of the well with dry gas flowing above the mist level. This is the condition that many high-flow wells flow in initially when they are high on the decline curve. Later the wells can move into Types 2 and 3 conditions.

A Type 2 well

In Fig. 2.9 a casing fluid level shot on a Type 2 well is shown with an expanded view near the end of the tubing. The well is operating in a stabilized condition with gas only flowing up the tubing, with the surface casing valve closed, and with no

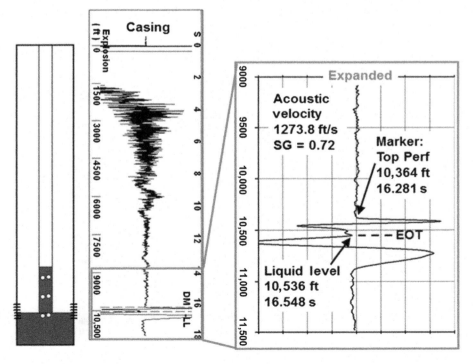

Figure 2.9 A "Type 2" well.

packer in the well. The fluid level is shot down the tubing immediately once the well is shut-in and then shot down the casing either after the tubing shot or before the well is shut-in. The casing fluid level shot confirms that the liquid level from the casing shot is located at the end of the tubing as expected. A troubleshooting technique is to shoot fluid level down the casing and confirm that the liquid level is located at the end of the tubing, because a high fluid level in the casing often indicates that there is a possible hole in the tubing.

The wellbore drawing displays the tubing gassy fluid level somewhere in the well below the surface. If the well is continued to be shut-in, then the fluid level will move downward as the gas flow into the tubing increases the tubing surface pressure. The casing pressure will gradually increase to support the increasing gas pressure in the tubing, thereby maintaining the liquid level at the end of the tubing.

The tubing gas velocity is not high enough to lift this well's 2087 ft of accumulated gassy tubing liquids to the surface because the flowing gas rate is below critical causing the well to be liquid loaded. Liquid loading accumulated in the bottom of the tubing causes 553 psi of tubing backpressure to act against the formation reducing inflow. Calculating the difference between casing minus tubing pressure, 429 psi, is a simplified way of estimating the liquid loading in the tubing. In this 10,536 ft deep well the fluid level shot shows there is 124 psi of additional liquid loading when compared to the difference between casing and tubing surface pressures. As the gas rate decreases, the concentration of liquid at the bottom of the well increases. Depending on the well as the casing pressure increases, the tubing liquid may unload from the bottom of the well. When the tubing is set deep and the gas flow decreases as the pressure increases, all flow from the formation can stop. When the tubing is set above the bottom of the perforations, the accumulated liquid in the tubing is usually pushed out by the increasing tubing pressure.

The simple picture of the well represents the flowing pressure gradient in the well bores. There is a light gas gradient above the gas/liquid interface (close to gradient of flowing gas), below the liquid level is a heavier gradient composed of gas flowing through liquid. The liquid is held up by the gas flow (zero net liquid flow) with gas bubbles or slugs percolating through the liquid. Below the end of the tubing the liquid gradient is heavier due to reduced gas velocity.

A Type 3 well

Fluid is at the bottom of the well. Gas flow has stopped. No liquids are transported to the surface. If the well is shut-in and a fluid shot is immediately made then the fluid level will be indicated. While the well is shut-in the fluid level will move down in the well as pressure builds up in the tubing. Fig. 2.10 shows the up-kick on the acoustic trace from the increase in cross-sectional area at the top of the perforations, and the liquid level at 6069 ft has been pushed to the bottom of the perforations.

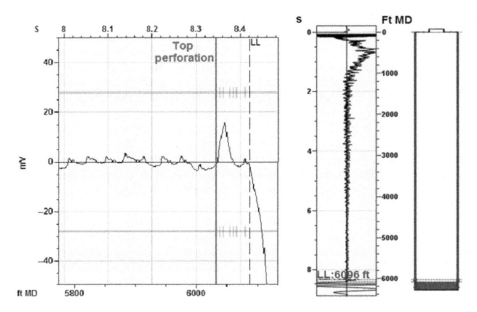

Figure 2.10 A "Type 3" well.

Summary of liquid level shots:

- Well must be in production until ready for shots.
  - Gaseous liquid column can move quickly and go below the end of tubing.
  - Requires good communication with operations for planning.
  - Start with relatively high pressure in gas gun for shots to have best chance of seeing deep into low-pressure wells.
- Complete well data important in further analysis.
- Flowing BHP (FBHP) calculation using measurements is reasonably accurate.

2.3.5 Determining well performance from a fluid shot

Fig. 2.11 depicts the analysis of a fluid shot showing pressure distribution in the well that has a gassy fluid level downhole. The surface pressure looks to be about 60 psi, the FBHP, if there was no fluid level, looks to be about 85 psi, and the actual FBHP with the gassy fluid level included is about 360 psi. The pressure added due to the fluid level (liquid loading) is 295 psi.

Determination of the well's performance:

- Combine FBHP with SBHP to estimate the IPR.
- Repeatable tests are an indication of stabilized conditions.
- Reduction of backpressure on formation by removal of liquid will increase the gas production (i.e., deliquify the well).

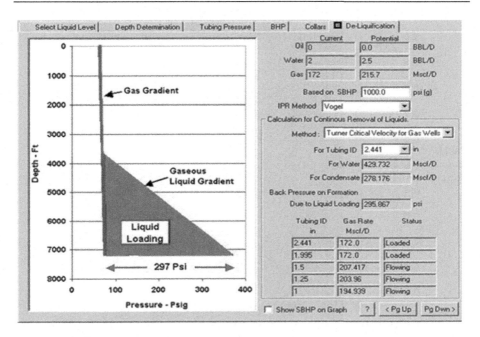

Figure 2.11 Analysis of gas well with fluid level.

Use of the Vogel expression for a gas well IPR.

- Vogel's IPR method can be used to describe inflow performance for gas wells (it approximates the backpressure equation commonly used for approximate IPRs in a gas well).
- The Vogel IPR is commonly used to describe a well's fluid inflow performance when both free gas and liquids flow simultaneously in the reservoir.
- Method is applicable for oil reservoirs with pressures below the bubble point pressure.

The formula for the Vogel expression is as follows:

$$Q/Q_{max} = (1 - 0.2(PRHP/SIBHP) - 0.8(PRHP/SIBHP)^2$$

where Q is the flow rate; SIBHP is the shut-in pressure; and FBHP is the flowing BHP.

It is seen that for many wells, due to tight permeability, often times the SIBHP is not known or tested for, that short-term tests can be misleading as the well is not stabilized, and that stabilization time can be very long. Therefore perhaps the recommendations here are better for a well with higher permeability or just used for understanding showing what the IPR would be if the stabilized data is available. However, SIBHP can be guessed or the value from adjoining well can be used or approximated by extrapolation if two producing conditions are known. Regardless, SIBHP is difficult to determine exactly for a tight formation well; many operators

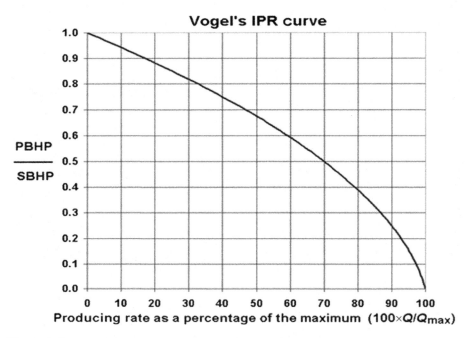

Figure 2.12 Example of using a Vogel curve for gas inflow.

of tight fractured wells do not know the SIBHP and rely on the decline curve for most analysis.

The PBHP should be approximately less than 10% of the SIBHP (Fig. 2.12) for good drawdown efficiency or for good operation of the well. If a high rate is indicated, then the well optimization is more important.

Improvements could be:

- lower WHP
- better dewatering technique
- is tubing right sized?
- look for other factors like as SV over perforations for plunger operation, etc.?

2.4 Summary

Several symptoms of wells suffering from liquid loading have been illustrated in this chapter. These indicators provide early warning of liquid loading problems that can hamper production and sometimes permanently damage the reservoir. These indicators should be monitored on a regular basis to prevent loss of production. Methods to analytically predict the loading problems and the subsequent remedial action will be discussed in more detail in later chapters.

Further reading

Gilbert WE. Flowing and gas-lift well performance. In: *Presented at the spring meeting of the Pacific Coast District, Division of Production*, Los Angeles, May 1954, Drilling and Production Practice, p. 126–57.

Thrasher TS. Well performance monitoring: case histories, SPE 26181. In: *Presented at the SPE gas technology symposium*, Calgary, Alberta, Canada; June 28–30, 1993.

Critical velocity

3

James F. Lea's experience includes about 20 years with Amoco Production Research, Tulsa, OK; 7 years as Head PE at Texas Tech; and the last 10 years or so teaching at Petroskills and working for PLTech LLC consulting company. Lea helped to start the ALRDC Gas Dewatering Forum, is the coauthor of two previous editions of this book, author of several technical papers, and recipient of the SPE Production Award, the SWPSC Slonneger Award, and the SPE Legends of Artificial Lift Award.

3.1 Introduction

To effectively plan and design the liquid loading problems of gas well, it is essential to accurately predict when a particular well might begin to experience excessive liquid loading. In the next chapter, Nodal Analysis (trademark of Macco—Schlumberger) techniques are presented, which can be used to predict under what conditions liquid loading problems and well flow stability occur. In this chapter, the relatively simple "critical velocity" method is presented to predict under what conditions the onset of liquid loading occurs.

The critical velocity/rate method to check the well liquid loading was developed by correlating a substantial accumulation of well data and has been shown to be reasonably accurate for near vertical wells. There are corrections for inclined wells. Calculation of critical velocity at any point in the well is applicable but graphically it is often shown at the well head conditions. It should be used in conjunction with methods of Nodal Analysis and by examining field symptoms if possible.

3.2 Critical flow concepts

The transport of liquids in near vertical wells is governed primarily by two complementing physical processes before liquid loading becomes more predominate and other flow regimes such as slug flow and then bubble flow begin. If the gas produced cannot lift the liquid upward then they accumulate in the well and cause liquid loading.

Gas Well Deliquification. DOI: https://doi.org/10.1016/B978-0-12-815897-5.00003-2

3.2.1 Turner droplet model

It is generally believed that the liquids are both lifted in the gas flow as individual particles and transported as a liquid film along the tubing wall by the shear stress at the interface between the gas and the liquid before the onset of severe liquid loading. These mechanisms were first investigated by Turner et al.[1] who evaluated two correlations developed on the basis of the two transport mechanisms using a large experimental data base as illustrated in Fig. 3.1. Turner discovered that he could best predicted by a droplet model that showed when droplets move up (gas flow above critical velocity) or down (gas flow below critical velocity).

Turner et al.[1] developed a simple correlation to predict the so-called "critical velocity" in near vertical gas wells assuming the droplet model. In this model, the droplet weight acts downward and the drag force from the gas acts upward (Fig. 3.1). When the drag is equal to the weight, the gas velocity is at "critical." Theoretically, at the critical velocity the droplet would be suspended in the gas stream, moving neither upward nor downward. Below the critical velocity the droplet falls and liquids accumulate in the wellbore.

In practice the Turner's critical velocity is generally defined as the minimum gas velocity in the production tubing required to move droplets upward. Methods to increase the gas velocity above the critical include a "velocity string" to reduce the tubing size until the critical velocity is obtained. Lowering the surface pressure (e.g., by compression) also increases the velocity. Foam will actually lower the velocity which can carry liquids upward. Pumps pump the liquids up the tubing and let gas flow up the casing. Gas lift adds gas to get the sum of produced gas and injected gas to be above critical.

Turner's correlation was tested against a large number of real well data having surface flowing pressures mostly higher than 1000 psi. Examination of Turner's data, however, indicates that the range of applicability for his correlation might be for surface pressures as low as 5−800 psi.

Liquid transport in a vertical gas well

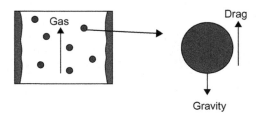

Figure 3.1 Illustration of concepts (film on wall and/or droplet model) investigated for defining "critical velocity."

Two variations of the correlation were developed; one for the transport of water and the other for condensate. The fundamental equations derived by Turner were found to under predict the critical velocity from the database of well data. To better match the collection of measured field data, Turner adjusted the theoretical equations for required velocity upward by 20%.

See Appendix A, Development of critical velocity equations, for the development of critical velocity and critical rate formulas of Turner. Later, Coleman (Exxon) used similar equations for critical velocity and critical rate, and these formulas were fit to lower well head pressure data. The formulas give a critical velocity and rate that is about 20% lower than Turner's correlation, so the use of Colemen's equations would suggest you continue to flow to lower rates than Turner's correlations, before you encounter critical flow. The development and statement of Coleman's equations are also shown in Appendix A, Development of critical velocity equations.

Since Turner shows a higher critical velocity and rate, it might be considered as being more conservative since it alerts you to do something about the encroachment of below critical flow and liquid loading at a higher rate (sooner) than Coleman equations do. Turner's equations is widely used in the industry. Even though other concepts for critical flow have come along, Turner equation is still one of, if not, the most used correlations for predicting critical flow rate and critical velocity, since it is based on comparing the data and is more conservative.

Fig. 3.2 shows the critical rate at the surface. However, one can use the same Turner's or Coleman's formulas at different pressures and temperatures to calculate a critical profile downhole.

3.3 Critical velocity at depth

Although the aforementioned formulas are developed using the surface pressure and temperature, their theoretical basis allows them to be applied anywhere in the wellbore if pressure and temperature are known. Gas wells can be designed with tapered tubing strings or with the tubing hung off in the well far above the perforations. In such cases, it is important to analyze gas well liquid loading tendencies at locations in the wellbore where the production velocities are lowest.

For example, in wells equipped with tapered strings, the bottom of each taper size would exhibit the lowest production velocity and thereby be first to load with liquids. For most lower presssure, lower temperture wells, it will first load at the bottom or a section of the well as production declines. However, if you calculate critical rate with depth it is found that for high pressure wells the onset of loading or dropping below critical can be near the surface of the well.

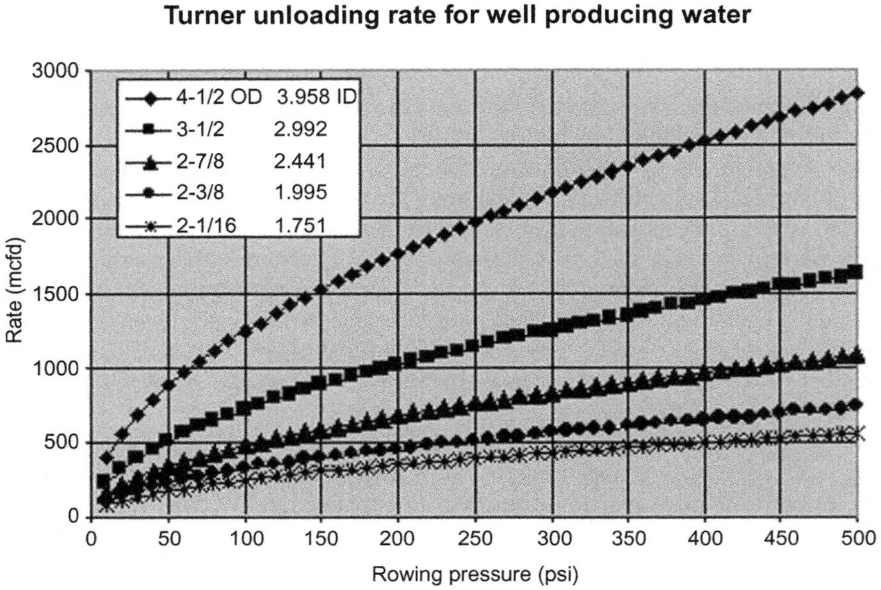

Turner unloading rate for well producing water

Figure 3.2 Simplified Turner critical rate chart (X-axis is psia).

When calculating critical velocities in downhole sections of the tubing or casing, downhole pressures and temperatures must be used. Minimum critical velocity calculations are less sensitive to temperature, which can be estimated using linear gradients. Downhole pressures, on the other hand, must be calculated by using flowing gradient routines (perhaps with Nodal Analysis, Macco-Sclumberger) or perhaps a gradient curve. Bear in mind that the accuracy of the critical velocity prediction depends on the accuracy of the predicted flowing gradient.

Example 3.1 Critical with depth: calculate critical profile in the well

Depth: 9450, casing ID: 5.1920, tubine OD/ID: 2.375/1.8530, roughness: 0.0018 in.
Separator P: 100 psi, separator T: 100 F, well head temperature (WHT): 100 F, bottom hole temperature (BHT): 245 F, gas rate: 555 Mscf/D
7.7 bbls condensate/MMscf, 111 bbls water/MMscf, gas gravity (GG).6, cond. gravity: 43.3, WG: 1.00
Use gray MPF correlation, tubing flow only.

Results:

From Fig. 3.3, one can see for this deep well that the critical calculated varies a lot from the top to the deep high BHT. For this relatively low-pressure well the critical increases with depth.

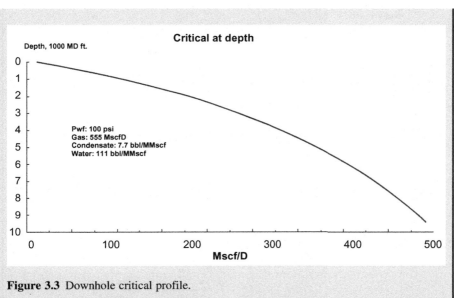

Figure 3.3 Downhole critical profile.

Example 3.2 Critical with depth: calculate critical profile in the well

Depth: 10,000 in., casing ID: 4.00, tubing OK: 2.375, tubing ID: 1.995, roughness: 0.0018

1000 ft. of casing flow below the end of tubing (EOT)

Separator: 100 psi, separator temp.: 100 °F, flowing WHT: 100, BHT: 230, Mscf/D: 550

No condensate, bblsW/MMscf: 20, well head pressure (WHP): 100 psi, GG: 0.65, condensate grav.: 50, water gravity: 1.03

Results:

You can see that the vertical straight line is the rate of 650 Mscf/D. The break in the horizontal line to the right shows the critical in the 1000 feet of casing flow is critical at 1900–2000 Mscf/D. It is liquid loaded at this time. The curve near vertical line to the left is the critical profile for tubing flow to 10,000 ft. It is nowhere liquid loaded (Fig. 3.4A).

Next put a dead string in the 1000 ft. of casing flow of 4½ casing. It gives an effective tubing flow with tubing of 2.875 OD for the 2 3/8's in the 4½ casing.

(Continued)

Example 3.2 (cont'd)

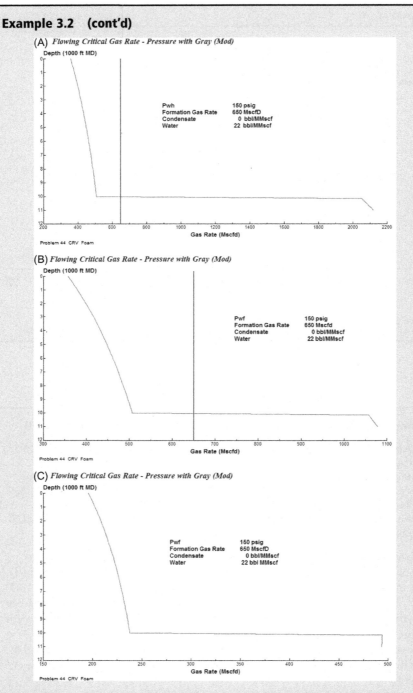

Figure 3.4 (A) Liquid loaded casing flow. (B) Liquid loaded casing with dead string inserted in the casing. (C) Simulating dead string in the casing and foamer in casing flow (and tubing flow).

Results (Fig. 3.4B):

Therefore now with dead string the ctitical in the casing has dropped from about 2000
to 1000 Mscf/D but is still more than the actual rate of 650 Mscf?D, so the casing flow is still loaded.
Next, simulate adding foamer by adjusting the gravity of the water in the program input.

Results with foam simulated from bottom of dead string:

Now (Fig. 3.4C) the tubing and the casing flows are both above critical with the dead string and foamer added. This simulates a protected commercial system called the CVR system (Weatherford).

Example 3.3 Critical flow at surface and bottom hole

Depth tubing: 9450', depth casing: 10,000', tubing OD: 2 3/8', tubing ID: 1.995, roughness 0.0018
Separator P: 100 psi, separator temp.: 100 F, flowing WHT: 100, BHT: 245, Mscf/D: 411
7.7 bblsc/MMscf, 307: bblsW/MMscf, WHP: 100 psi, GG: 0.65, condensate grav.: 43.3, water gravity: 1.03

Fig. 3.5 shows critical rate calculated using the Gray correlation. The vertical line is the actual rate. The curved and then horizontal well is the critical rate.

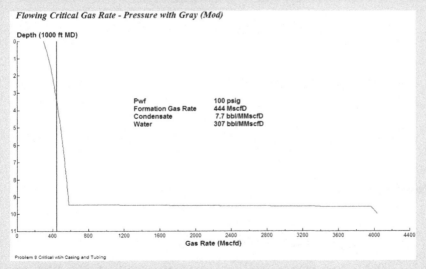

Figure 3.5 Critical velocity with depth.

(Continued)

Example 3.3   (cont'd)

Note the well is predicted to be just above critical rate at the surface but the rest of the tubing is below critical and, as usual, well below critical for the casing flow. Normally the required rate is maximum at the bottom of the tubing but for high pressure, high temperature (unusual for most loaded) gas wells the critical can be calculated to be maximum at surface conditions. This shows that even if the well is calculated to be above critical at surface, much of tubing flow can be below critical or liquid loaded.

Example 3.4 Critical flow downhole: tapered tubing string

Fig. 3.6 shows a tapered tubing string for wells where critical starts at bottom. This type of a string will allow a well to flow longer before loading. It is not used much as it prevents the use of plunger or use of rod string but the example is informative showing how critical behaves downhole and with different tubing size.

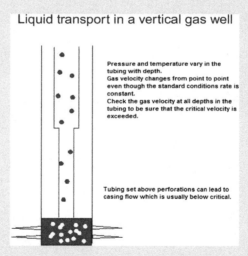

Figure 3.6 View of tapered tubing string.

Data for example 3.4: tapered tubing string:

Initially 2 3/8's tubing (1.853 in. ID) in 5 in. casing to depth of 9450 ft., roughness: 0.0018

Separator P: 100 psi, separator temp.: 100 F, flowing WHT: 100, BHT: 245, Mscf/D: 375

7.7 bblsc/MMscf, 111: bblsW/MMscf: WHP: 100 psi, GG: 0.65, condensate grav.: 43.3, water gravity: 1.03 Fig. 3.7A shows results with one size tubing. Note the bottom ~1/3 of the tubing is shown to be flowing below critical.

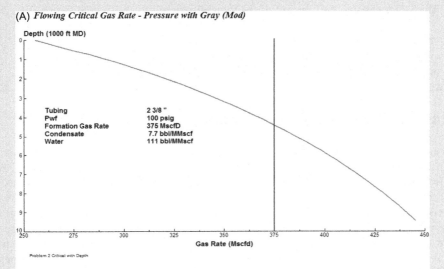

(A) *Flowing Critical Gas Rate - Pressure with Gray (Mod)*

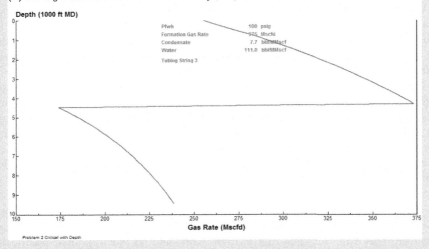

(B) *Flowing Critical Gas Rate - Pressure with Gray (Mod)*

Figure 3.7 (A) Critical velocity with depth. (B) Critical velocity with depth: tapered tubing.

(Continued)

Example 3.4   (cont'd)

Now with the smaller tubing at the bottom, the smaller string reduces the critical velocity/rate and now higher of the two strings show liquid loading. This demonstrates the concept of the tapered string.

Guo et al.[2] presented a kinetic energy model and show critical rate and velocity at downhole conditions. They mention that Turner underpredicts the critical rate. They mention that the controlling conditions are downhole. There are several models of critical velocity published in the industry.

3.4   Critical velocity with deviation

In inclined or horizontal wells the earlier correlations for critical velocity cannot be used. In deviated wellbores the liquid droplets have shorter distances to fall before contacting the flow conduit rendering the mist flow analysis less effective. Due to this phenomenon, calculating gas rates to keep liquid droplets suspended and maintain mist flow in horizontal sections are different situations for tubing or slanted tubing. Fortunately, hydrostatic pressure losses are minimal along the lateral section of the well and only begin to come into play as the well turns vertical where critical flow analysis again becomes applicable.

Another, less understood effect that liquids could have on the performance of a horizontal well has to do with the geometry of the lateral section of the wellbore. Horizontal laterals are rarely straight. Typically the wellbores "undulate" up and down throughout the entire lateral section. These undulations tend to trap liquid, causing restrictions that add pressure drop within the lateral. A number of two-phase flow correlations that calculate the flow characteristics within undulating pipe have been developed over the years and, in general, have been met with good acceptance. One of such correlations is the Beggs and Brill method.[3] These correlations have the ability to account for elevation changes, pipe roughness and dimensions, liquid holdup, and fluid properties. Several commercially available Nodal Analysis programs now have this ability.

A rule of thumb developed from gas distribution studies suggests that when the superficial gas velocity (superficial gas velocity = total in situ gas rate/total flow area) is in excess of ≈ 14 fps, the liquids are swept from low-lying section. However, this rule requires fairly high rates and as such is not practically applied.

Fig. 3.8 shows the equation and the results of the changes in critical. At about 40 degrees the critical has increased from vertical to about 40% more. Then as deviation continues to increase, the percent of increase begins to decrease again.

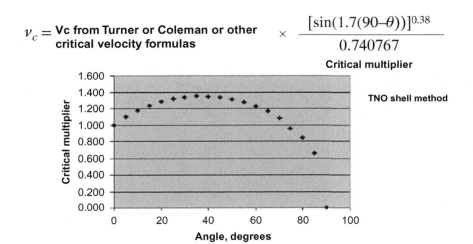

$$v_c = \text{Vc from Turner or Coleman or other critical velocity formulas} \times \frac{[\sin(1.7(90-\theta))]^{0.38}}{0.740767}$$

Figure 3.8 "Critical velocity" changes with deviation angle. Belfroid et al.[4]

References

1. Turner RG, Hubbard MG, Dukler AE. Analysis and prediction of minimum flow rate for the continuous removal of liquids from gas wells. *J Petrol Technol* 1969;1475–82.
2. Guo B, Ghalambor A, Xu C. A systematic approach ot predicting liquid loading in gas wells, SPE 94081. In: *Presented at the 2995 SPE production and operations symposium*, Oklahoma City, OK, 17–19 April 2005.
3. Beggs HD, Brill JP. A study of two-phase flow in inclined pipes. *J Petrol Technol* 1973;607.
4. Analyzing the Effect of Well Parameters on Critical Velocity, Sept17, 2008 Presented by S. P. C. Belfroid, W. Schiferli, G. J. N. alberts, TNO Science and Industry and C. A. M. Veeken, E. Biezen, Shell- NAM.

Further reading

Bizanti MS, Moonesan A. How to determine minimum flowrate for liquid removal. World Oil; 1989, pp. 71–73.
Coleman SB, Clay HB, McCurdy DG, Norris III HL. A new look at predicting gas-well load up. *J Petrol Technol* 1991;329–33.
Nosseir MA, et al. A new approach for accurate predication of loading in gas wells under different flowing conditions, SPE 37408. In: *Presented at the 1997 middle east oil show in Bahrain*, March 15–18; 1997.
Trammel P, Praisnar A. *Continuous removal of liquids from gas wells by use of gas lift*. Lubbock, TX: SWPSC; 1976. p. 139.

Nodal Analysis*

4

Analyzing loaded wells

James F. Lea's experience includes about 20 years with Amoco Production Research, Tulsa, OK; 7 years as Head of PE Department at Texas Tech; and the last 10 years or so teaching at Petroskills and working for PLTech LLC consulting company. Lea helped to start the ALRDC Gas Dewatering Forum, is the coauthor of two previous editions of this book, author of several technical papers, and recipient of the SPE Production Award, the SWPSC Slonneger Award, and the SPE Legends of Artificial Lift Award.

4.1   Introduction

Nodal analysis can be a model of a gas or an oil well. It can simulate a single well's performance including such components flow as the flowline, a surface or down-hole choke, tubing performance, completion effects such as perforations, and inflow performance (vertical flow and approximations to horizontal well). If simulated approximately well, the user can then see what impacts in the variations of in tubing size, choke size, surface pressure, and inflow can have on the performance of the well.

This chapter emphasizes the modeling of a liquid loaded well and solutions (using nodal) to the loading situation. The details of nodal analysis are relegated to Appendix B, Nodal concepts and stability concerns, describing whether or not how Nodal can show stable flow. Also there are discussions on the tubing performance curve (TPC) and on inflow expressions and how nodal modeling can still be beneficial even if no inflow expression is available by analyzing the tubing performance curve.

Fig. 4.1 is repeated here to depict how Nodal shows the intersection of an inflow and outflow curve to indicate a predicted flow rate (gas or liquid). See Appendix B for details and concepts for Nodal Analysis. Examples using Nodal Analysis are presented in this chapter.

---

* NODAL (production system analysis) is a mark of Schlumberger.

Gas Well Deliquification. DOI: https://doi.org/10.1016/B978-0-12-815897-5.00004-4

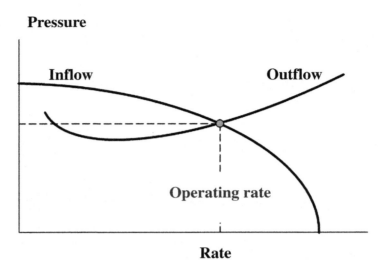

Figure 4.1 System of Nodal Analysis.

4.2 Nodal example showing liquid loading and solutions

4.2.1 Liquid-loaded well

To begin with, an example of a liquid-loaded well is input into a Nodal program and the output is discussed. A commercial Nodal program is used for most examples in this chapter (program SNAP; System Nodal Analysis Program by permission).

Input data:

Gas well: Use back pressure equation for inflow expressions: well head temperature (WHT): 110 F, well head pressure (WHP): 220 psi, well head pressure (BHT): 170 F, gas gravity (GG): 0.65, liquids: 22 bbls/MMscf, API: 35, WG: 1.0.

Inflow: Pr: 1111 psi, WC: 0.4, C (tested coefficient from back pressure equation): 0.000236 Mscf/d/(psi^{2n}), n value for back pressure equation: 1.0. Back pressure equation: $Q = C(Pr^2 - Pwf^2)^n$ where Pr is well shut-in presssure, Pwf is downhole producing prssure, Q is gas rate, and C and n are constants to be determined from test data.

Depth: 6000' with 2 3/8's tubing. Use Gray correlation for tubing multiphase flow pressure drop. Tubing roughness: 0.0018 in.

Fig. 4.2 is the calculated Nodal plot at bottom hole conditions.

Discussion:

The TPC in Fig. 4.2 slants downward and to the right at the point of intersection. This tubing curve intersects the IPR (inflow performance curve) to the left of the minimum in the tubing performance curve. As per discussion in Appendix B, Nodal concepts and stability concerns, this indicates unstable flow or essentially a Nodal way of saying it is unstable because of liquid accumulation in the tubing.

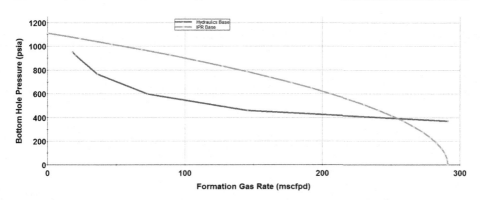

Figure 4.2 Loaded well.

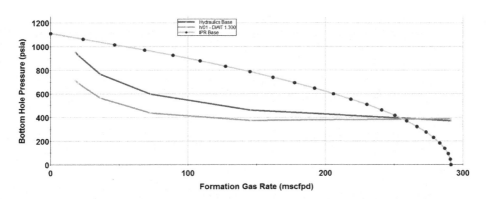

Figure 4.3 Smaller tubing as solution.

Even without the inflow curve, the downward and right slanting tubing performance curve (TPC) indicates unstable flow over the entire range of flows indicated in Fig. 4.2. Since the well is essentially liquid loaded, what can be done (using Nodal) to indicate a solution to the liquid loading problem? Note, for the information, that Turner calculated critical rate is about 500 Mscf/D, so the critical rate also says that this well is liquid loaded over the plotted range of flow rates.

4.2.2 Solutions to the loading situation

Smaller tubing as solution

For smaller tubing try 1.5 in. tubing with ID of 1.3 in. (Fig. 4.3).

Discussion:

The tubing chosen is small enough such that the TPC for this tubing size slants upward and to the right at the point of intersection so this tubing size results in stable flow where the flow is indicted, i.e., the point of intersection. Note if the rate

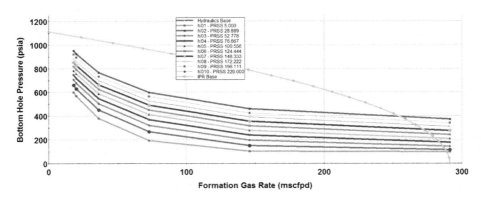

Figure 4.4 Compression as a solution.

declined to about 150 Mscf/D then this smaller tubing would also be unstable but for higher rates it is a stable situation. Therefore one solution to the loading situation can be a smaller tubing to get stable flow at the point of intersection the tubing and inflow curves.

Compression as a solution

Discussion:

Compression is not too good of a solution for this particular problem. From Fig. 4.4, WHP need to be brought down to about 5 psi before the TPC just flattens out with a very little upward slant indicating a solution. Putting the suction of a compressor on the well seems to give a solution if the pressure is brought to the very low well head pressure of 5 psi. For other situations, the lower presssure to stablize the tubing curve may occur at a higher tubing surface pressure which requires less Hp and is more easily obtainable by several types of compressors. This shows that compression (brought on by using the suction of a compressor on the surface of the tubing) can indeed be a solution to loading. The low pressure at the wellhead increases velocity of the gas and will also put some of the liquid back in the gasseous state.

Using chokes as solution

Discussion: using chokes

Several chokes were introduced at the surface for conditions that were used to generate the plots in Fig. 4.5. The chokes input were 24, 20, 18, 16, and 14 64's, and the corresponding choke IDs are 0.375, 0.3225, 0.28125, 0.25, and 0.21875 in. Looking at the results, the combination of the TPCs and choke give outflow curves. When the choke is downsized to a size of 14 choke, stable flow is indicated while the other choke sizes still indicate unstable flow. A well can be stabilized by using a choke but the flow is still below critical. This is a point of confusion as to how to add pressure (with a choke) to the surface of a well and expect a stabilized flow?

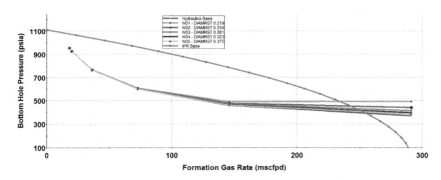

Figure 4.5 Using chokes.

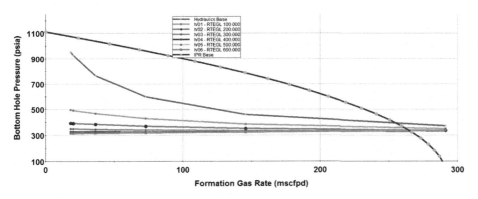

Figure 4.6 Inject gas to stabilize.

The answer is that if you just add a constant pressure (like into a higher surface pressure collection system), it will still be unstable and at a lower rate. However, if you add a pressure-dependent restriction at the surface of the well (i.e., a choke) then smaller chokes show the combination of the small tubing and the choke will show a stable outflow curve.

Inject gas to stabilize

Discussion:

From the results in Fig. 4.6, one can see that you must inject 300−400 Mscf/D to get the TPCs to slant upward at the intersection point with the IPR. So gas lift works but requires compressed gas to get the job done to unload this well.

Use foam to stabilize

Discussion: Adding foam (as simulated) just fattens the TPC at the point of intersection, so the use of foam seems marginal. However, this base case has only 40% water cut. We know that foaming (see Chapter 7) works better in case of water.

Therefore for this case change the WC from 0.4 to 0.9 and repeat the calculation as shown in Fig. 4.8.

Discussion:

Adding foamer as in Fig. 4.7 shows a marginal application but this application has only 40% water. With foamer and 90% water in Fig. 4.8 the application looks better and is good over a wider range of flow rates. Many operators will not run foam as a lift method unless the percent water is greater than 50%. In general foam is more likely to stabilize the flow (unoad the well) if there is a high percent of water in the production. If most of production is with water and if it is foamed effectively, the initial critical can be lowered using surftants by as much as two-third. For a high oil percent age the critical canot be lowered as much and in fact if a high percent age of condenstes are produced, surfactants may not be able to foam the liquids.

Plunger to unload

Here plunger is approximated by using another program, PDA (Fig. 4.9).

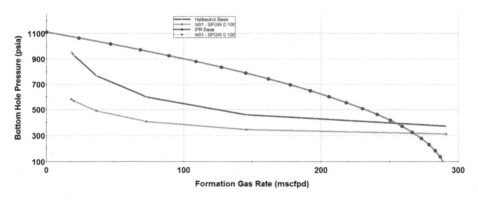

Figure 4.7 Add foamer.

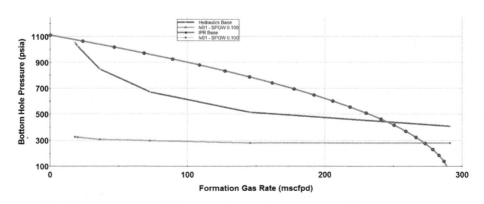

Figure 4.8 Foam 90% WC and foam.

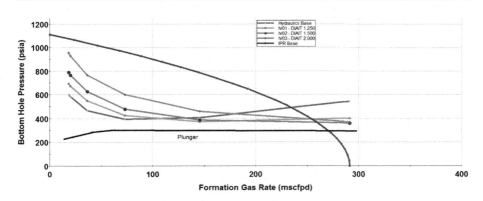

Figure 4.9 Nodal plunger compared to tubing sizes.

Discussion:

This example shows some tubing sizes compared to plunger over a range of flows defined by the input IPR. Even with no IPR you can make a comparison over a range of anticipated flows. Note that the base case (top TPC curve) is loaded as before and the next one with the 1.5 in. tubing shows this size tubing is not loaded at the IPR intersection. Some smaller tubings are shown but the smallest has a lot of friction near the IPR. Note that plunger is in this case predicted to give a lower BHP at intersection and for lower rates than any of the smaller tubings or velocity strings. Plunger does not stabilize the previously flowing tubing performance curves. Instead plunger helps the gas to lift an amount of liquid above the plunger. The plunger prevents a slug of liquid from falling back as it rises up the tubing. See Chapter 6 for details of the plunger lift method.

Pumped-off pumping well to unload- Use of pumps to lift the liquids

The gas flows up the casing, and if the well is pumped off, there is no liquid at the bottom of the casing. Pumping the height of fluid over the pump to lower levels, lowers the pressure on the well formation as well as any dewatering method unless the pump has gas or solids or other problems. By setting the nodal program to casing flow of gas and no liquids coming up the casing, you get a Nodal plot that shows how low the producing pressure or at least the pressure at the bottom of the tubing can be for pumped-off well with 220 psi on the CHP.

This is same (since friction low in the casing) as:

Casing BHP = surf CHP
$$\times EXP(.01875 \times GG \times depth/(avg\ Z\ factor \times (BHT + 460))$$
$$= 220 \times (EXP(.010875 \times 6000 \times .6/(.9 \times (1870 + 460)) = 247\ psi$$

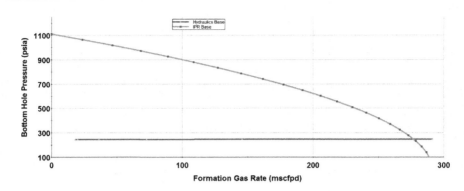

Figure 4.10 Nodal simulation of pumped-off pumping well.

The pressure increases with depth in the casing due to the weight of gas. This is included in Fig. 4.10 Nodal plot showing where on the IPR you would operate with a pumped-off pumping unit by pumping fast enough to lower the fluid in the casing to the bottom of the tubing. Acutually it is difficult to keep the liquid pumped to the bottom of the tubing but if the fluid level is low (on the average) in the casing, the pumping system has done it's job.

Discussion:

The pumped-off well gives a low BHP and solves liquid loading as all of the liquid is pumped up the tubing. However, it is difficult to keep a well completely pumped off over time so this result is somewhat optimistic.

4.3 Summary

In this chapter an example loaded well is simulated and the results are plotted on a Nodal plot showing the inflow and TPC curve.

Then using Nodal the same well loading is solved by (Nodal simulation) using:

1. Smaller tubing (velocity string)
2. Compression (lower WHP)
3. Use of a surface choke on the tubing
4. Gas lift (inject gas at depth into the tubing)
5. Add surfactants to the production at depth to foam the liquids
6. Plunger lift: use of a rising and falling piston to more effectively lift the liquds (nodal not used for ths simulation)
7. Use of pumping systems to lift liquids and let the gas flow up the casing

The results of each simulation are plotted on a Nodal plot. This should give the user a good idea of what can be done with simulating loading and some solutions to loading with respect to using Nodal. However, Nodal is not used as much as decline curve analysis and use of simple critical calculations since Nodal requires more data and possibly well tests to get data in order to input the Nodal program/s.

Note if the well is pumped off, it does not matter what kind of pumping system is used. However, some are able to operate to mitigate the effects of ineffective gas separation, (gas through pump, etc.) than other systems.

Nodal has other advantages having components such as chokes, completion effects, casing/tubing flow, and other options available to the user. Also the techniques presented do not have to be applied individually. They can be applied in tandem or as higher combinations of techniques.

Further reading

Duns H, Jr., Ros NCJ. Vertical flow of gas and liquid mixtures in wells. In: *Proc. sixth world pet. congress*; 1963, p. 451.

Fetkovitch MJ. The Isochronal testing of oil wells, SPE Paper No. 4529. In: *48th annual fall meeting of SPE of AIME*, Las Vegas, Nevada, September 30–October 3; 1973.

Gray HE. Vertical flow correlations in gas wells. API user's manual for API 14, Subsurface controlled subsurface safety valve sizing computer program. Appendix B; June 1974.

Greene WR. Analyzing the performance of gas wells. In: *Presented the annual SWPSC*, Lubbock, Texas, April 21; 1978.

Lea J, Brock M, Kannan S. *Artificial lift with declining production*. Lubbock, TX: SWPSC; 2016.

Rawlins EL, Schellhardt MA. 7 *Back pressure data on natural gas wells and their application to production practices*. Baltimore: Bureau of Mines Monograph; 1935.

Shahamot S, Tabatabale SH, Matter L, Motomed E. Inflow performance relationship for unconventional reservoirs (Transient IPR), SPE 175975 MS. In: *Presented at the SPE/CUR unconventional resources conference*, Calgary, Alberta, October 20–22, 2015.

Compression

5

Larry Harms started his consulting company, Optimization Harmsway LLC, after retiring from ConocoPhillips in 2015. He has over 35 years of experience in the application of compression to optimize production. He has conducted training courses for hundreds of operations, maintenance, and engineering personnel on compression, production optimization, systems Nodal Analysis, artificial lift, and gas well deliquification.

5.1 Introduction

Compression is crucial to all gas well productions as it is the primary means to transport gas to market. Compression is also vital to deliquification, lowering wellhead pressure, and increasing gas velocity. The lower bottom hole producing pressure from deliquifying wells and lowering surface pressures with compression can result in substantial production and reserves increases. These increases can range from a few percent to many times the current production. Increases in rates and reserves from lower surface pressure require investment in the compressor and associated equipment as well as operating costs for the maintenance and power to continue running the compressor. In the end, all decisions on deliquification are based on economics. However, many times compression can be the most economical way to keep the wells deliquified, providing higher production rates, and additional recovery at lower pressures.

The process of choosing how to apply compression and the proper equipment to achieve the desired pressures and rates is important in optimizing results. Fortunately, Systems Nodal Analysis and Integrated Production Modeling (IPM) can be used effectively to help in the process of evaluating the effect of compression on rate and recovery as well as providing insight on when, where, and what size compression equipment should be installed to maximize profits.

Compression and reduced surface pressure is usually the first and sometimes the only tool used in the life of a gas well to keep it deliquified. Compression can also increase the effectiveness of other artificial lift deliquification methods including plunger lift, foamers, gas lift, beam pumping, electric submersible pumps (ESPs), and velocity strings. When applying compression or any deliquification method, it is important to ensure that the system upstream and downstream have sufficient capacity to enable uplift to the overall production.

There are many different types of compressors, each of which has its own operating ranges, efficiencies, strengths, and weaknesses. Most of the applications for

Gas Well Deliquification. DOI: https://doi.org/10.1016/B978-0-12-815897-5.00005-6

gas well deliquification involve the use of reciprocating or screw compressors, with reciprocating compressors being the more common application of the two.

5.2   Compression horsepower and critical velocity

As noted in Chapter 3, critical velocity is directly proportional to the surface pressure, so lowering surface pressure can always help to keep a well deliquified. However, reduction in surface pressure requires energy (horsepower) and cost. Compression horsepower is related to the ratio of the discharge and suction pressures in psia commonly known as the compression ratio. In general, the best way to calculate compression horsepower is to use a compressor manufacturer's software.

Table 5.1 shows the horsepower required to compress gas at different suction pressures using a multistage reciprocating compressor to discharge at pipeline conditions of 1000 psig. Note that as suction pressure drops, compression ratio increases, and the amount of horsepower increases substantially. Also note in Table 5.1 that the amount of fuel gas that would be required to drive a natural gas engine to power the compressor is almost 6% of the gas being compressed at 0 psig (calculations are from a compressor vendor's software and are based on the most efficient engine and compressor that are used in normal gas field operations).

By combining the amount of horsepower required at a given pressure with the critical velocity or rate (see Chapter 3 and Appendix B: Critical velocity) required to keep a well deliquified, it is possible to identify the minimum amount of horsepower required to keep any well deliquified. This is important because if there is not enough horsepower to keep the well deliquified, the desired results in compression will not occur. Required horsepower is shown in Fig. 5.1 for different tubing sizes assuming again a 1000 psig pipeline pressure and a reciprocating compressor.

A similar evaluation can be done for any specific well to determine the horsepower that will be required to keep the well deliquified based solely on well fluids/tubulars, surface pressures, and compression horsepower per unit of flow at the given surface pressures. However, this evaluation shows only the minimum horsepower required and neglects the well's performance. Also, neglected are the limits on the specific compressor's performance. In order to assure that the well will

Table 5.1 Compression horsepower and fuel gas

Suction (psig)	Suction (psia)	Discharge (psia)	Compression ratio	Horsepower (MMCFD)	%Fuel gas required
0	14.7	1014.7	69.0	309	5.9
10	24.7	1014.7	41.1	253	4.9
25	39.7	1014.7	25.6	216	4.2
50	64.7	1014.7	15.7	181	3.5
125	139.7	1014.7	7.3	130	2.5
300	314.7	1014.7	3.2	75	1.4

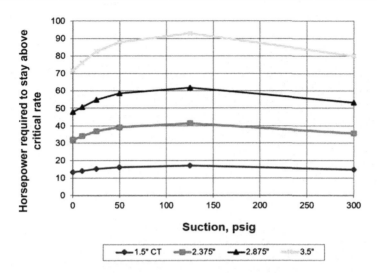

Figure 5.1 Compression horsepower required for different tubing sizes to stay above the critical rate.

respond as desired, it is necessary to include the well's performance characteristics as well as the specific compressor's performance.

Considering the fuel gas used as well as the increased size and cost of the compression equipment needed to keep gas flowing above the critical rate, it is important to closely scrutinize the economics of using compression to lower surface pressure.

It is therefore very useful to apply Systems Nodal Analysis to evaluate the current potential uplift and future results expected from compression.

5.3 Systems Nodal Analysis and compression

Systems Nodal Analysis tools (see Chapter 4: Systems Nodal Analysis) are ideally suited to evaluating the effect of reducing the surface tubing pressure using compression. The following is an example of how Systems Nodal Analysis can be used to evaluate a specific well and possible compressor options.

Example 5.1 Wellhead Compressor Option Evaluation Using System Nodal Analysis

Well depth—7000 ft to mid-perf
   Well tubulars—2.375 in. (1.995 in. ID) Tubing set at mid-perf
   Bottomhole temperature—180°F

(Continued)

Example 5.1   (cont'd)

Surface flowing temperature—80°F
Surface flowing pressure—125 psig
Gas gravity—0.65
Water gravity—1.03
Condensate gravity—57 API
Liquid—20 BBL/MMSCFD, 50% water
Reservoir pressure—600 psia
Reservoir packpressure, n—0.97
Reservoir backpressure, C—0.0015 Mscf/D/psi2
C and n used in "backpressure equation for gas flow," q, gas $= C$ $(P_r^2 - P_{wf}^2)^n$
P_r is reservoir shut in or average pressure, psia
P_{wf} is producing pressure at perforation mid-point, psia
Tubing flow correlation—Gray

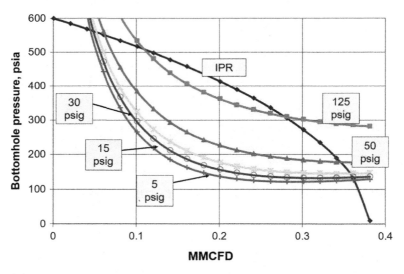

Figure 5.2 Systems Nodal Analysis for well in Example 5.1.

The example well is loaded up and will not flow continuously but averages about 200 MCFD from intermittent flow periods. Systems Nodal Analysis for this well is shown in Fig. 5.2. The solution point indeed falls to the left of the minimum in the tubing performance curve (inflection point), which represents the minimum rate for stable flow (see Appendix B).

In addition, Fig. 5.2 shows that the well can become stable with reduced surface pressure. One way to more easily see this is to plot the solution points from the

Systems Nodal Analysis well prediction and the critical rate calculated for the tubing size and fluid in the well per Coleman (some may prefer using Turner) as shown in Fig. 5.3. This check on solution points above critical rate is always recommended as in general critical rate calculations (hopefully tuned by real well performance), such as Coleman, do a better job of estimating stability than multiphase flow correlations.

It can be seen in Fig. 5.3 that the wellhead pressure must be reduced to about 108 psig to result in stable flow at the current reservoir pressure of 600 psia, however, a compressor sized about 350 MCFD at 50 psig suction would give an additional 50 MCFD of uplift and provide additional stability over time as reservoir pressure and well productivity drops. Also included in Fig. 6.2 is the well's performance with 450 psia reservoir pressure. This shows that the well can be produced down to 450 psia reservoir pressure if a flow rate of 170 MCFD at 25 psig suction pressure can be achieved (compressor capacity is reduced as suction pressure drops).

From these calculations, rates and pressures are generated which start informing the capacity curve of the compressor that might be chosen. Fig. 5.4 shows the

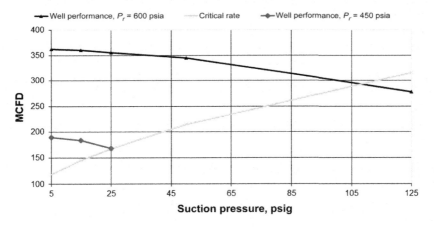

Figure 5.3 Well prediction and critical rate comparison for Example 5.1.

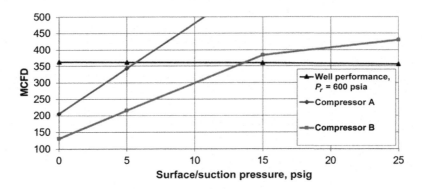

Figure 5.4 Compressor performance comparison for Example 5.1.

well's performance at 600 psia reservoir pressure along with the performance curves for two different wellhead compression units (two of the smaller reciprocating compressors, screw, and possibly some other compressor types could be used). It was assumed that the compressor would have a 125 psig discharge pressure into the current gathering system and no pressure drop in the suction piping.

Inspection of these curves shows that both the units would have the ability to deliquify the well with an increase in rate. Compressor A, the larger unit would result in a wellhead pressure of about 6 psig at a rate of 362 MCFD while compressor B would provide a rate of 360 MCFD at a wellhead pressure of 13 psig.

As Section 5.2 showed, the increase of fuel gas must be considered at the lower pressures and indeed in this case *reducing the pressure to 6 psig instead of 13 psig results would burn more fuel gas than the 2 MCFD production increase* resulting in a net reduction of rate from reducing the wellhead pressure by an additional 7 psig. In practice, the natural gas engine driving the compressor can be slowed down or changes to the compressor can be made (Section 5.2) to control the capacity and optimize the performance and fuel gas use; however, this requires active monitoring and optimization from knowledgeable compression personnel or automation.

Since either compressor will work to deliquify the well and increase production, and the smaller compressor B should cost less and consume less fuel gas, it seems to be the obvious choice. Indeed, as seen from Fig. 5.3, an even smaller unit than Compressor B might be considered which will move 350 MCFD at 25 psig. However, future performance must be considered. As the reservoir depletes will compressor B still be the best choice to keep the well unloaded, and what reduction in reservoir pressure can be expected before the well begins to have loading problems again?

We know that both of these units have the capacity to move more than the 170 MCFD at 25 psig suction pressure required to pull the reservoir down to 450 psia, but how low can we go? In order to investigate this additional nodal analysis runs can be done with lower reservoir pressures and other compressor options can be considered. In this example, additional runs (not shown here) dropping the reservoir pressure to 380 psia show that this is the last point where there is enough productivity for the well to stay unloaded with Compressor B. The well can be expected to stop flowing steadily when the reservoir pressure drops to this point and will require shut-in periods to prevent liquid loading. This fluctuating flow situation can be particularly difficult for gas engine—driven wellhead compressors unless they have sufficient bypass capacity and fuel gas supply to run through the low/no flow periods (see Section 5.11) or can be switched off and on. At this reservoir pressure, a different compressor or some other form of artificial lift could be applied, if economically feasible.

In order to do a more thorough evaluation of the options and generate rates over time which can be used for economic analysis, an IPM,[1] which incorporates a material balance model of the reservoir, as well as the wellbore, compressor performance curve, and surface facilities can be used.

The well in Example 5.1 can be considered a fairly tight (low permeability) well so the main value from the compression is in keeping the well unloaded with small uplifts seen from the incremental lowering of wellhead pressure past this point. There can be substantial differences when evaluating higher permeability wells.

5.4   The effect of permeability on compression

The goal of compression can be expanded from keeping the well deliquified to including a significant acceleration component in higher permeability/productivity wells. Also, high-productivity wells do not need compression to stay deliquified until significantly lower reservoir pressures. These differences can be very important to optimizing the effect of compression on different productivity gas wells.

This is shown in Table 5.2, where Wells L and H are identical except for the difference in permeability and completion type (Well L has been fracture stimulated).

Systems Nodal Analysis shows that lower permeability Well L can be expected to experience liquid loading at a substantially higher reservoir pressure. Also, if compression is put on Well L just as the well reaches the critical rate (preventing liquid loading), a modest accompanying uplift of about 200 MCFD is the expected result. Well H can be expected to provide a much higher 1100 MCFD uplift, even though it is at a much lower reservoir pressure. This indicates that it might be worthwhile to install compression sooner on this well depending on the economics of accelerating the production.

Since higher permeability wells in any given field flow reliably down to lower reservoir pressures, ironically, many times they do not receive as much attention as they deserve to be sure they do not load up. Also these type of wells are generally the most valuable assets in any field and they should be the focus of operations and engineering personnel to make sure they are optimized and get the appropriate deliquification solutions in a timely manner.

A dependable rule of thumb is that high productivity, high cumulative production, low reservoir pressure wells make the best compression candidates.

That is not to say that compression is unimportant for tighter gas wells. Especially on fairly tight gas wells that have been allowed to liquid load at higher

Table 5.2 Comparison of well's permeability and compression uplift

	Well L	Well H
Perm. (md)	0.2	2
Reservoir thickness (ft)	100	100
Skin	−3	0
Depth (ft)	7000	7000
Tubing diameter (in.)	2.875	2.875
Surface pressure (psig)	500	500
Critical rate (MCFD)	900	900
Reservoir pressure at critical rate (psia)	1500	870
Increase from drop to 100 psig surface pressure (MCFD)	200	1100

surface pressures/rates and produced intermittently and/or to flow in bubble flow at low average flow rates for a long period of time, large changes in producing bottomhole pressure can occur with compression (assuming the well can be unloaded) making it attractive to reduce surface pressures on these wells also.[2,3]

In all cases IPM can be a very useful tool to determine the increased reserves and acceleration component over time for different compression options in order to optimize the value of compression.

5.5   Pressure drop in compression suction

Because the goal of compression is to transmit the suction pressure of the compressor all the way to the bottom of the well, anything that causes a pressure drop between the wellhead and the compressor is undesirable. Surface restrictions increase the horsepower used and/or result in a reduction in uplift or quicker liquid loading because the pressure reduction at the wellhead is reduced. In extreme cases, surface restrictions may completely choke the flow resulting in a very small reduction of pressure at the wellhead even though the compressor has a low suction pressure.

Examples of surface restrictions include small diameter piping, use of multiple pipe elbows, chokes, orifice meters, and suction control valves. Although all these restrictions may not be eliminated, they can be minimized.

As an example, a well with a 2 in. flow line produces 390 MCFD up 2.875 in. tubing. This well is below the critical rate and analysis shows the rate can be increased to 650 MCFD if the tubing pressure is dropped to 17 psig. The pressure drop through the flow line was measured at 21 psi for the current average rate conditions. Calculations show that it would be impossible to reach the target rate with the current flow line used as the compressor suction due to excessive friction pressure drop. Increasing the pipe size to 3 in. would result in 6 psi pressure drop at the expected 650 MCFD and 17 psig compressor suction, allowing the well to approach the target rate. However, in this evaluation the best economic decision identified was to install a 4 in. flow line with an expected pressure drop less than 2 psi, essentially meeting the target rate. Systems Nodal Analysis as well as IPM can be helpful in modeling these situations. Using the current actual pressure drop as a basis for tuning the model can improve the accuracy of the results.

A suction control valve is typically installed upstream of compressors to prevent horsepower and mechanical limits from being reached. Unfortunately, good control requires some pressure drop to occur. When choosing these control valves in applications where the suction pressure is below 50 psig, lower pressure drop designs such as V-Ball or butterfly valves may be appropriate to achieve good control with minimal pressure drop.

Discharge piping restrictions are not as large a factor as the suction piping because they have smaller effects on the compression ratio and thus horsepower. However, discharge piping systems should always be analyzed to determine the expected effect and whether changes to reduce pressure loss can be economically justified.

5.6 Wellhead versus centralized compression

Because there will always be some pressure drop in piping, no matter the size, between a well and the compressor suction, minimizing the distance helps minimize the pressure drop. The ability to have low losses on the suction and the higher efficiency of flowing at higher pressures on the discharge explain why wellhead compression is always potentially more efficient on a hydraulic basis than the centralized compression.

Offsetting this is the increased fuel and mechanical efficiencies along with reduced capital cost per horsepower of installed compression that can be obtained with a few larger horsepower centralized units when compared with many smaller wellhead units. Also, cost to operate and maintain larger centralized units is less per unit of compression capacity.

However, as has been shown, when wells in one operating area with significant contrast in productivity are involved, customizing the compression to the individual well can be attractive to optimize rates, fuel, and operating cost.

In most cases with multiple gas wells being gathered in the same system, a combination of wellhead and centralized compressions will provide optimum economic results. A good rule of thumb is to install central compression sufficient to produce the "average" well in the field optimally and then to use wellhead compression (and other costly deliquification techniques) on the best wells. Again, IPM can be a great tool to use when developing this compression "strategy."

Compressors are many times used in series, for example, a wellhead screw compressor discharging to a centrally located reciprocating gathering compressor. In all cases of compressors being operated in series, the reliability depends on the product of the reliability of the two machines, that is, if both units have 95% reliability then the overall reliability can be expected to be slightly over 90%, so this effect should be included in economics for developing a compression and maintenance strategy.

5.7 Developing a compression strategy using Integrated Production Modeling

The questions of when to install compression in a field, what suction pressures to use, where to locate the compressors, what size gathering and flow lines to use, should there be more than one pressure system in the field, etc., are technically challenging.

One documented study[4] shows how this was approached in a major gas producing field in South Texas. This field had 1800 producing wells so constructing an overall field model would be daunting, therefore typical wells were chosen and an IPM model was used to determine the effect of compression on these wells.

Many options were analyzed and Table 5.3 is a summary of the cases where wells were produced until they reached the critical rate and then switched into a lower pressure system, from high pressure to intermediate pressure to low pressure, and then "Ultralow" (wellhead compression).

Table 5.3 Results of IPM modeling with expected production and recovery on different type wells

Well type (OGIP)	HP (950 psig), MMCF/% rec.	IP (200 psig) MMCF/% rec.	LP (50 psig) MMCF/% rec.	Ultralow (0 psig) MMCF/% rec.
1 bcf	426 (52%)	215 (26%)	168 (20%)	17 (2%)
3 bcf	1683 (65%)	557 (22%)	291 (11%)	53 (2%)
6 bcf	3850 (74%)	974 (18%)	385 (7%)	78 (1%)

These results validate what has been presented earlier in that lower permeability and limited reservoir wells still show substantial increases in recovery with compression; however, the best compression candidates are the higher permeability/larger reservoir wells (6 BCF).

As a result of economic runs with the well mix in this field, the following conclusions were drawn:

1. Only the areas where there were expected to be many of the three BCF + wells would be considered for an intermediate pressure system, in most normal areas there would only be a high- and a low-pressure system.
2. Only six BCF + wells would be considered for wellhead compression with 50 psig pressure being sufficient for the other wells.

This information is presented to show that reliable results can be produced with IPM which can enable economic evaluations of options to allow a system strategy to be generated. Results will vary widely between different fields and in different gas price and cost environments.

5.8 Downstream gathering and compression's effect on uplift from deliquifying individual gas wells

A frequent subject of debate when deliquifying gas wells in a field (actually when any production-increasing project is done) is whether there has been a true uplift in not only the individual well but also the overall field's production. This is an area of considerable concern when using any of the deliquification methods as economics are usually based primarily on the uplift. Particularly important is knowing if the wells are producing from the same reservoir as offset wells as this might mean losses in these wells which could offset much of the uplift (in rate and reserves) in the deliquified well. More in-depth reservoir analysis, reservoir models, and/or IPM modeling should be used in these cases.

On the surface system, in order to ensure that true uplift is seen from deliquifying an individual well, there must be adequate gathering piping size and compression capacity to result in minimal pressure increases on the other wells in the system. This can be confirmed using a Systems Nodal Analysis of the complete system but should be examined with at least a simple piping and compressor analysis to assure that the method of deliquifying the well will result in the expected overall uplift demanded to provide economic success.

Consider the following case study: a well has loaded up in the 1000 psig gathering system and needs to be put into the low-pressure system to unload and increase flow. A check of the compressor shows that there is adequate compression horsepower/capacity to put the well on compression at 85 psig, which will unload the well. Upon switching the well to the low-pressure system, it is found that wellhead pressure dropped only to 180 psig but the well unloaded with an uplift of 500 MCFD. Unfortunately, the increase in pressure on the downstream gathering system in this part of the field which was not adequate to handle the increased gas flow resulted in more than a 500 MCFD loss from other wells on this same system so that overall production was actually decreased. In this case, the well was put back into the high-pressure system until an additional gathering line could be repurposed from the high- to the low-pressure system to reduce the increase in the gathering system pressure (and subsequent loss of production from other wells).

5.9   Compression alone as a form of artificial lift

Sometimes compression is the most economical, lowest risk choice as the sole artificial lift method. This may be true in wells that are sand producers, making them prone to operating problems with capillary strings, plungers, or pumps. Also, mechanical problems in the tubulars such as holes or restrictions may make it very expensive and risky to install other artificial lift types which might be effective like gas lift. In some cases, the available expertise in the area to operate plunger lift or pumping systems may be insufficient dictating only compression be used.

The significant risk of reducing the well's productivity (for example by killing the well) during any work-over operations that may be required to run other artificial lift methods may also swing the best choice to stay with compression only. This can be particularly true in low pressure, high-productivity reservoirs which struggle to produce back work-over fluids and to reverse relative permeability damage from kill fluids. Also, in these types of wells (which were previously pointed out as top deliquification candidates), the reduction of flow area from the insertion of additional equipment in the well may limit the rate due to increased friction pressure drop.

Lower pressures from compression can substantially increase the water vapor-carrying capacity of the gas which may be helpful for deliquification. However, in wells with produced water that has significant total dissolved solids (high chlorides) with low water production rates, this may cause the water to evaporate completely leaving salt/dissolved solids downhole and eventually plugging up the well. In these cases, fresh water may need to be circulated to prevent these solids from forming.

In many cases, lower wellhead pressures with compression is an aid to the other types of artificial lift/ deliquification. With an effective and optimized artificial lift system, lowering surface pressure ultimately is the limiting factor on reducing producing bottomhole pressure.

There will be a time in the life of all gas wells when the only option will be to lower surface pressure or pump cement.

5.10   Compression with foamers

Compression can aid foamers because reducing the wellhead pressure also reduces the density of the gas and increases agitation (see Chapter 7: Foam).

Foamer can be very useful for aiding the effectiveness of compression and vice versa. In order to optimize and preserve uplift in one field case,[3] 37 of the 54 wells, which were being produced with wellhead compression, were using continuous foamer. In this field, it was found to be more economical to first apply foamer than wellhead compression because of lower costs.

Foamer can cause liquid (and sometimes solids) carryover into downstream compressors causing serious damage if it is not properly designed and applied. In cases where central compression pressures are lowered as well as foamer is used in the wells during a deliquification campaign, there is a high probability that frac sand or other solids along with slugs of liquid may carry into and through the primary separation. In wellhead compressors which are installed on wells that are loaded up, large liquid slugs should be expected and prepared for.

When installing a wellhead compressor or significantly reducing wellhead pressures with central compression, checks on the separators/scrubbers should be done to ensure they are clean, properly sized and working effectively before foamer application. It is also important to be careful to check foamer efficacy, half-life, and dosage to ensure the foamer does not carry over. In some cases, defoamer must be used.

5.11   Compression and gas lift

Gas lift and compression are almost always used concurrently as the high-pressure gas used for lift gas injection usually is provided by the discharge of a compressor (see Chapter 11: Gas lift). There is an optimum combination of wellhead pressure, gas lift injection pressure, gas lift injection rate, and compression requirements that can be found for any specific well. Again, Systems Nodal Analysis can be a useful tool to determine this optimum. Often this optimum is at a low wellhead pressure (below 100 psig), which allows reduced lift gas injection rates. In general, optimizing existing gas lift compressors to provide lower suction pressures (not more lift gas) will deliver higher production rates.

A limit to the advantage of lower wellhead pressures in gas lift is seen when the friction pressure drop upstream of the compressor becomes too large a factor. Systems Nodal Analysis can be very helpful in these evaluations.

5.12   Compression with plunger lift systems

Plunger lift is an artificial lift system that can benefit greatly from compression as lower surface pressure reduces critical rate increasing after flow time and decreasing shut-in time (see Chapter 6: Plunger lift). Both gas-powered and electric compressors have been shown to have application in plunger lift installations.

Fig. 5.5 shows a simple schematic of a plunger lift installation equipped with a surface compressor. The compressor is switched on to lower the wellhead pressure when the well is opened while the plunger is arriving and during the after-flow period. The compressor is then shut down or in full recycle while the well is shut-in to build up pressure to bring the plunger up for the next cycle.

Plunger lift is sometimes thought of as an intermittent gas lift system where the lift gas is provided by the reservoir and stored in the annulus. Sometimes the reservoir gas is supplemented taking gas from the compressor discharge or other external gas and injecting into the well annulus allowing the plunger to be cycled more frequently. This is called gas-assisted plunger lift.

Fig. 5.6 shows production data for a plunger lift installation equipped with compression. The initial production followed a fairly steep decline until it was put on compression, at which time the oil and gas increased markedly then fell off. A sustained uplift was achieved after the plunger lift was installed 4 months later and operated in conjunction with the compressor. When plunger lift wells begin having trouble cycling, it is possible that the life of the plunger wells can be extended considerably if the WHP is lowered. Again economics will dictate if this solution is viable.

Morrow[6] further discusses compression with plunger lift.

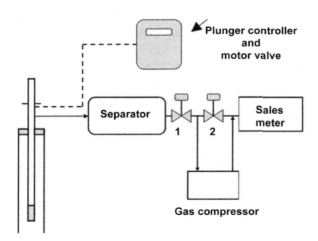

Figure 5.5 Compression system installation with plunger lift (Phillips and Listiak[5]).

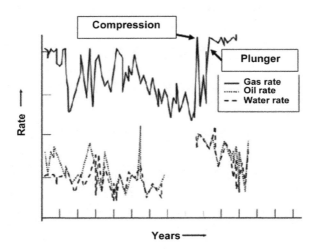

Figure 5.6 Performance improvement using plunger lift and compression (Phillips and Listiak[5]).

5.13 Compression with beam pumping systems

In a beam pump system, the liquid production is governed primarily by the down-hole stroke length, the number of strokes per minute, and the pump size. Lowering the casing head pressure (CHP) on a pumping well producing liquids up the tubing and gas up the annulus will bring up the liquid level in the annulus and allow you to pump faster if desired. If the pump is set below the perforations and there is adequate liquid over the pump, but all liquid is below the perforations, then the producing pressure is a function of the CHP. This emphasizes that pumping liquids lower the producing pressure, but if the CHP remains high the producing pressure on the perforations is always a little higher than the CHP. This shows that *pumping* and *compression* are needed for low producing pressures on the formation.

Fig. 5.7 compares the effect of three wellhead pressures on the liquid level in the tubing casing annulus. Note that these assume a constant production rate and flowing bottom pressure. The figure shows that as the CHP is reduced, the liquid level in the casing/tubing annulus is raised substantially. Conversely, high wellhead pressure puts the liquid level in the annulus low in the well near the pump intake. If the CHP is too high and annulus fluid level is too low, there is a distinct possibility of incomplete pump fillage and lower pump efficiency. This can be overcome by putting the pump below the perforations if possible. Otherwise, a low but adequate fluid level is needed. As reservoir pressures deplete, lower annulus fluid levels over time are expected with a fixed wellhead surface pressure.

Therefore one way to ensure proper pump fillage and more efficient pump operation is to lower the surface casing pressure by compression. If the liquid level was initially low and pump fillage was not complete, compression will have the effect of increasing the well's production simply from higher pump efficiency.

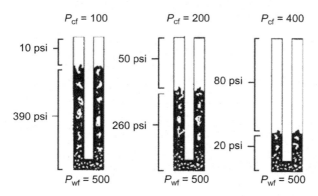

Figure 5.7 Pressure relations on a pumping well with a gaseous fluid column (McCoy et al.[7]).

In addition, since there is a higher liquid level in the annulus and better pump fillage with lower surface pressure, the unit can also be run at higher speeds and/or longer stroke lengths to thus lower the fluid level, which in turn lowers the producing bottomhole pressure, resulting in higher production rates of liquids and gas. As shown in Section 5.4, this increase can be considerable if the well's productivity is high. Bottomhole producing pressure is composed of mostly contributions from the fluid height in the annulus and surface casing pressure. The fluid height can be controlled (within limits) by the pumping capacity and the casing head pressure can be modified by compression.

Increased pump fillage can also reduce pump failures and other associated failures, especially if "fluid pound" is avoided. See Chapter 10, Use of beam pumps to deliquify gas wells, for a complete discussion on gas separation and effects on the beam pump system.

5.14 Compression with electric submersible pump systems

In general, ESPs operate with a fixed pressure between the pump intake and the pump discharge (see Chapter 12: Electrical submersible pumps). This translates to a fixed pressure increase between the well's bottomhole flowing pressure and the wellhead pressure. Therefore lowering the surface wellhead pressure on a typical ESP installation proportionally lowers the flowing pressure. The lower bottomhole flowing pressure increases production and/or lowers the power demand of the unit.

However, again if the CHP is still high, even with all the fluid off the perforations (pump below perfs. with motor cooling options), the producing pressure will never be lower than the CHP (plus the contribution of the weight of gas in the casing), emphasizing the need for compression on the casing. Artificial lift can control the fluid height (within limits) in the casing but compression is needed to control the CHP.

5.15   Types of compressors

There are a number of compressor types that are used to lower the pressure on entire fields of gas wells, or to lower the pressure on individual gas wells.

For single well applications, the following compressor types may be used[8]:

- Liquid injected rotary screw
- Reciprocating
- Liquid ring
- Sliding vane
- Rotary lobe
- Reinjected rotary lobe

Over 95% of the units used in onshore gas fields are either reciprocating compressors or liquid injected rotary screw compressors and further information on these types is given below.

5.15.1   Liquid injected rotary screw compressor

Refer to Fig. 5.8.

- Higher cost per cfm.
- Liquid injected.
- Approximately 10−20 compression ratio possible.
- Medium displacement.
- Power frame required.
- Requires seal oil cooling system.
- Liquid ingestion dilutes seal oil and solids ingestion can be catastrophic. Good suction scrubber is required.
- Requires gas/oil separator.

Figure 5.8 Elements of a screw compressor.

- Can handle very high compression ratios in one stage of compression as the oil absorbs the much of the heat of compression. Excellent for very low suction pressure even down to vacuum. Oil cooling system required.
  - Except for gear amplification, very few wearing parts, which provides very high reliability.
  - Mechanical and adiabatic efficiency is high if unit is run at design conditions.
  - Efficiency suffers if unit is run too far off of design conditions (a slide stop and slide valve can help).
  - Discharge pressure limited, usually the maximum discharge pressure is less than 400 psig.
- Oil used in screw compressors is hydrophilic, so the discharge temperature of screw compressors must be maintained above 205°F (adjust oil flow, cooling or discharge pressure to achieve this temperature) to prevent water buildup in the oil but no more than 215°F to avoid degrading the oil.
  - Oil can become contaminated with heavy hydrocarbons and other liquids causing operational problems. Selection of proper oil type is critical. Test oil frequently for fines content.

5.15.2 Reciprocating compressor

Refer to Fig. 5.9.

- High cost per cfm.
- Air or liquid cooled.
- Approximately 4.0 maximum compression ratio per stage limit.
- Low displacement/power frame.
- No amount of liquid ingestion allowed.
- Valve losses affect compression ratio and volumetric efficiency but can have the highest efficiency.
- Most flexible of all compressors in that it potentially can handle varying suction and discharge pressures and still maintain high mechanical and adiabatic efficiency within temperature and mechanical design limits.

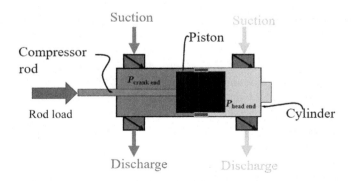

Figure 5.9 Elements of a dual acting reciprocating compressor cylinder.

- Overall compression ratio is limited only by discharge temperature and rod load rating of frame. Units can be 2 staged (or even 3 + staged) to produce very high discharge pressures with low suction pressure.
- Level of knowledge required for maintaining unit can easily be obtained. Good engine mechanics can be good reciprocating compressor mechanics.
- Potentially high operating expense and downtime due to compressor valve maintenance. This valve maintenance is highly dependent on gas quality (solid and liquid contamination), which can be a problem with wellhead compression.
- Good separation upstream of the compressor and the proper choice of compressor valves can effectively eliminate almost all valve problems.
- Not as efficient with very low suction pressures (vacuum).
- Max. rod load (force on piston rod) should never be exceeded.

In all compressor applications, as wells/fields decline the existing compressors will become oversized resulting in inefficient operations. If the cost of inefficient operations (excessive maintenance, fuel gas, and downtime) and/or if profits from lowering the surface pressure with a more efficient operating unit exceed the cost of changing out the existing compressor, then a different compressor should be installed.

As an alternate to installing a different compressor, the oversized compressor capacity might be useful (if proper rate and pressure available) to implement single point gas lift (see Section 11.8) or an ejector might be used to provide an additional stage of compression as shown in the next section.

5.16   Gas jet compressors or ejectors

Gas jet compressors, or ejectors, are classified as thermocompressors and are in the same family as jet pumps, sand blasters, and air ejectors. They use a high-pressure gas for motive power. Ejectors using gas can impart up to two compression ratios; using liquid they can generate higher ratios if cavitation can be avoided

The ejector, or gas jet compressor, operates on the Bernoulli principle as illustrated in Fig. 5.10. The high-pressure motive fluid enters the nozzle and is accelerated to a high velocity/low pressure at the nozzle exit. The wellhead is exposed to the low pressure at the nozzle exit through the suction ports and is mixed with the motive fluid at the entrance to the throat. Momentum transfer between the motive and produced fluids in the throat and velocity decrease in the diffuser increases the pressure to the discharge pressure.

Eductors have potential advantages including:

- No moving parts.
- Low maintenance/high reliability.
- Easy to install, operate, and control.
- Can handle liquid slugs.
- Low initial cost/payback time usually short.
- Nozzle sizes can be changed to meet changing well conditions.

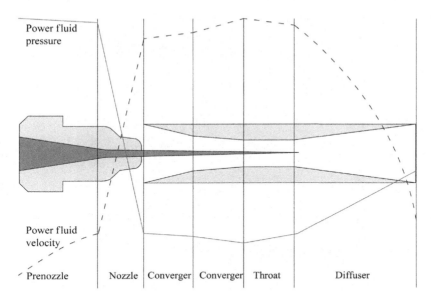

Figure 5.10 Principles of gas jet compressor (ejector).

Fig. 5.10 shows an ejector in actual field service. One successful configuration uses a flooded screw compressor to pull the tubing/casing annulus down to 8−10 psig. A portion of the gas discharged by the compressor is used to drive an ejector to pull the tubing down to 1−5 psig. The exhaust of the ejector is combined with the casing gas and sent to the compressor (Fig. 5.11).

If high-pressure fluid is available (e.g., from a nearby high-pressure gas well) to power the ejector, then it is advantageous to utilize this wasted energy with an eductor to lower surface pressure on a lower rate well to prevent liquid loading.

The principle disadvantage of eductors is that they have a higher hp/MMCF requirement than other technologies (i.e., they have lower mechanical efficiency). This lower efficiency can often be offset by extremely low capital cost. For example, a well was limited to 600 MCF/d with 9 psig wellhead pressure by the two-stage reciprocating compressor installed. The compressor had plenty of horsepower to move more gas but the piping and cylinder configuration did not allow lower pressures. Replacing the 300 hp compressor with a different machine would be very expensive so an ejector was added between the wellhead and the compressor to compress the full stream—basically adding a compression stage. This ejector lowered the wellhead pressure to −5 psig with 9 psig discharge pressure (atmospheric pressure at this sited is 11 psia so the ejector developed 3.3 compression ratios) and increased the well's production to 900 MCF/d. The efficiency of the ejector is only 46%, but it reduced the capital outlay required by more than an order of magnitude.

Figure 5.11 Ejector installed on wellhead from Gas Well Deliquification 2nd Edition, Compression Chapter by D Simpson.

5.17   Other compressors

Other types of compressors continue to be developed or adapted for application on gas wells including multiphase pumps, which can act as compressors and scroll compressors. Attempts continue to develop an economically feasible downhole compressor.

5.18   Centrifugal compressors

Discussion of centrifugal compressors, which are installed on all the largest gas lift, gas plant, and transmission applications, have not been included as these are used infrequently in normal gas field deliquification service. However, large central centrifugal compressors have been installed in Australian coal bed methane (CBM) applications and these are working successfully including four stage units which are compressing gas from 30 to 1900 psig with the capability to go down to 10 psig suction. Although centrifugals are in general less efficient than the positive displacement compressors such as reciprocating and screw compressors, large centrifugals can have quite good efficiencies, higher reliability, and so may be good candidates for fields with high gas rates.

5.19   Natural gas engine versus electric compressor drivers

Although much has been presented on compressors, the drivers that provide power to the compressor are very important parts of any compression system. Most of the units in onshore gas fields use natural gas engines as drivers but electric motors can also be very good choices.

In general, electrics should be seriously considered if the electricity is readily available (no long expensive power lines to run) and if electricity is available at a competitive price compared to natural gas. Also, noise requirements may make electric drivers much more desirable.

Natural gas engines are substantially more costly to buy, operate, and maintain than electric motors. They are usually responsible for 70%–80% of the downtime that occurs on compression units. It is important to consider the fuel economy and expected loss of revenue from increased downtime in any driver decision.

5.20   Optimizing compressor operations

The in-depth details of how compressors can best be applied and optimized are beyond the scope of this book; however, the following items are presented as useful tips for optimizing compressor operations.[9]

1. Compressor profitability depends mostly on the gas throughput of the unit and not the operating and maintenance costs so most of the focus should be on be on the revenue side (optimizing throughput) and not the cost side.
2. Compressor capacity and operating ranges must be monitored and adjusted[10] to maintain optimal performance by matching compression with field/well performance which is changing over time. For reciprocating compressors this includes setting the unit up to have the proper desired suction and discharge pressure ranges, speed, clearance of each cylinder (volume left after every compression stroke) and settings for suction and recycle controllers.
3. Recycling gas from the discharge to the suction of the compressor is the most inefficient capacity control and should only be used as the method of last resort/temporarily. The most efficient way to reduce compressor capacity is to change the speed. Automating this function can be particularly helpful; however, care should be taken to ensure emissions requirements (if any) continue to be met in the RPM adjustment range.
4. Variable volume clearance pockets can be used to adjust the capacity of reciprocating compressors and should be considered for installation on all compressor stages to increase the flexibility and useful life of the units in any application.
5. Wells that are loaded up will not "automatically" unload once wellhead pressure is lowered with compression, especially if they have been flowed below critical rate for a long period of time. On most of these type wells extended shut in periods, foamer, swabbing or jetting must be used to get them unloaded and to see the desired production uplift.
6. Monitor and track compressor run times and causes of downtime to eliminate/minimize the most common causes of losses. This includes wellhead compressors.
7. Reduce increases in the gathering system operating pressures which occur when there are multiple central compressors on a common suction system and one goes down by shutting in wells (preferably automatically) to keep other wells from loading up with higher pressures.[11]
8. All wells without another artificial lift method (downhole pumps, plunger lift, etc.) that allows them to stay unloaded at higher pressures should be shut in automatically when the compressor goes down to keep the well from loading up.

9. Production separators and scrubbers upstream of compressors must be properly maintained (mist extractors cleaned, solids removed, etc.) to keep undesirable liquids and solids out of the compressor.

10. Many of the compressors in deliquification service are provided by compressor rental companies. Teamwork between the operator and the compressor rental company as well as alignment of goals on focusing on optimizing throughput, not downtime, will provide the most profitable operations.

5.21   Unconventional wells

Although the examples used in this chapter have referred to conventional wells, the same principles apply to unconventional wells.

On the productivity side, it is more challenging to model these well; however, they have very low permeability so compression serves mainly to keep them above critical rate, to lower the vapor pressure/recover vapors from condensate as well, and to assist other deliquification techniques.

In operational terms, very fast declines and slugging that can be expected affect the compression/separation choices.

5.22   Summary

Compression can help a liquid loading well by increasing the gas velocity to equal or exceed the critical unload velocity and also lowers pressure on the formation for more production/reserves by lowering the wellhead flowing pressure.

Because of the differing response that can be expected from different types of wells, it is important that the compressor type and size be matched to the well. Systems Nodal Analysis and IPM can be very helpful tools to accomplish this.

Compression often is used on a field-wide basis to lower the gathering system pressure; however, for any compressor the amount of pressure reduction that can be transmitted back to the wellhead must be taken into account for optimal results.

Compression can be used as a primary artificial lift method or to aid the other types of artificial lift to different degrees.

There are many types of compressors that can be successfully applied to help deliquify gas wells. The key to attaining the best economic success in deliquifying gas wells is to pick the best areas, wells, match the compressor to the well's performance, and to operate efficiently.

References

1. Harms LK. Better results using integrated production models for gas wells. In: *Paper SPE 93648 presented at the 2005 SPE production and operations symposium*, Oklahoma City, OK, April 17–19, 2005.

2. Harms LK. Installing low-cost, low-pressure wellhead compression on tight Lobo Wilcox wells in South Texas: a case history. In: *Paper SPE 90550 presented at the 2004 SPE annual technical conference and exhibition*, Houston, TX, USA, September 26–29, 2004.
3. Harms LK. SPE 138488 — Wellhead compression on tight gas wells in the long run: a follow-up case history on seven years of success in Lobo. In: *Presented at the SPE tight gas completions conference in San Antonio*, TX, November 2–3, 2010.
4. Bui Q, Harms L, Munoz E, Becnel J. Low pressure system for gas wells: do we need it? How low should we go? A compression strategy for tight gas wells in South Texas. In: *Paper SPE 124869 presented at the 2009 SPE annual technical conference*, New Orleans, LA, October 4–7, 2009.
5. Phillips D, Listiak S. Plunger lifting wells with single wellhead compression. In: *Presented at the 43rd southwestern petroleum short course*, Lubbock, TX, April 23–25, 1996.
6. Morrow SJ, Aversante OL. Plunger-lift, gas assisted. In: 42nd annual southwestern petroleum short course, Lubbock, TX, April 19–20, 1995.
7. McCoy JN, Podio AL, Huddleston KL. Acoustic determination of producing bottomhole pressure. In: *SPE formation evaluation*, September 1988, p. 617–21.
8. Thomas FA. Low pressure compressor applications. In: *Presentation at the 49th annual liberal gas compressor institute*, April 4, 2001.
9. Harms L. Optimizing compressor operations. In: *Presented at the 15th annual gas well deliquification conference*, Denver, CO, February 20–22, 2017.
10. Harms L, Garza M. SPE 142293 — Stay in the "Box": a consistent method for configuring (loading) reciprocating compressors to optimize performance. In: *Presented at the SPE production operations symposium*, Oklahoma City, OK, March 27–29, 2011.
11. Schulz R, Harms L. SPE 117433 — An unconventional but definitive analysis of a field's production improvement. In: *Presented at the SPE eastern regional/AAPG eastern section joint meeting*, Pittsburgh, PA, October 11–15, 2008.

Further reading

Simpson DA. *Practical onshore gas field engineering*. Gulf Professional Publishing; 2017.

Plunger lift

6

James F. Lea's experience includes about 20 years with Amoco Production Research, Tulsa, OK; 7 years as Head PE at Texas Tech; and the last 10 years or so teaching at Petroskills and working for PLTech LLC consulting company. Lea helped to start the ALRDC Gas Dewatering Forum, is the coauthor of two previous editions of this book, author of several technical papers, and recipient of the SPE Production Award, the SWPSC Slonneger Award, and the SPE Legends of Artificial Lift Award.

Lynn Rowlan, BSCE, 1975, Oklahoma State University, was the recipient of the 2000 J.C. Slonneger Award bestowed by the Southwestern Petroleum Short Course Association, Inc. He has authored numerous papers for the Southwestern Petroleum Short Course, Canadian Petroleum Society, and Society of Petroleum Engineers. Rowlan works as an Engineer for Echometer Company in Wichita Falls, Texas. His primary interest is to advance the technology used in the Echometer Portable Well Analyzer to analyze and optimize the real-time operation of all artificial lift production systems. He also provides training and consultation for performing well analysis to increase oil and gas production, reduce failures, and reduce power consumption. He presents many seminars and gives numerous talks on the efficient operation of oil and gas wells.

6.1 Introduction

Plunger lift is a system that assists natural flow by helping a slug of liquid to be carried to the surface using gas production. The plunger assists flow by reducing the amount of liquid that would fall back in the well as the liquid rises. By reducing or eliminating the liquid fall back from a slug of liquid as it rises, the efficiency of the production of liquids and gas is greatly increased. Plunger lift is normally not applied until the gas flow drops to near or below the critical velocity (as determined by the Turner or Coleman critical velocity techniques). At or below the gas critical flow rate, liquids are no longer brought to the surface as in mist flow. As slug and later bubble flow occurs in the well, liquids occupy a greater portion of the tubing volume. These accumulated liquids add pressure to the formation and less or no gas is produced. The plunger lift system intermittently carries slugs of liquid to the surface, allowing gas to be produced with less pressure from accumulated liquids (Fig. 6.1).

Most plunger lifts are applied without using external sources of energy; however, in some cases, additional gas can be injected [gas-assisted plunger (GAPL)], and

Gas Well Deliquification. DOI: https://doi.org/10.1016/B978-0-12-815897-5.00006-8
© 2019 Elsevier Inc. All rights reserved.

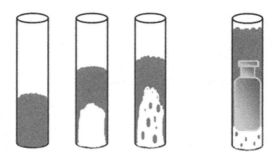

Figure 6.1 Slug rise with no plunger on left: carrying slug upward with plunger on the right (Listiak and Phillips, SWPSC).

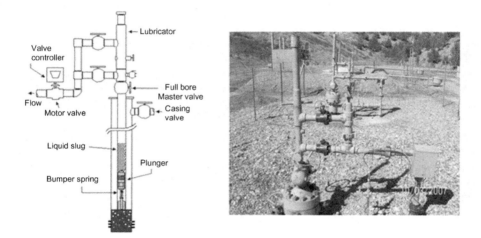

Figure 6.2 Plunger well.

foam can also be applied. In addition, there is a technique of using more than one plunger in one well (progressive plunger), which will be discussed later.

Most plunger lift wells use no external energy. External systems needed in the plunger lift system are inexpensive compared to other gas well deliquification systems, both from a Capex and Opex standpoint. The top four methods of deliquification of gas wells are plunger, surfactants, gas lift, and pumps. Compared to other systems, plunger is applied most often but cannot move the most liquids due to certain feasibility requirements, which will be discussed later.

A plunger lift system is relatively simple and requires few components. A typical plunger lift installation, as shown in Fig. 6.2, would include the following components:

- A downhole bumper spring which is wire-lined into the well to allow the plunger to land more softly downhole.
- A plunger free to travel the length of the tubing.
- A well head designed to catch the plunger and allow flow around the plunger.
- A controlled motor valve that can open and close the tubing to the flowline.

- A sensor on the tubing to sense arrival of the plunger.
- An electronic controller that contains logic to decide when to open and close the well.

There are degrees of sophistication for the controller which is discussed later in the following.

Plunger lift is an intermittent artificial lift method that uses only the energy of the reservoir to produce the liquids. A plunger is a free-traveling piston that fits within the production tubing and depends on well pressure to rise and solely on gravity to return to the bottom of the well. Fig. 6.2 illustrates a typical plunger lift installation.

Plunger can produce perhaps a few hundred bpd at very shallow depths but has less liquid production capacity for deeper wells. However, many plunger lift wells produce only about 5 bpd or less. If these wells producing below critical do not have the small amounts of liquid artificially lifted, the wells will load up and in some cases stop producing.

6.2 Plunger cycles

Two types of cycles are designed for plungers. The cycles depend on using certain types of plungers and available well producing characteristics.

6.2.1 The continuous plunger cycle

The continuous plunger cycle is shown in Fig. 6.3. It is illustrated using the ball/sleeve plunger and can also be used with a rod actuated plunger to adjust the deal mechanism automatically.

The events shown are:

A: The ball/sleeve makes a seal with the ball at the bottom of the hollow seal and lifts a slug of liquid to the surface.
B: The ball/sleeve arrives at the surface and the sleeve slides over a downward protruding rod. The rod hits the ball and begins to fall downward (if appropriate weight for the flow) and the sleeve is held on the rod with pressure/flow. The slug is produced out of the well.
C: The sleeve is still held on the rod by pressure/flow, and the ball is allowed to fall near or at the bottom of the well. Also during this time the well is allowed to produce a slug of liquid into the bottom of the well if the flow is low enough. Then the well is given only a short shut-in of perhaps 10—20 seconds. This allows the sleeve to drop off of the rod and begin to travel downward.
D: The sleeve is pictured to have fallen through the gas and the accumulated liquid and hits the ball at the bottom. This automatically makes a seal and the liquid, sleeve, and the ball begin to travel upward.
E: The liquid slug, the sleeve, and the ball are shown moving upward and the cycle is repeated.

The sleeve and the ball are designed to fall against the flow of the well or while the well is flowing. When the ball is up against the sleeve, a seal is implemented, allowing the plunger to lift the liquids efficiently to the surface. No shut-in time is

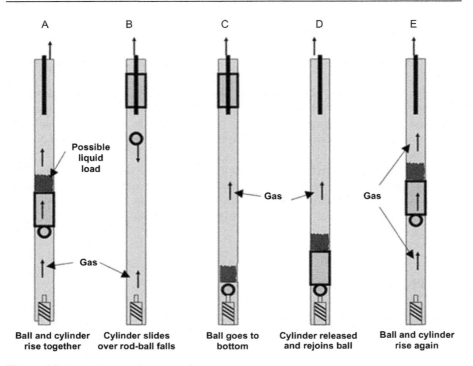

Figure 6.3 A continuous plunger cycle.

needed except the short time to allow the sleeve to fall and clear the rod. This same cycle can be completed using a rod actuated bypass plunger where on fall the seal is not active and on rising, the rod actuates the plunger to have seal.

Fig. 6.4 shows supervisory control and data acquisition (SCADA) data for continuous plunger cycle. This is particular for ball/sleeve plunger in shallow well with higher than usual liquid production.

The first rise of casing pressure (CP) (and other data) indicates the arrival of the liquid slug and plunger. The ball begins to fall and the sleeve remains on the rod at the surface. When the ball falls, liquid accumulates up to the point that a 10-second shut-in is shown on the tubing pressure. This allows the sleeve to fall and then the production resumes after the short shut-in. When the sleeve hits the ball, the well picks up the fluid load and the CP rises again. The effects of slug size are shown but are not accurate, as the pressure indication is dynamic and not static. Following the shut-in data from a conventional cycle can more confidentially predict the slug size from static pressure measurements.

Plungers for the continuous cycle are shown in Fig. 6.5. Continuous plunger cycles or bypass plungers have been able to fall with the seal cabability, either

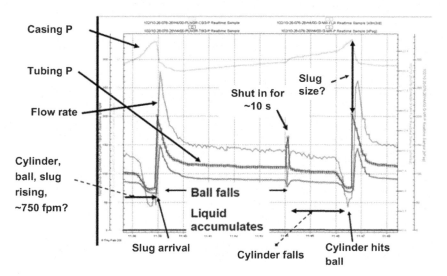

Figure 6.4 Continuous plunger cycle system: SCADA traces (ball and sleeve in this case).

Open/close seal mechanisms

Rod actuated   Ball and sleeve

Figure 6.5 Some continuous plunger cycles.

eliminated or somewhat disabled so that the plunger can fall. Then it automatically must establish the seal again once it falls under the liquid at the bottom in order to lift the liquid. The ball and sleeve do this by being open to fall when the ball is at the bottom of the sleeve and then it makes a seal when the sleeve hits the ball. The rod actuated plunger opens/closes an internal valve that either opens to allow fall or closes to make a seal. There are other versions available for these plungers.

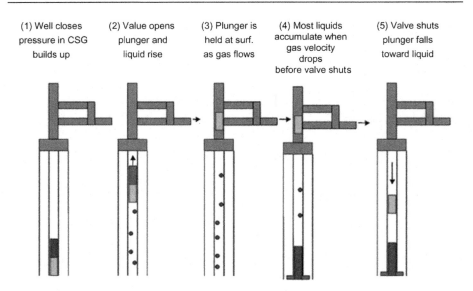

Figure 6.6 Conventional plunger cycle events.

6.2.2 The conventional plunger cycle

The conventional plunger lift operates on a relatively simple cycle, as illustrated in Fig. 6.6. Fig. 6.7 shows in more detail the casing, and tubing and bottom-hole pressures throughout one complete plunger cycle. The numbers on top of Fig. 6.6 labeling the steps of the cycle are also provided on the figure for clarity.

1. The well is closed and pressure in the casing is building. When the pressure is enough to lift the plunger and the liquids to the surface at a reasonable velocity (≈ 750 fpm) against the surface pressure, the surface tubing valve will open.
2. The valve opens and the plunger and liquid slug rise. The gas in the annulus expands into the tubing to provide the lifting pressure. During the rise time the well produces some energy required to lift the plunger and liquid.
3. The liquid reaches the surface and travels down the flowline. The plunger is held at the surface by pressure and flow. The gas is allowed to flow for some time.
4. The flow velocity begins to decrease and liquids begin to accumulate at the bottom of the well. The casing pressure begins to rise, some indicating a larger pressure drop in the tubing. If the flow is allowed to continue too long, a "too large" liquid slug will be accumulated at the bottom of the well, requiring a high casing build-up pressure to lift it.
5. The valve is shut. The plunger falls. The liquids are at the bottom of the well for the most part. The plunger hits the bottom and the cycle repeats.

The cycles continue and may be adjusted according to different schemes that may be programmed into various controllers available. The well must be closed in order for the continuous type plungers to fall as they have good deal (better than

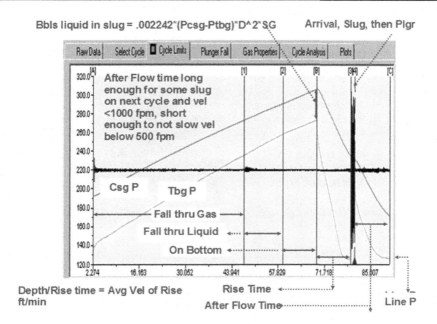

Figure 6.7 SCADA data for a conventional plunger cycle (annotated).

continuous of bypass seals) but also good seal on the way down preventing fall or making fall very slow.

The data in Fig. 6.7 show the casing and tubing pressures during the conventional cycle. From Fig. 6.7 the well is shut in from A to B. Here the CP (top trace, shut-in casing pressure at surface) and TP (shut-in tubing surface presssure) rise during shut-in. Additional data are shown, indicating the plunger falling through gas, liquid, and then sitting at the bottom before the well is opened. The plunger is rising from B to 3. At 3 the liquid arrives and at 4 the plunger arrives. The plunger is in the well head or lubricator from 4-C during a gas after-flow period. The after-flow period must be long enough for some liquid to be accumulated in the well for the next cycle to lift.

The well should be shut in long enough for the plunger to fall through the gas and liquid at the bottom and also long enough for the CP to reach a value that will bring up the liquid/plunger at a reasonable speed. Desired rise times are from 500 to 1000 fpm as determined from industry experience.

Some conventional plungers are shown in Fig. 6.8. The pad plunger is very common and as a seal second only to a new brush plunger (one that is not cut back). However, the pads can stick if sand is present. The brush plunger is flexible and is less prone to stick with sand but it wears out very quickly when sand is present. There are some other sand-resistant plungers that do not wear out as fast as the

Pad plungers:
Common, spring loaded
seal

Turbulent seal plungers:
Poor seal, for stronger
wells

Brush plunger:
Flexible, used for some
sand present

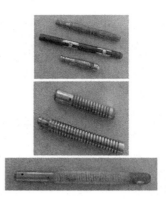

Figure 6.8 Some conventional plungers. The seal is active for both rise and fall.

brush plunger. The turbulent seal (also called barstock or spiral grooved plunger) is a cheap plunger with a seal that is not very good. It can be used when the well is strong and leakage is not as critical. Also it can break loose some scale deposits. Other conventional plungers will be discussed later.

6.2.3 When to use the continuous/conventional plunger cycle

Fig. 6.9 shows that the continuous plunger should be installed early and after the first onset of liquid loading below the critical flow. Since the continuous plunger system does not have an extended shut-in time, the continuous cycle will allow more production early on. This can continue until the continuous plunger does not automatically rise to the surface when the seal is made at bottom-hole. Then shut-in time must be added to the cycle so you are back to types of conventional plunger initially using ones that fall fast to conventional plungers that have very good seals as the well weakens. Different plungers are discussed in the following sections.

However, if the sand is present then the chart shown in Fig. 6.9 may not be able to be implemented and a sand-resistant plunger may be needed over the life of plunger lift.

6.2.4 Additional plunger types

Fig. 6.10 shows some typical plungers that were tested to provide data for developing plunger lift system models.[1] The types shown are typical but do not include all types of plungers available to the industry.

In this figure the plungers are identified from left to right as:

1. Capillary plunger, which has a hole and orifice through it to allow gas to "lighten the liquid slug above the plunger."

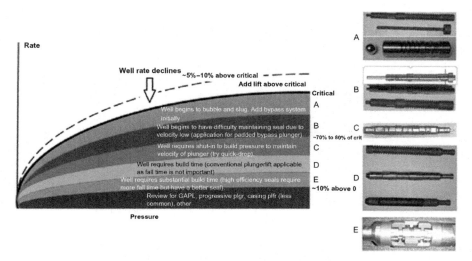

Figure 6.9 Progression of what plungers are best as the well declines below the onset of below-critical flow (or liquid loading threshold).

Figure 6.10 Various types of plungers.

2. Turbulent seal plunger with grooves to promote the "turbulent seal."
3. Brush plunger used especially when some solids or sand is present.
4. Another type of brush plunger.
5. Combination grooved plunger with a section of "wobble washers" to promote sealing.
6. Plunger with a section of turbulent seal grooves and a section of spring-loaded expandable blades. Also, a rod can be seen that will open and close a flow-through path through the plunger depending if it is traveling down or up.

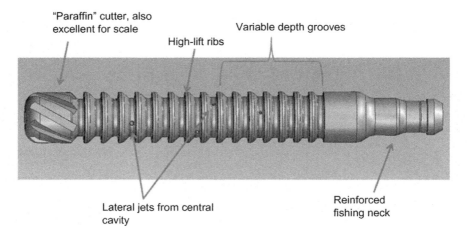

"Paraffin" cutter, also
excellent for scale

Variable depth grooves

High-lift ribs

Lateral jets from central
cavity

Reinforced
fishing neck

Figure 6.11 Metal plunger with swirler, hard metal, hollow with hole to jet solid form
goove from inside out. Designed to replace fast wearing brush plunger if desired when sand
is present.

7. Plunger with two sections of expandable blades with rod to open flow-through plunger
   on down stroke.
8. Mini-plunger with expandable blades.
9. Another with two sections of expandable blades and rod to open flow-through passage
   during plunger fall.
10. Another with expandable blades and a rod to open a flow-through passage during the
    plunger fall and close it during the plunger rise.
11. Wobble washer plunger and rod to open flow-through passage during the plunger fall.
12. Plunger with expandable blades with a rod to open a flow-through passage on the
    plunger fall which could fall against the flow and operate as "continuous" flow.

Several of the above plungers have a push rod to open a flow-through passage
during the plunger fall to increase the fall velocity. When the plunger arrives at the
surface, the push rod forces the flow-through passage open for the next fall cycle.
When the plunger hits at the bottom, the rod is pushed upward to close the flow-
through passage for the next upward cycle.

The brush plunger was found in testing to be the best seal for gas and liquids, but
it typically wears out quickly, compared to others. Though brush plunger is designed
for sand, it wears out very fast. There are other metal plungers that are designed for
sand with a bigger clearance and hollow with vent holes in the grooves. Plungers
with the spring-loaded expandable blades show the second best sealing mechanism
and they do not wear out nearly as fast as the brush plunger (Figs. 6.11 and 6.12).

There are many plunger types available to design for special purposes. One thing
to mention about plungers is that the usually the liquid weighs much more than the
plunger, so a somewhat lighter plunger may show no gain in performance.
However, lighter plungers may create less damage if they hit up/down with no
cushioning liquids present.

**Dynamic turbulent seal
(solid)**

- **Tangential jets create
  rotation**

- **Fluid-film bearing effect**

- **Designed for even wear,
  consistent behavior
  with depth in inclined
  tubing**

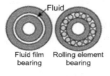

Fluid film   Rolling element
bearing   bearing

Non rotating   Rotating
(no or radial jets)   (tangential jets)

Figure 6.12 The concept of "jetted plunger" which is hollow and provides straight or tangential jets from inside out to help plunger be more centered and create less wear out in a deviated well.

6.3 Plunger lift feasibility

Field testing of various artificial lift methods to determine their applicability can be costly. Although plunger lift is a relatively inexpensive technique (possibly $4000 for a "minimum" installation), additional equipment options can increase the initial costs. Also, downtime for installation, adjustments to see if the plunger installation will perform, and adjustments to optimize production will all add to the costs.

To alleviate these costs, methods have been developed to predict whether plunger lift will work in advance of the installation, under particular well operational conditions. Although these methods vary in rigor and accuracy, they have historically proved useful tools when predicting the feasibility of the plunger lift method.

Several screening procedures can be used to determine if plunger will work for a particular set of well conditions.

6.3.1 Gas/liquid ratio rule of thumb

The simplest of these is a simple rule of thumb that states that the well must have a gas/liquid ratio (GLR) of 400 scf/bbl for every 1000 ft of lift or some value that is fairly close to the 400 approximate value. This is a rule for conventional plunger lift. (This corresponds to approximately 233 m³ gas/m³ liquid for every 1000 m depth.)

Example 6.1

Will plunger lift work for a 5000 ft well producing a GLR of 500 scf/bbl?
    Applying the rule of thumb of 400 scf/bbl for each 1000 ft of lift, the required GLR is:

    GLR, required = 400 scf/ (bbl-1000 ft) × 5 = 2000 scf/bbl

However, the actual producing GLR is 500 scf/bbl, so this well is not a candidate for plunger lift, according to this rule.

Although useful, this approximate method can give false indications when the well conditions are close to that predicted by the rule of thumb. Due to its simplicity the simple rule method neglects several important considerations that can determine plunger lift's applicability. This rule of thumb, for instance, does not consider the reservoir pressure and resultant casing build-up operating pressure that can play a pivotal role in determining the feasibility of plunger lift. Well geometry, specifically whether or not a packer is installed, can also determine if plunger lift is feasible.

An industry rule of thumb for pressure is that a well should build to 1.5 times the line pressure to be a candidate. But this rough rule does not say how long should be allowed for the pressure to build. Probably, it should be around an hour to be on the order of the time required to reach the operating casing pressure.

6.3.2   Feasibility charts

To get around some of the shortcomings of the GLR rule-of-thumb requirement, charts from Beeson et al.[2] have been developed which provide a more accurate means for determining the applicability of plunger lift. These are shown in Figs. 6.13 and 6.14 that examine the feasibility of plunger lift for 2 3/8 in. and 2 7/8 in. tubing, respectively.

With reference to the charts, the horizontal X-axis lists the "net operating pressure." The net operating pressure is the difference between the casing build-up operating pressure and the separator or line pressure to which the well flows when opened.

The casing build-up pressure represents a casing pressure to which the well builds to within a reasonable "operating" period of time. Since this time dictates the time permitted for each plunger cycle, reasonable time suggests a matter of a few hours rather than days or weeks.

Although the line pressure used in the net operating pressure is more straightforward, some special considerations deserve mentioning. The line pressure used to enter the chart must be the flowing well head pressure. Often, if the separator is located a significant distance from the well, and particularly if the two are connected through a small diameter flowline, the line pressure might build when the

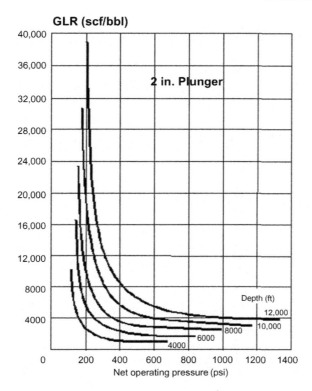

Figure 6.13 Feasibility of plunger lift for 2 3/8 in. tubing.[2]

well is allowed to flow. For example, if the separator pressure is 100 psi, the line pressure might build to 200 psi at the well head when the well comes on as the liquid slug is forced into the small diameter line. Therefore the proper use of Figs. 6.13 and 6.14 requires some judgment on the part of the design engineer. The vertical Y-axis of the charts is simply the required minimum produced gas/liquid ratio in scf/bbl.

Use the figures by entering the X-axis with the net operating pressure. Track vertically upward to the intersection with the well depth. Then track horizontally to the Y-axis and read the minimum produced GLR required to support plunger lift.

If the well producing measured GLR is greater than or equal to that given by the chart, then plunger lift will likely work for the well. If the measured GLR of the well is close to the value given by the charts, the well may or may not be a candidate for plunger lift. Under those conditions, the accuracy of the charts requires that other means be employed to determine the applicability of plunger lift. The following example illustrates the use of the charts as shown in Figs. 6.13 and 6.14.

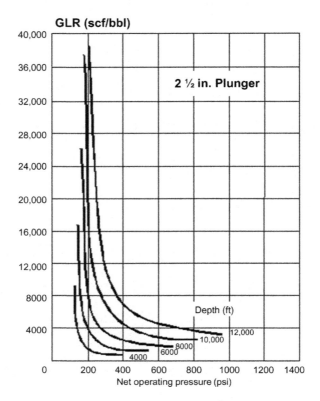

Figure 6.14 Feasibility of plunger lift for 2 7/8 in. tubing.[2]

Example 6.2

A given well is equipped with 2 3/8 in. tubing (a 2 in. plunger approximately).
Is this well a good candidate for plunger lift?

Operational data:

Casing build-up pressure	350 psi
Line or separator pressure	110 psi
Well gas/liquid ratio (GLR)	8500 scf/bbl
Well depth	8000 ft

Use Fig. 6.13 to determine whether plunger lift will work for this well.

$$\text{Net operating pressure} = (\text{casing build-up pressure} - \text{line pressure})$$
$$= 350 - 110 = 240 \text{ psi}$$

> Entering Fig. 6.13 shows that at a depth of 8000 ft, the well is required to produce a GLR (gas−liquid ratio) of about 8000 scf/bbl to maintain plunger lift.
>
> The example well has a measured GLR of 8500 scf/bbl and therefore is a likely plunger lift candidate. Note that pressure, gas rate, and depth are accounted for from this chart.

Comparing Fig. 6.13 with Fig. 6.14 suggests that there is an advantage in using the larger diameter tubing. As the tubing diameter increases, however, the likelihood is that the plunger may lose more of the liquid on the upstroke (as liquid falls back around the plunger) and the dryness of the plungers increases. If the plunger comes up dry, it (a large metal object) will impact the well head with great force, possibly causing damage. Because of this and other reasons, plunger lift is not as common with 3½ in. tubing and especially larger tubing sizes. On the way up the well liquid is lost from above to below and gas is lost from below to above the plunger.

Also note that the casing size is not listed with either Fig. 6.13 or Fig. 6.14. Since the casing volume is used to store the pressured gas used to bring the plunger to the surface, the casing size is important. In general the bigger the casing size, the smaller the required casing build-up pressure to lift the plunger and liquid. From Ref. [2], it is unclear if the figures were developed using 5½ in. or 7 in. casing data, or both.

6.3.3   Maximum liquid production with plunger lift

According to the approximate chart as depicted in Fig. 6.15 a few hundred bpd are possible with plunger at shallow depths. However, the well has to have enough pressure build-up and frequent cycles. As depths increase the possible rates decrease. This is because the rise and fall time increase and during these times no production takes place until the end of travels.

Plunger lift with packer installed

Although some installations have employed plunger lift systems successfully in wells having packers installed, packer-less completions are highly preferred. In the event that the well does have a packer installed, perforation of the tubing above and near the packer, allowing the casing annulus to accommodate gas storage, can drastically improve the efficiency of the plunger lift system. However, packer liquid may have to be drained from the well annulus before going on to the production, perhaps by setting a plug below and bailing the liquid out of the well.

Some wells, however, have sufficient reservoir pressure and gas flow to produce liquids with plunger lift even with a packer. The rough industry rule is that to operate with a packer the well has to have about 1000−2000 scf/(bbl-1000 ft) where only about 400 is needed for operation with no packer in place.

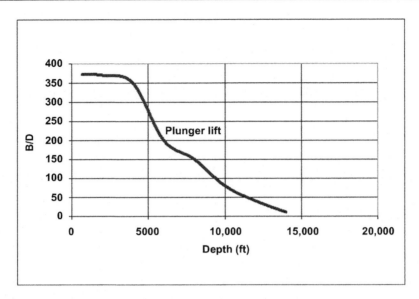

Figure 6.15 Approximate liquid production from plunger with depth.

Thus for a 5000 ft well the well should have at least a 2000 scf/bbl GLR and for operation with a packer the well should produce at least 5000–10,000 scf/bbl GLR.

Plunger lift nodal analysis

Ref. [3] describes how to estimate the average bottom-hole pressure for all portions of the cycle. The average pressure includes the rise, the flow period, the fall period, and the build-up period. This is compared to various sizes of tubing and what pressure is required to flow up the tubings are of various rates. Then the plunger lift performance can be compared to flowing up various sizes of tubings. The results of this type of analysis are shown in Fig. 6.16. For the case in Fig. 6.16, plunger does better than all of the velocity strings. However, in some cases at higher rates, some velocity strings can achieve lower bhp than using plunger. But in general as rates get lower, usually all velocity strings are predicted to load up while plunger is predicted to continue to some lower rate.

6.4 Plunger system line-out procedure

The following section outlines hints and suggestions to incorporate into the procedures used to bring a plunger lift system online. The section covers procedures covering all aspects of plunger lift from the initial start-up, considerations before and during the first kickoff of the plunger, methods to adjust the plunger cycle, and techniques to optimize the plunger cycle to maximize production. The following

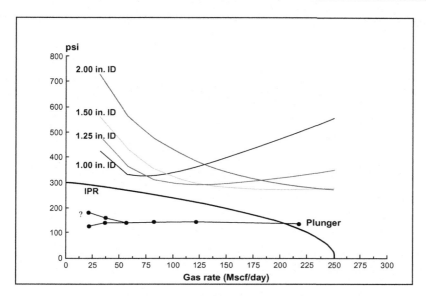

Figure 6.16 Plunger versus velocity string performance with declining rates.[3]

material on system operation and maintenance follows the *Ferguson Beauregard Plunger Operation Handbook*[4] with some updates and alterations. Although most of these functions may be done with computer controlled algorithms, the precautions are listed here so one can compare to computer control.

6.4.1 Considerations before kickoff

Several parameters must be considered before kicking off a plunger lift well.

The most important is the casing pressure. Remember the rough rule of thumb is that casing pressure should build to 1.5 times the line pressure for feasibility. If little CP build-up pressure is available then plunger may not work and energy may have to be added in the form of injected gas or a pump, or gas lift might be necessary.

Another key factor to consider is the liquid load or the amount of liquid accumulated in both the casing and the tubing. Remember the rule of 400 scf/(bbl-1000 ft) and if the well produces a value lower than this there may be too much liquid compared to gas. Before starting the well, if it is shut in for a while, some of the initial liquid may be pushed back into the formation, making it easier to start the well. If there is a very tall column of liquid in the well to start with and some shut-in time does not push much liquid back into the formation, it may be that gas list or pumps will be required to unload the well. Swabbing could be the initial step.

A third major factor to be considered is back-pressure. Lower well head pressure helps plunger lift last long and also makes it easier to initiate.

Load factor

It is extremely important to properly prepare the well before you open it to flow. First, it should be as "clean," or as free of liquid, as possible. This may mean swabbing the well until it is ready to flow or it may mean leaving it shut in for several days to allow the well pressure to build high and to push liquids back into the formation. The effects can be seen by calculating the load factor.

The load factor can be used to see if the well is ready to be opened. This may be automated or could be used in manual operation. The definition is as follows:

$$\text{Load factor} = 100 \times \frac{\text{shut-in casing pressure} - \text{shut-in tubing pressure}}{\text{shut-in casing pressure} - \text{line pressure}}\%$$

A good rule of thumb is to ensure that the load factor does not exceed 40%−50% before opening the well to let the plunger and liquids rise.

Example 6.3

A given well has been shut in until the following conditions prevail. Determine whether the conditions are sufficient to start the plunger cycle.

Casing pressure:	600 psi
Tubing pressure:	500 psi
Sales line pressure:	100 psi

$$\text{Load factor} = 100 \times \frac{600 - 500}{600 - 100}\% = 20\%$$

Since the load factor is less than the maximum limit of 40%−50%, the plunger and liquid slug are predicted to rise when the well is opened. Continue to shut in the well and the LF will decrease.

Conditions are predicted to be acceptable to start the plunger cycle.

In many cases, it is desirable to vent the gas above the liquid in the tubing on the initial cycle to a lower pressure. This creates more differential pressure across the slug and plunger, pushing the slug to surface. Regardless, every effort should be made to remove as many restrictions in the flowline as possible. If a flowline choke is required, the largest possible choke for the system should be used. It is also a good practice to put large trims in the dump valves of the separator. A slug traveling at 1000 ft/min corresponds to a producing rate of 5760 bpd in 2 3/8 in. tubing. Frequently a larger orifice plate in the sales meter is used to measure the peak flow of the head gas.

Kickoff

Once adequate casing and tubing pressures have been reached, the well is ready to bring the plunger to the surface. The casing and tubing pressures required to kick off the well are obtained from the methods outlined previously.

The motor valve should be opened quickly so that the tubing pressure is bled off quickly. If done, this quickly establishes the maximum pressure difference across the plunger and the liquid slug to move them to the surface.

Record the time required for the plunger to reach the surface. The current thinking is that the plunger should travel between 500 and 1000 ft/min for optimum efficiency with a midpoint of ~750 fpm being best unless plunger specific rules are developed. Experience has shown that plunger speeds in excess of 1000 ft/min tend to excessively wear the equipment and waste energy and plunger speeds lower than 500 ft/min will allow gas to slip past the plunger and liquid slug lowering the system efficiency. The plunger travel speed is controlled by the casing build-up pressure and the size of the liquid slug that is produced with the plunger. Note that a plunger could be run slower if it had a very good sealing mechanism.

With the plunger at the surface initially, close the motor valve and allow the plunger to fall. This is called a "swab" cycle where the well is shut in once the liquid is produced. Gas begins to pressurize the casing and tubing for the next cycle. The plunger must also be allowed to reach the bumper spring. New data from Echometer can help indicate when the plunger hits at the bottom or you use the Echometer system or a plunger like the PCS (Denver) smart plunger to measure when the plunger hits at the bottom. Once the casing pressure has regained its initial value, the cycle can be set for automatic operation if a few of these manual cycles are used to start the plunger operation.

Some controllers will do the starting procedure without manual intervention.

Cycle adjustment

Liquid loading can occur not only in the tubing but also in the reservoir immediately surrounding the wellbore. Liquid accumulation in the reservoir near the wellbore can reduce the reservoir's permeability. To partially compensate for this, it is recommended to run the plunger on a "conservative" cycle for the first several days. A conservative cycle implies that only small liquid slugs are allowed to accumulate in the wellbore and that the cycle is operated with high casing operating pressures. Also that the after-flow is set to zero of small time after liquid production.

In summary, the kickoff procedure is outlined as follows:

- Check (and record) both the casing and tubing pressures. Apply the rule of thumb demonstrated in Example 6.3.
- Open the well and allow the head gas to bleed off quickly. Record the time required for the plunger to surface (plunger travel time).
- Once the plunger surfaces and production turns gassy shut the well in and let the plunger fall back to the bottom.
- Leave the well shut in until the casing pressure recovers to the pressure it had on the previous cycle. Better is to return to the casing pressure in excess of line pressure.
- Open the well and bring the plunger back to surface and again record the plunger travel time. Shut the well in.

- If this cycle has been operated manually, then set the timer and sensors to the recorded travel time and pressures.
- If you have no casing pressure sensor or magnetic shut-off switch, then it is necessary to use time alone for the cycle control. Allow enough time for adequate casing build-up and enough flow time to get the plunger to the surface. A two-pen pressure recorder can be a valuable asset under these conditions. By monitoring the charts, you can quickly compare the recovery time of the casing and adjust the cycle accordingly.
- Whichever approach you use, once you see the cycle is operating consistently, leave it alone and allow the well to clean up for 1 or 2 days until the liquids in the reservoir wellbore area have been somewhat cleared.

Although many new controllers will take care of the earlier steps, the steps are listed to show what physically should be considered to start a plunger installation, and also for when newer controllers are not being employed.

Stabilization period

Once "swab" cycles are repeated, then some after-flow can be gradually added or the operation can be turned over to hopefully a somewhat sophisticated controller.

If the plunger ran smoothly during the initial installation, then it is unlikely that tubing is either crimped or mashed or has scale buildup. If damaged tubing is suspected then a wireline gauge ring should be run in the tubing having an OD corresponding to the tubing's manufactured drift diameter. It is also a good practice to run a gauge ring with a gauge length equal to at least the length of the plunger. Care should be exercised while running the gauge ring, however, to prevent the ring from becoming stuck in the event that there is foreign debris in the tubing.

6.5 Optimization

You can optimize using arrival velocity. If the arrival velocity is too low (well below 750 ft/min or below a 500 ft/min threshold) then you can reduce the flow time on the next cycle and/or increase the shut-in time.

If the velocity is too high then you can either decrease the shut-in time or increase the flow time. However, caution should be exercised on reducing the shut-in time as there must be time allotted for the plunger to fall to the bottom bumper spring.

However, only looking at arrival velocities can cause problems. A high CP and a large slug of liquid might travel the same velocity upward as that of a low CP and a small liquid slug. In general, it will be shown that more production occurs with more cycles. This is saying you should trend to lower liquid slugs but still you must lift a big enough slug each cycle in order to meet your desired liquid production. A common production decreasing control is to shut in much longer than necessary.

6.5.1 Oil well optimization

To fully optimize the flow time for an *oil well*, it is necessary to install a magnetic shut-off switch in the lubricator at the surface to shut the well in upon plunger arrival. Any reliable arrival transducer would serve the purpose. The switch activates the motor valve shutting the well in immediately upon plunger arrival, which saves the needed tail gas for the next cycle. The plunger then starts its return to bottom with only a small hesitation at the surface, shortening the cycle time and increasing liquid production.

This prevents the well from depleting the vital gas supply stored in the casing. Depleting this stored gas would require longer shut-in periods to rebuild pressure and in most case would lower the overall liquid production.

In the event that the casing pressure remains too high after plunger arrival, rather than allowing the well to flow gas after the plunger has surfaced, the recommended practice is to lower the casing operating pressure. Lowering the operating pressure generally prompts an increase in production since the pressure against the formation is reduced. This is the type of cycle described in the Foss and Gaul[5] paper. The authors have witnessed oil wells on plunger making as much as 300 bpd from about 4000 ft.

6.5.2 Gas well optimization

Optimizing the flow time for a *gas well* requires more effort, if done manually, since the time that the gas is allowed to flow after plunger arrival is considerably longer than that of an oil well.

An older method of optimization is as follows: The flow time for a gas well is optimized by continually adding small increments to the amount of time allotted to gas flow while recording the plunger travel time. These small increments should be added over the period of several days to allow the well to regain stability after each change. As the flow time is increased the plunger travel time will decrease. Once the plunger travel time drops to approximately 750 ft/min or at least >500 fpm, the flow time used to be considered optimized. However, now the velocities mentioned are achieved but attention is given to the average pressure on the formation during the cycle and this is minimized by trending to smaller liquid slugs per cycle but still large enough to sum to the desired daily liquid production.

6.5.3 Optimizing cycle time

The above mentioned methods to examine rise velocity only did work to establish cycles but do not optimize production. For instance, a large slug can be brought in to the well during the flow period and then a large casing build-up pressure will allow the plunger and liquid to be lifted to the surface at 750 fpm. This would exert a high average casing pressure and pressure on the formation. As such the production would be reduced.

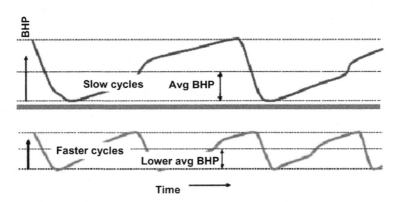

Figure 6.17 Faster cycles with a smaller liquid slug of liquid result in a lower average casing and flowing bottom-hole well pressure. If the well is producing a small slug size then it may not be advisable to reduce the flow time. If the fall time is known or measured, then one must provide an off-time for the plunger to fall through the gas and the liquids at the bottom of the well.

It would be better if a small slug of liquid is accumulated in the tubing during a brief flow period, and then only a small casing build-up pressure would be required to lift the slug, at an average rise velocity of about 750 fpm or at least >500 fpm. This would result in a smaller average pressure on the formation and the production would be higher (Fig. 6.17).

6.6   Monitoring and troubleshooting

There are several common methods of monitoring.

6.6.1   Decline curve

One of the most common is to see whether or not well is on the decline curve or the target. Fig. 6.18 is a right example on the decline curve where production dropped off and plunger was adjusted/repaired and production returned to the target or goal for this well.

6.6.2   Supervisory control and data acquisition data

Another common technique is to observe the SCADA traces if a system is present. Fig. 6.19 is a SCADA trace with CP, TP, Line P, and flow rate plotted.

The CP and TP are the top two traces. The well is shut in when the flow is flat and at zero. The line under the CP and TP is the line pressure.

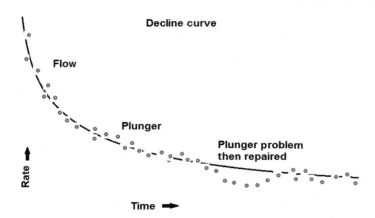

Figures 6.18 A decline curve extended into plunger operations.

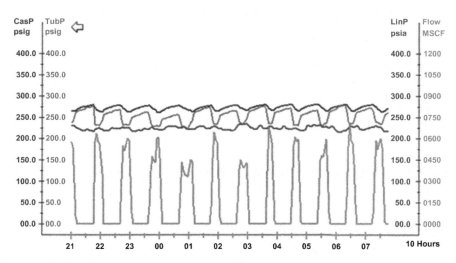

Figure 6.19 SCADA plot for plunger lift well. Top to bottom: casing pressure, tubing pressure, line pressure, production.

Indications:

- The bigger the vertical spread between CP and TP, the more liquid is at the bottom ready to be lifted.
- If CP and TP come together during off cycle, then there is either large tubing hole or no liquid downhole.
- If CP and TP gradually come together during shut-in, then there is either tubing leak or liquid falling out at the end of the tubing (no standing valve).
- If there are flow indications during shut-in, the valve needs new trim.
- If the line pressure spikes when the well is on, then there is not enough capacity in the flowline.

- If TP falls during the off cycle, then there is leak near well head or in flowline.
- If the plots are erratic, then this is an indication that the plunger is worn.
- If TP and CP always come together, then there could be parted tubing.
- There could be other rules as well.

6.6.3   Some common monitoring rules

If you understand the following rules, then you know a lot about plunger lift operation.

If the plunger ascends too fast, it is assumed that:
    Casing pressure is too high
    Slug size is too small
    Line pressure has fallen
The controller may respond by:
    Increasing flow time
    Decreasing shut-in time
If the plunger ascends too slowly, it is assumed that:
    Casing pressure is too low
    Slug size is too large
    Line pressure has increased
    Plunger seal is becoming less efficient
The controller may respond by:
    Decreasing flow time
    Increasing shut-in time

6.6.4   Tracking plunger fall and rise velocities in well

Plunger fall velocity

Plunger fall velocity is important and needs to be known for good operation of a plunger. If the well is opened before the plunger gets to the bottom then it will reverse directions and return to the surface dry. This upsets the plunger cycles and coming up dry and fast can cause damage. On the other hand, if the well is kept shut in much longer than it takes for the plunger to get to the bottom of the well (falling through gas and liquid) then production can be reduced.

1. Data is used to correlate construction features of plungers to fall velocity.
2. Some features cause a plunger to fall rapidly, while other features cause a plunger to have a slower fall velocity.
3. Well conditions (gas flow rate and pressure) have significant impact on plunger fall velocity.
4. Use plunger fall velocities to determine shut-in time.
   a. Use of rule of thumb for fall velocity is not accurate.
   b. Impacted by many parameters.
5. Setting controller to the shortest shut-in time will maximize oil and gas production.

Plungers to use as well declines shown on the critical plot:

Initially, plungers that fall faster are needed as the well can build pressure quickly and be ready for the plunger to rise. Later as the well declines slow falling plungers are sufficient as the well needs more time to build pressure and a good seal is needed to conserve gas. This was shown in Fig. 6.9.

Therefore the question becomes how to measure the fall velocity of the plungers and also a method for predicting fall velocity will be presented.

Methods to determine plunger fall velocity

Echometer has researched various methods to track falling plungers.
Active methods:

1. Manually shoot down tubing.
2. Automatically shoot down tubing.
      Passive methods:
3. Determine round trip travel time (RTTT) of acoustic pulse created by plunger.
4. Determine elapsed time from beginning to end of fall.
5. *Count each collar using a high-speed acoustic signal.*

Currently the last method listed here is the standard to measure plunger falling in gas and gassy liquid. This passive method requires a sensitive microphone at the well surface and as the plunger falls past a collar recess, a pressure pulse (about 0.1 psi) can be measured at the surface. Then by counting joints versus time the position and velocity of a falling plunger can be measured and recorded.

Since method (5) is the most used at present it is described in more detail.
Passive method #5:
Count each collar in high-speed acoustic signal.

1. Acoustic liquid level instrument used records the acoustic signal as plunger falls during shut-in.
2. Manually count each acoustic pulse created by the plunger passing through a tubing collar recess.
3. Depth to the collar reflection can be found by multiplying the average tubing joint length with the count of each tubing collar recess's acoustic pulse.
4. This method is best suited for computer processing of the acoustic data (Fig. 6.20).

Instrumentation required for earlier measurements and data collection includes:

1. Three-channel high-frequency (30 Hz or greater) data acquisition,
2. tubing
   a. pressure and
   b. acoustic signal.
3. Casing pressure.

Count signals from plunger in collar OR RTTT: acoustic signal during shut-in (1 minute) (Figs. 6.21−6.23).

As the signals are weak, a high-resolution measurement of data is required (Fig. 6.24).

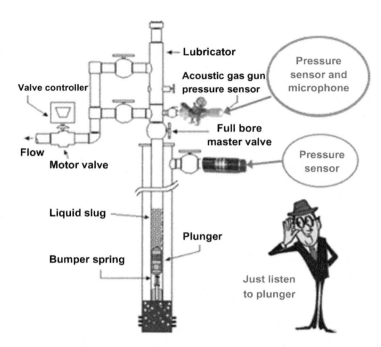

Figure 6.20 Sensitive recorder at surface collects collar recess data as plunger falls and transmits the data to a computer for storage and analysis (Echometer).

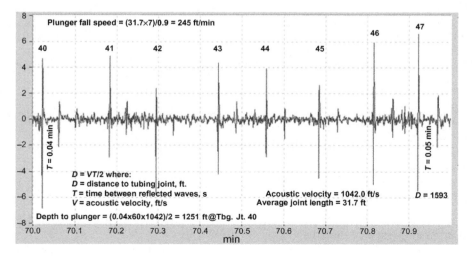

Figure 6.21 Count individual collar reflections to track falling plunger.

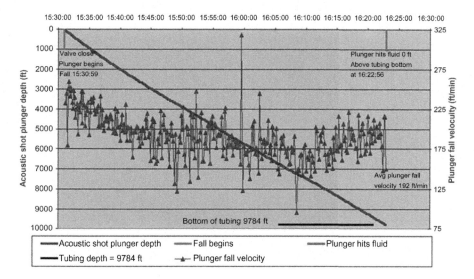

Figure 6.22 Count collars of 2.875 in. dual pad (192 ft/m average velocity).

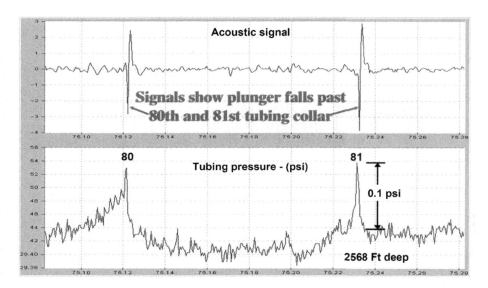

Figure 6.23 Passive monitoring requires high-resolution pressure and acoustic data.

The earlier occurs when the weight of the plunger is transferred to the gas (Figs. 6.25−6.27).

Count collars to get plunger position and calculate the velocity (distance/$(T_{i-1} - T_i)$).

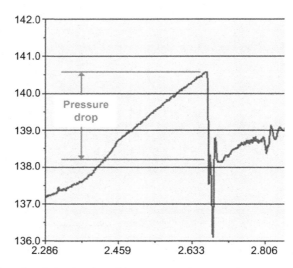

Figure 6.24 When shut-in begins the tubing pressure drops as plunger starts to fall. (Data for above: pressure drop = plunger weight (8 lb)/area of 2 3/8 in.).

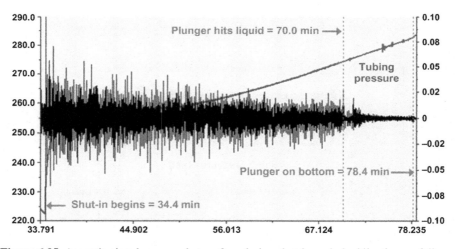

Figure 6.25 Acoustic signal measured at surface during shut-in period while plunger falls.

Some measured fall velocity data through gas at about 230 psi WHP (Tables 6.1 and 6.2).

What do we know?

1. Measured plunger fall velocities for grooved, ultra-seal, dual pad, and brush type are much less than 1000 ft/min.
2. Two-piece and bypass plungers are fast! (> 1000 ft/min).

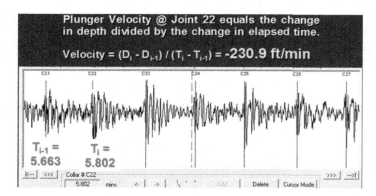

Figure 6.26 Plunger velocity calculated between to consecutive counted collars.

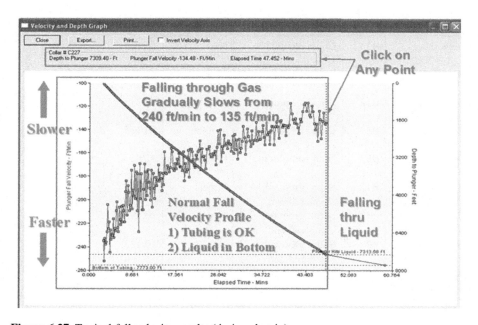

Figure 6.27 Typical fall velocity results (during shut-in).

3. Worn 2 3/8 in. brush type plungers (408−477 ft/min). New brush plungers fall slowly. Fall velocity changes w/wear.
4. 2 3/8 in. dual pad type plungers (259−265 ft/min).
5. Increasing the diameter from 2.375 to 2.875 in. resulted in the pad type plunger falling slower (> 200 ft/min).
6. Improving the seal on a dual pad plunger (ultra-seal) results in even slower fall velocities (159 ft/min).
7. Solid plungers are "fast," that is, 300−400 ft/min.

Table 6.1 Comparison of plungers fall velocities measure in 2 3/8 in. tubing

Field test	Acoustic velocity (ft/s)	Plunger type	Tubing depth (ft)	Height of liquid(ft)	Average fall velocity (ft/min)
1	1424	Brush	7400	260	477
2	1269	Pad 2 3/8 in.	4008	608	265
3	1216	Pad 2 3/8 in.	5800	232	259
4	1280	Ultra-seal	7485	41	159
5	1242	Grooved	3042	13	408
6	1320	Clean out	9896	1400	326

Table 6.2 Comparison of plunger fall velocity 2 7/8 in. tubing in similar wells

Date	Well	Description	Condition	Tubing length (ft)	Fall speed (ft/min)
September 11, 2001	B3	Brush plunger	New	10,123	150
September 11, 2001	B3	Dual pad plunger	New	10,123	162
September 13, 2001	B12	Dual pad plunger	Existing	10,235	179
September 11, 2001	B3	Dual pad plunger	4-month old	10,123	187
June 05, 2001	B9	Dual pad plunger	Existing	9,784	200
September 13, 2001	B12	Solid, grooved, tapered end	Repeat	10,235	364
September 13, 2001	B12	Solid, grooved, tapered end	New	10,235	368
September 11, 2001	B3	Solid, grooved, tapered end	New	10,123	423
September 13, 2001	B12	Bypass valve, dual pad	New	10,235	1690

8. In the same well new plungers fall slower when compared to the same type of older/worn out plunger.

Effects of inclination on plunger fall:

Williams tracked plungers in their S-shaped wells. Their configurations are shown in Fig. 6.28.

Williams' published results shed some light on how plunger construction affects fall in an inclined well (Fig. 6.29).

When angle is over about 20 degrees, the plunger slowed down (as expected). Also other styles of plungers (except the pad plunger) exhibit the same trends (Fig. 6.30).

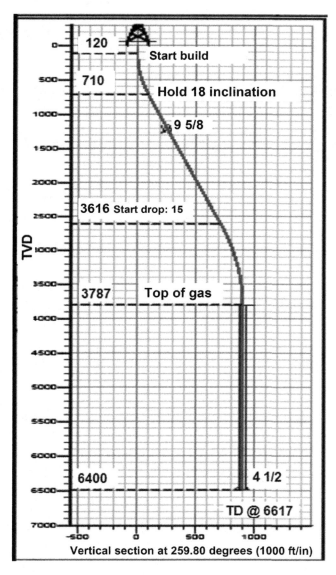

Figure 6.28 Typical Williams' S-shaped well.

When angle exceeded about 20 degrees, the pad plunger fall velocity increased. The concept is that when the plunger is tilted over, it compresses the pad seal and the plunger is free to fall faster as it looses its seal. The inference is that if a pad plunger in an inclined section looses its seal falling, then it would also loose seal on the way up and part of the liquid load could be lost.

Therefore one should use pad plungers in a deviated well with caution.

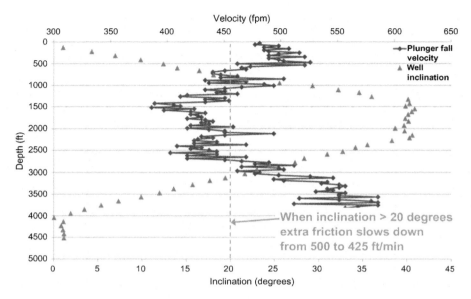

Figure 6.29 Brush plunger falling in S-shaped well.

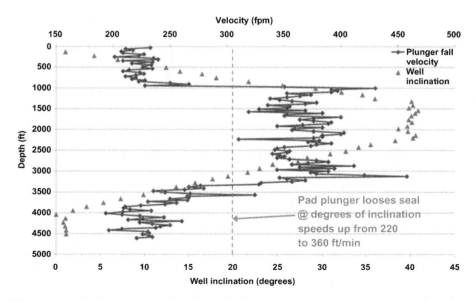

Figure 6.30 Pad plunger falling in S-shaped well.

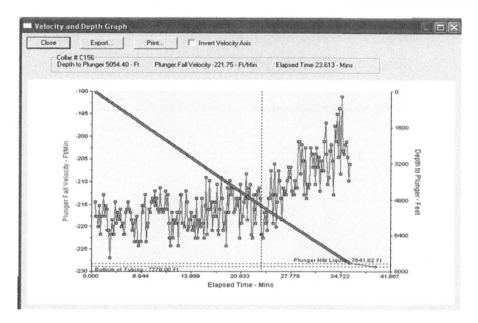

Figure 6.31 Results show plunger slows as it passes through the hole.

Troubleshooting using fall velocity (Fig. 6.31):

When there is a hole above the tubing, the gas can escape through the hole but once the plunger passes through the hole, the gas underneath is trapped above the liquid and slows down.

Effects of falling through gas versus gassy fluid (Fig. 6.32):

A good plunger seal will give slower fall velocity through both gas and liquid and a poor seal will give faster fall velocity through both gas and liquid (as explained in Echometer analysis).

The plunger fall velocity is affected by the following factors:

1. Diameter of plunger—larger diameter falls slower.
2. Effectiveness of seal between plunger and tubing—better seal plunger falls slower.
3. Brush stiffness—if the bristles do not provide an effective seal then the plunger falls faster.
4. Increased friction due to contact with the tubing—plunger falls slower.
5. Old age/increased wear—as the plunger wears out the worn plunger falls faster.
6. If gas can pass through plunger (i.e., bypass)—then a plunger falls faster.
7. When the plunger becomes stuck and stops—usually indicated by a 3 psi increase in pressure.
8. If the tubing is sticky—the plunger falls slower wellbore deviation—more than 20 degrees of deviation impacts plunger fall velocity
   a. Padded plungers faster due to loss of seal.
   b. Solid plungers slower due to increased friction.
9. Gas flow rate into the tubing—gas flow into tubing reduces plunger fall velocity.

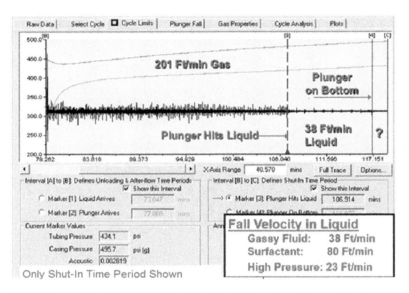

Figure 6.32 Plunger speed drops considerable when falling through gassy fluid versus falling through gas.

10. Pressure or density of gas
    a. High pressure when plunger fall is slow.
    b. Low pressure when plunger fall is fast.
11. Liquids increase density—plunger falls slow
    a. Surfactant lightens gradient and plunger falls faster, but more time may be required.
    b. High pressure also causes plunger to fall more slowly through liquid.

New plunger fall velocity model:

- The model predicts fall velocity in well at any pressure conditions and temperature.
- All plungers generally fall fast at low pressure and slow at high pressure.
- Use known fall velocity at a specific pressure and temperature to calculate fall velocity at other pressures and temperatures.
- Published fall velocities can be used for each plunger type, but may not be accurate for other conditions. The model overcomes this shortcoming.
- Data from many wells will be used to compare the measured fall velocity of different types of plungers to the predicted fall velocity (Fig. 6.33).

Possible plunger fall velocity models and equations for two possible fall models are shown in Figs. 6.34–6.37.

Results seem accurate for model compared to data at other well pressures (Fig. 6.38).

More results from model using a low lab pressure fall velocity to project fall velocities for wells at much higher pressures. The data group compares very well to the model (Fig. 6.39).

Elapsed time (min)
0 25 50 75 100 125 150
0 -100 -200 -300 -400 -500 -600
Fall velocity (ft/min)
240 200 160 120 80 40 0
Tubing pressure (psig)
Fall velocity for each cycle
-255.8 ft./min 163.0 Tbg-psig
-319.8 ft./min 130.7 Tbg-psig
-355.9 ft./min 102.0 Tbg-psig
-394.8 ft./min 73.3 Tbg-psig
-478.1 ft./min 41.6 Tbg-psig
Tubing pressure-psig
Dual pad plunger fall velocity faster 1.75 ft/min with each 1 psi decrease in pressure

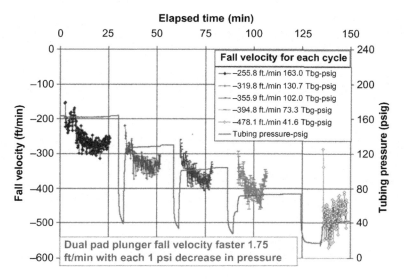

Figure 6.33 Results show that plungers fall slower through gas when the well pressure is higher.

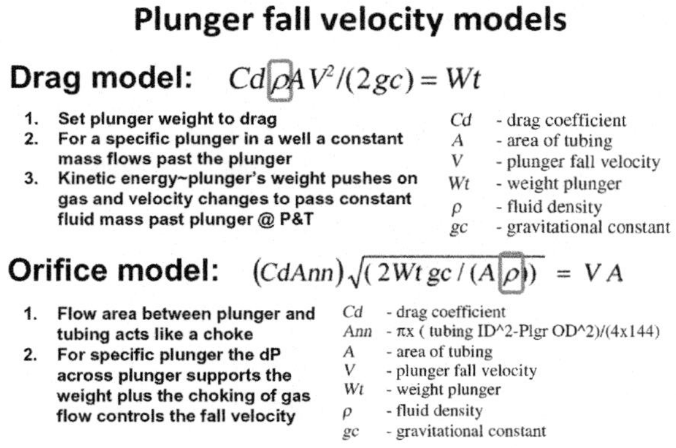

Plunger fall velocity models
Drag model: $Cd\,\rho A V^2/(2gc) = Wt$
1. Set plunger weight to drag
2. For a specific plunger in a well a constant mass flows past the plunger
3. Kinetic energy~plunger's weight pushes on gas and velocity changes to pass constant fluid mass past plunger @ P&T
Cd - drag coefficient
A - area of tubing
V - plunger fall velocity
Wt - weight plunger
ρ - fluid density
gc - gravitational constant
Orifice model: $(CdAnn)\sqrt{(2Wt\,gc/(A\rho))} = VA$
1. Flow area between plunger and tubing acts like a choke
2. For specific plunger the dP across plunger supports the weight plus the choking of gas flow controls the fall velocity
Cd - drag coefficient
Ann - πx (tubing ID^2-Plgr OD^2)/(4x144)
A - area of tubing
V - plunger fall velocity
Wt - weight plunger
ρ - fluid density
gc - gravitational constant

Figure 6.34 Equations that can be used to predict fall velocity.

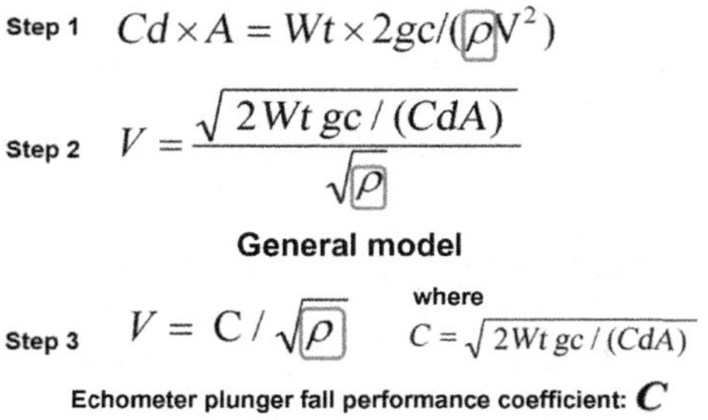

Step 1 $Cd \times A = Wt \times 2gc/(\rho V^2)$
Step 2 $V = \frac{\sqrt{2Wt\,gc/(CdA)}}{\sqrt{\rho}}$
General model
Step 3 $V = C/\sqrt{\rho}$
where $C = \sqrt{2Wt\,gc/(CdA)}$
Echometer plunger fall performance coefficient: C

Figure 6.35 Both models show an inverse fall velocity relationship to the square root of gas density.

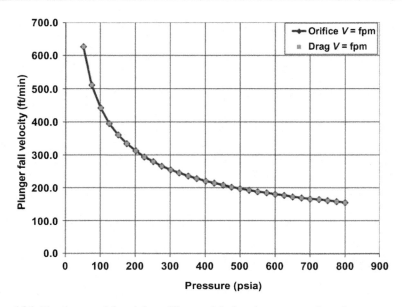

Figure 6.36 The drag model and the orifice model give the same results using pressure (density) as a parameter for comparison.

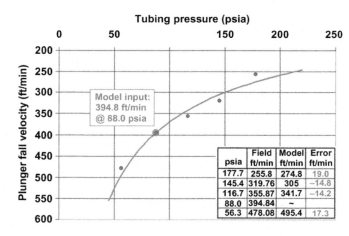

Figure 6.37 For one plunger use measured fall velocity and well conditions to predict the C value and then one can predict the fall velocity at other well conditions (pressures) as earlier.

Optimize plunger performance in well:

- Published fall velocity for a plunger type is NOT accurate, because fall velocity is significantly impacted by gas density (at higher pressure the density of the gas is more and the fall velocity is less).

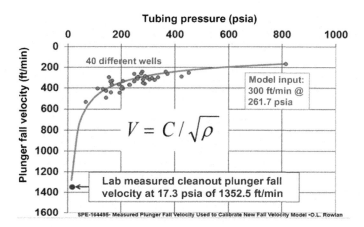

Figure 6.38 New general plunger fall model predicts plunger's fall velocity at density for pressure and temperature (example shown earlier).

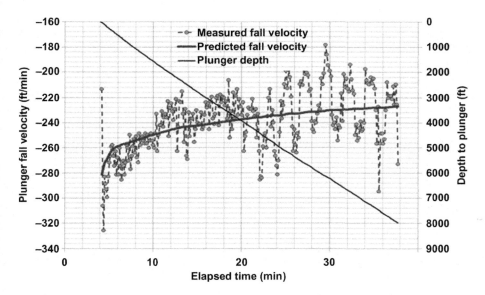

Figure 6.39 Plunger fall can be erratic as the SV opens/closes but using this model and the C factor from model can give average conditions.

- From measured plunger fall velocity use model to predict plunger fall velocity at other P&T.
- Each plunger type has its own performance coefficient.
- Changing the plunger cycle impacts operating pressure, and the model calculates new velocity.
- When changes to the well cycle impact the operating pressure, use model to determine the change in time required for the plunger to fall to the bottom during shut-in.

- Predicting accurate fall velocity will ensure that the plunger will reach the bottom by the end of the shut-in period.

Conclusion:

- All plungers have faster fall velocity at low pressure versus slower fall velocity at high pressure (gas density).
- Use C *(determined from the fall velocity equation(s) and with the use of measure data at one P&T)* to predict plunger behavior at other conditions.
- Gas density and C are used to predict plunger fall velocity through gas at depth in the tubing.
- But, plunger velocity relative to the tubing slows down as the upward velocity of the gas increases—due to increase of gas flow rate into the tubing.
- Plunger fall velocity is directly impacted by gas density when gas flow is laminar and gas flow rates are slow.
- The concepts presented are general plunger fall model, theory of relative to gas flow, and Echometer plunger fall performance coefficient C.

Plunger rise velocity in well

The industry has adapted rules that if the rise velocity is between 500 and 1000 fpm, then better performance can be expected. If the rise velocity is too high, damage can occur. If too low, the plunger has more time to leak gas up past the plunger and may not even surface.

How to control plunger rise velocity:

Typical ascent velocities should be between 500 and 1000 fpm. If the rise velocity is in this range, it has been found that there is a good balance between the casing buildup pressure and the amount of liquid to be lifted.

- Plunger rises too fast, because:
    - Casing pressure is too high.
    - Slug size is too small.
    - Line pressure has fallen.
- Decrease rise speed by:
    - Increasing flow time (lower FBHP and increase slug size).
    - Decreasing shut-in time (less energy/gas in system).
- Plunger rises too slow, because:
    - Casing pressure is too low.
    - Slug size is too large.
    - Line pressure has increased.
    - Plunger seal is becoming less efficient.
- Increase rise speed by:
    - Decreasing flow time (raise FBHP and decrease slug size).
    - Increasing shut-in time (more energy/gas in system).

Earlier points are the main rules that the industry would follow to adjust the plunger rise velocity (Fig. 6.40).

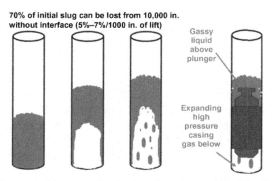

Figure 6.40 Plunger acts to prevent liquid fall-back. Velocity affects leakage, damage, and efficiency of lifting liquids.

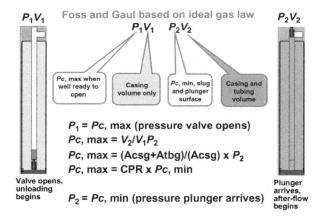

Figure 6.41 Illustration of basic concept of Foss and Gaul model (earlier).

Use of Foss and Gaul model allows the prediction of rise velocity if liquid load, well conditions, initial tubing, and casing pressures, etc. are known:

1. Foss DL, Gaul RB. Plunger-lift performance criteria with operating experience—ventura field. Drilling and Production Practice, API (1965), 124−140.
2. Modified Foss and Gaul model used in this study.
3. Show impact of low-line pressure when producing gas wells.
4. Prediction accounts for liquid load, frictional effects, tubular sizes and lengths, surface line pressure, and fluid properties (Fig. 6.41).

The original Foss and Gaul model used only the expansion of gas from the casing to the tubing to lift the plunger and liquids. A modified model later can include the produced gas and effects of gas slipping past the plunger while rising.

Modified Foss and Gaul equations:

$$P_{c,\min} = (14.7 + P_P + P_{wh} + P_c \times \text{slug})(1 + \text{depth}/K)$$

where $P_{c,\min}$ is the casing pressure as the slug and plunger just reach the surface. The casing pressure will drop as the well flows; P_p is the pressure to lift the plunger; P_{wh} is the well head pressure when well opened; P_c is the psi/bbl to lift weight of slug and overcome friction of liquid slug moving upward; K accounts for the gas friction in the tubing below the plunger; slug is slug size in barrels.

For the following tubing sizes, P_c and K have the following approximate values according to the original Foss and Gaul model:

Tubing	K	P_c
2 3/8 in.	33,500	165
2 7/8 in.	45,000	102
3	57,600	63

A_{ann} is the annulus cross-sectional area between casing and tubing; A_{tbg} is the tubing inside the cross-sectional area.

$$\text{CPR} = \frac{A_{\text{ann}} + A_{\text{tbg}}}{A_{\text{ann}}}$$

$$P_{c,\max} = \text{CPR} \times P_{c,\min}$$

This approach assumes conservatively that all energy comes from expansion of the gas from the casing to the tubing as the plunger comes up. It is modified here to account for the gas that is produced as the plunger is coming up to the surface.

K (gas friction) can be calculated using the following formula:

$$K = \frac{(T_{\text{avg}} + 460)(Z)(Tbg_{\text{OD}}/12)(2 \times 32.2 \times 144 \times 3600)}{(144/53.3 \times Sg \times f_{\text{gas}} \times V^2)}$$

P_c (slug, psi/bbls) can be calculated from:

$$P_{\text{weight}} = \text{Slug}_{\text{length}} \times 0.433 \times \text{SG}_{\text{liquid}}$$

$$P_{\text{friction}} = \frac{62.4 \times \text{SG}_{\text{liquid}} \times f_{\text{liquid}} \times \text{Slug}_{\text{length}} \times V^2}{\left\{ (Tbg_{\text{id}}/12) \times 2 \times 32.2 \times 144 \times 3600 \right\}}$$

$$P_c = P_{\text{weight}} + P_{\text{friction}}$$

To account for production and also for leakage of gas up and past the plunger, the following equation can be used:

$$P_{c,\max} = P_{c,\min}\text{CPR} - \frac{14.7(T_{R,\text{avg}})(z)(1000)(\text{Mscd}_p - \text{Mscfd}_l)}{(520)(24)(60)(V)}$$

where $\text{Mscfd}_p - \text{Mscfd}_l$ is the difference of the flow rate of formation produced gas minus the flow rate of gas leaking past the plunger during plunger rise. The original Foss and Gaul model did not account for the production of gas from the formation. The gas to lift the plunger was modeled to come entirely from the expansion of the gas from the casing into the tubing.

Following are the graphical results (using the modified Foss and Gaul model):

- Here bpd is the maximum rate for conditions.
- Casing pressure is the operating pressure at the end of shut-in time period.
- Tubing pressure is the corresponding tubing pressure just before opening the well.
- Liquid slug lifted each time is related to (casing − tubing) pressure.
  - 1 bbl in 2.375 tubing is 108 psi or 250 ft.
  - 1 bbl in 2.875 tubing is 74 psi or 171 ft.
- Plunger fall velocity is assumed to be 250 ft/min in gas and 40 ft/min in gaseous liquid (Fig. 6.42).

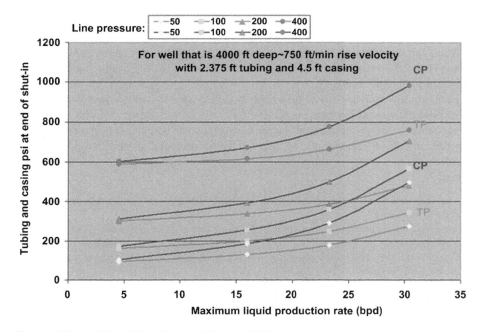

Figure 6.42 1 bbl in 2.375 tubing is 108 psi or 250 ft.

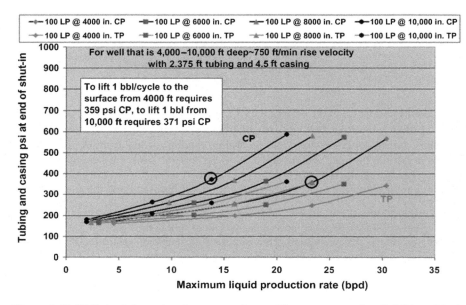

Figure 6.43 Well depth has minor impact on plunger lift system operation (2 3/8 in. tubing).

Well depth does not affect the casing pressure required to lift the liquid substantially but it does affect the production. Deeper wells make less liquid because more of cycle time is used to travel and less trips are possible (see Figs. 6.43–6.46).

Since larger tubing allows the same amount of liquid to not stand as high in the tubing (less hydrostatic) the casing pressure required to lift the liquid can be less as shown in Figs. 6.47 and 6.48.

Plungers that fall fast spend less time traveling and this allows more cycles and more production possible (see Fig. 6.49).

Since more energy is stored in larger casing, the predicted casing pressure to lift a given amount of liquid can be less in larger pressure as shown in Fig. 6.50 (Table 6.3).

Summary:

- The Foss and Gaul model modified and has been used and has been compared successfully to field data.
- Improved model predicts casing pressure build-up for specified rise velocity with input of well conditions.
- Use of result can show when plunger will begin to cease to perform requiring compression or other lift methods.
- Use of results can assist in optimization of on-going cycle performance.

Conclusion:

1. Modified Foss and Gaul model has been presented.
2. Data from 10 wells compared favorably to the model (NOT enough data to statistically decide on best model options).

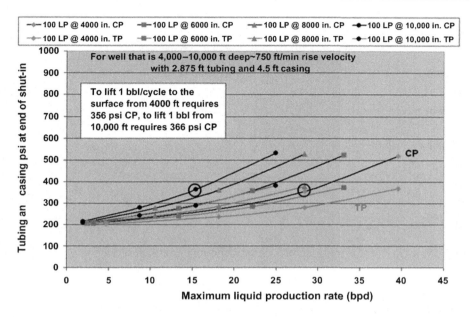

Figure 6.44 Well depth has minor impact on plunger lift system operation (2 7/8 in. tubing).

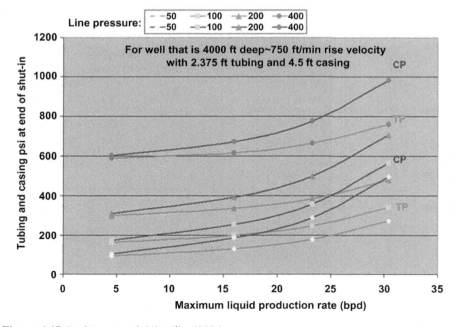

Figure 6.45 Performance: 2 3/8–4½–4000 in.

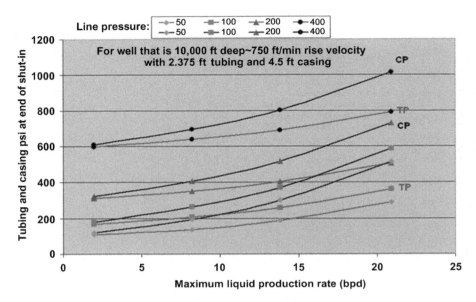

Figure 6.46 Performance: 2 3/8–4½ in. @ 800 psi casing 10,000 in. = 13 bpd at 4000 in. = 24 bpd.

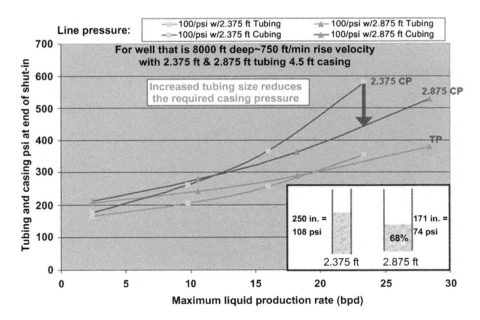

Figure 6.47 Required casing pressure is reduced if tubing size is increased.

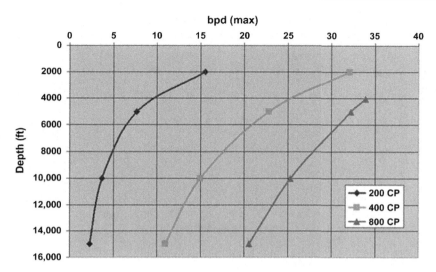

Figure 6.48 Plunger bpd with depth, 2 3/8 in., 4.5 ft casing, WHP: 100 psi, fall velocity 250 fpm in gas and 40 fpm in liquid (CP is casing operating pressure available, psi) bpd max.

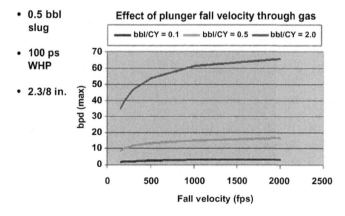

Figure 6.49 Fast falling plunger allows more production.

3. Predicted minimum casing pressure compares favorably to within an average of 1% error and 14.4% absolute average error.
4. $P_{c,max}$ from modified Foss and Gaul predicted build casing pressure to unload the plunger at the measured rise velocity.
5. Prediction accounts for liquid load, frictional effects, tubular sizes and lengths, surface line pressure, and fluid properties.
6. Option to consider or not, the influx of gas and leakage past the plunger is available.

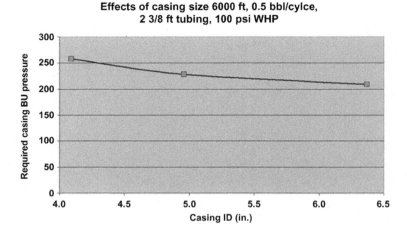

Figure 6.50 Larger casing reduces casing pressure requirement.

7. Results indicate that original Foss and Gaul model uses too high of friction and rise velocities would be too fast.

Measurement of rise velocity profiles

Measurement systems allow measurement of velocity and position by monitoring pressure and temperature as the plunger falls and also rises. Also there are systems that magnetically sense the collars and store the sensed data. This requires an instrumented plunger with a module to collect data for a time period and the then plunger is pulled and the collected data is downloaded and analyzed. Fig. 6.51 shows equipment with a data collection module that measures temperature and pressure. Temperature and pressure rise during downfall, so the time the plunger is falling is indicated by P and T rising and vice versa for falling. The disadvantage is that measurements are made on the instrumented plungers rising and falling and not the operator's plunger that is in the well.

6.7 Controllers

The controller decides when to open and close the well. There are simple controllers but they require more trial and error to get productive cycles in operations. There are more sophisticated controllers that take some of the guess work out of setting up the plunger cycle that will repeat and also tend to maximizing production.

Well is shut-in. Controller opens well when:

1. Tubing pressure $\geq$ *on* pressure system input.
   Operator has input a value and when SI TP reaches this value, the controller tells the motor valve to open.

Table 6.3 Calculated required shut-in casing pressure @ measured rise velocity

Well ID	Tbg ID, (in.)	P_{wh} (psi)	P_p (psi)	Depth (ft)	Slug (bbl)	Slug length (bbl, ft)	Avg. temp. (°F)	SG liquid	SG gas	Rise velocity (ft/min)
Normal cycle	1.995	126.3	2.6	7773.0	0.6	258.8	130.0	0.9	0.7	679.3
EFM flow rate well	1.995	68.3	3.35	6276.0	0.5	258.8	120.0	1.0	0.8	499.8
EFM flow rate well	1.995	167.6	3.5	7728.0	0.1	258.8	125.0	0.9	0.8	933.3
EFM flow rate well	1.995	159.3	0	6082.0	0.2	258.8	120.0	0.9	0.8	1730.9
Optimize before	1.995	65.3	2	8687.0	0.7	258.8	160.0	0.8	0.8	610.7
Optimize after	1.995	69.6	7.5	8687.0	0.1	258.8	160.0	0.8	0.8	1396.2
SV before	1.995	79.2	2.5	6273.0	0.1	258.8	120.0	0.9	0.8	792.4
SV after	1.995	64	2	6273.0	0.4	258.8	120.0	0.9	0.8	636.5
Horiz 0 load	2.441	110	5	4173.6	0.0	172.9	79.0	1.0	0.7	1923.8
	2.441	110	5	4173.6	1.5	172.9	79.0	1.0	0.7	332.8

Well ID	Rise velocity (ft/min)	Actual P_c (psi)	$P_{c,max}$ (psi)	% Error	Abs % error	$P_{c,max}$ new	% Error	Abs % error	$P_{c,max}$ psia orig.	% Error	Abs % error	$P_{c,max}$ new orig	% Error	Abs % error
Normal cycle	679.3	320.7	324.0	1.0	1.0	311.0	−3.0	3.0	380.2	18.6	18.6	367.7	14.6	14.6
EFM flow rate well 1	499.8	163.1	204.0	25.1	25.1	184.0	12.8	12.8	211.1	29.4	29.4	191.0	17.1	17.1
EFM flow rate well 2	933.3	303.1	277.5	−8.4	8.4	270.4	−10.8	10.8	323.0	6.6	6.6	317.0	4.6	4.6
EFM flow rate well 3	1730.9	390.0	388.5	−0.4	0.4	285.4	−26.8	26.8	625.0	60.3	60.3	622.0	59.5	59.5
Optimize before	610.7	268.7	246.0	−8.4	8.4	216.0	−19.6	19.6	304.0	13.1	13.1	274.0	2.0	2.0
Optimize after	1396.2	308.7	192.0	−37.8	37.8	174.0	−43.6	43.6	311.0	0.7	0.7	293.0	−5.1	5.1
SV before	792.4	144.0	151.0	4.9	4.9	147.7	2.6	2.6	170.4	18.3	18.3	167.1	16.0	16.0
SV after	636.5	131.7	171.6	30.3	30.3	167.9	27.5	27.5	192.5	46.2	46.2	188.7	43.3	43.3
Horiz 0 load	1923.8	211.7	177.5	−16.2	16.2	177.4	−16.2	16.2	220.0	3.9	3.9	217.0	2.5	2.5
Horiz 100 load	332.8	275.6	306.8	11.3	11.3	289.9	5.2	5.2	307.0	11.4	11.4	289.0	4.9	4.9
			Average	0.1	14.4		−7.2	16.8		20.8	20.8		15.9	17.0

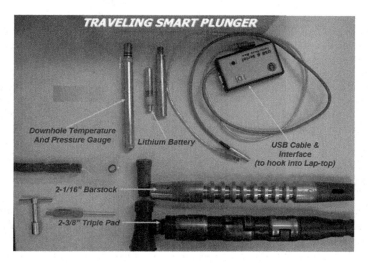

Figure 6.51 An instrument plunger system with a module to fit in the plunger and collect data while the plunger is rising and falling. Vendor's plunger performance is measured.

2. Casing pressure ≥ *on* pressure limit: casing pressure (CP) > a value.

Operator has input a value so that when SI CP reaches the value, the controller tells the motor valve to open.

3. Tubing line ≥ *on* pressure limit: TP exceeds value over LP.

Operator has input a value of TP − LP. When value is exceeded the well opens. This compensates for changing LP during shut-in.

4. Casing line ≥ *on* pressure limit: CP exceeds a value over LP.

Operator has input a value of CP-LP. When the value is exceeded the well opens. This compensates for changing LP curing shut-in.

5. Foss and Gaul calculations

This model needs the data for how much liquid is in the well and then the model can tell the user exactly what CP is needed so plunger and liquid will surface at a specified velocity (i.e., between 500 and 1000 fpm).

6. Load factor

This algorithm will open the well when the value of (CP − TP)/(CP − LP) < ~40%−50%.

This is a formula form experience but works well in many situations.

Of the earlier algorithms, the last two consider more of the important well operational variables and should require less trial and error from the operator although many of the high TP and CP controllers still operate.

Well is open and producing: Controllers signals shut in when the below happens:

1. Plunger has arrived (used on oil wells).

2. Casing pressure ≤ off pressure limit (looks for casing to fall below a set pressure).

3. Delta pressure across restriction (lower means lower flow rate tending to low critical or below).

4. Flow rate (calculate the flow and when it is below, say a Turner critical, then shut well in).

5. Casing–tubing ≥ off pressure limit (looks for differential between casing and tubing to increase while producing).

Item 5 does indicate that liquid has formed during the low period but many times if you wait for the casing pressure to show an increase at the surface, you may have too large a slug of liquid to lift on the next cycle. Of the earlier the flow below critical seems to be a good control algorithm but many are used successfully.

6.8 Problem analysis

The following section outlines solutions to some of the more common problems encountered with plunger lift systems. These items are grouped with respect to the system components and particular malfunctions.

Table 6.4 can be used as a quick reference for some general points. Many of the table entries are field specific but the user might develop a similar field-specific table for a particular operation.

See Appendix C, Plunger troubleshooting procedures, for discussion of troubleshooting procedures.

6.9 Operation with weak wells

When conventional plunger ceases to work, then one can add expensive gas lift or sucker rod pumping but since the operational staff is familiar with plunger lift, it would be beneficial if a modified form of plunger operations could maintain production.

6.9.1 Progressive/staged plunger system

In wells where conventional plunger system's struggle due to low gas, liquid ratios, high line pressure, packer or slim hole completions without sufficient GLR, or low reservoir pressure, there is an application for progressive(4) or multi-staged (5) plunger lift. A staged plunger system essentially operates as two separate, yet intricately linked systems within the same well (Fig. 6.20). In a conventional system the gas above the plunger and fluid level is rapidly produced down the line in order to create a differential. This differential allows the casing or near wellbore gas to expand into the tubing, forcing the plunger and its fluid to surface by attempting to drive gas past the plunger at a high velocity that forces a turbulent seal. In staging plunger lift the gas above the top system produces the same as in a conventional system. However, the gas that sits below the top and above the fluid in the bottom system is used as the energy source for the top plunger. This allows the top plunger to require only the energy to travel ¼ to ½ up the well depending on

Symptom	Check/change plunger	Optimize programs settings	More off time	More after flow	Less off time	Less after flow	Check well TBG (restrictor office)	Check well head (design)	Clean screen/check writing	Check module/writing	Change (+) load from link	Power down and restart/module	Set sensitivity of sensor	Change supply gas filter	Adjust supply gas preware (20–30 pad)	Clean control flood perb	Change o-rings under batch valve	Check charge rattory	Check solar panel	Repair motor vatro trine	Ginate flow restrictions	Check catcher	Change module	Change latch valve	Check special settings	Check motor valve diaphragm	Impact plunger
No plunger arrival	6	3	2			1	8	7	5	7			4								10		8				
Slow plunger arrival	4	3	2		2	1	6													9	6	4					
Fast plunger arrival		3		1			4						2							5		3					5
Fast plunger arrival @ all settings or plunger will not fall	4	1	2			1	7																				
Slow plunger arrival @ all settings or plunger will not come to surface		3						6													5						
Short lubricator spring life		4		2	3		5		3	4												1	5				
Short plunger life		3		1	2		5	4	5	7													8				
Sensor error	6	3	2			1			3	4			2									1	6				
plunger error		1					12	11		1	5		4						9	9	10						
Good trip, no count (plug-in sensor)		1									4	3	2														
Good trip, no count (strap-on sensor)		1							5	10		4	2														
Fatal error code@LED									5																2		
LED control screen blank																											
Sales valve will not open/close													4	3	6	7	2	2	3				5	12		13	
Tank valve will not open/close													4	3	6	7	2	8	9				11	11		12	
Latch valve will not switch													4	3	5	6	1	8	9				10	7			
Motor valves will not close or close slowly													4	3	1	2	5	2					9	7			
Short battery life		2								1							2	6	8				4				
Will not go to after flow										3								3					4		1		1

the design. This makes the plunger system significantly more efficient. It also reduces the necessary GLR to drive the plunger system and decreases the amount of pressure necessary. For this reason by staging a plunger well significantly increases the operating window and decreases the necessary pressure to produce the plunger that extends the wells' life.

From the bottom or near the end of tubing to surface, the composite of a staged plunger system usually requires the following: a bottom-hole bumper spring with hold down, a conventional plunger (usually a spiral but dependent on conditions), an Intermediate Landing Assembly (6) or Multi-Stage Tool, another conventional or quick-trip plunger and a typical surface lubricator with control system.

Typically a staged plunger system begins with a conventional plunger system evaluation due to low GLRs or pressure built that require significant shut-in time that the original system does not optimally lift the well. At this point the well is evaluated using two or more systems. The results of the evaluation process will typically mean a decrease in necessary GLR between 1 and 3 Mcf/barrel. The decrease in necessary pressure may be significant in a typical installation. Thus the necessary casing pressure to drive the plunger system decreases between 50 to in some extreme cases 500 psi or greater. This results in lower average flowing bottom-hole pressures and increased inflow. In most cases the staged plunger system will normally mean two to three times more trips with smaller controlled load sizes due to the decreased gas usage per cycle. Usually the necessary casing pressure will decrease which provides significant value in high line pressure applications where the initial casing pressure necessary is quite high.

While staging plunger lift is a significant advantage to wells performing beyond conventional applications, it does not necessarily improve production on typical conventional candidates. This is due to the bottom plunger system functioning as a choke during after-flow causing it to run dry if the well cycled at a high frequency. The staged plunger system, unlike the conventional, also requires slick line work in order to inspect the bottom plunger due to the fact that it does not travel to surface (Fig. 6.52).

6.9.2   Casing plunger for weak wells

The casing plunger travels in the casing only and there is no tubing in the well. The plunger senses when a head of liquid appears above the plunger, the internal bypass valve closes, and the well gas production lifts the plunger and slug of liquid to the surface. Then at the surface the plunger internal valve will open and the plunger will drop. The casing plunger rises and falls slowly. It has rubber cups that fit the casing. The cups will not last long enough for satisfactory service if the casing is very rough. Fig. 6.53 shows a casing plunger.

In summary the casing plunger can be used for shallow low-pressure wells with relatively good casing condition to extend the life of the well to very low pressures. The casing plunger can also be made for 5½ in. casing. Multi-Systems, IPS, and others can supply a casing plunger.

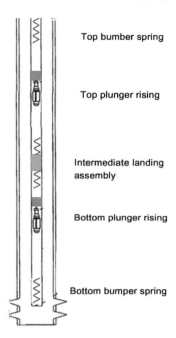

Top bumber spring

Top plunger rising

Intermediate landing assembly

Bottom plunger rising

Bottom bumper spring

Figure 6.52 Progressive plunger system.

Figure 6.53 Casing plunger and casing plunger well head.

6.9.3 Gas-assisted plunger

This is a system where gas is injected down the casing from the surface to help build pressure in the well to lift the plunger and the liquid slug. It can be done with no packer where the gas goes down the casing and turns into the bottom of the tubing. It can be done with a packer, but above the packer there is a gas lift valve to allow casing gas to enter the tubing. Gas is injected so the pressure at the time of opening the well is the same as it would be in an operating conventional plunger well before opening the well. Injection can begin from the surface as soon as the plunger begins to fall. It requires no extra hardware downhole if there is no packer but it does require a source (compressor) of high-pressure gas.

6.9.4 Plunger with side string: low-pressure well production

Plunger lift with a side string can be used to produce gas or oil wells with low bottom-hole pressures where a source of higher pressure makeup gas is available at the well head.

A plunger lift system in combination with a side string for injecting makeup gas and pressure for lift is used for this system.

The plunger lift system with side string injection requires that the tubing be removed from the well. As the tubing is run back in the hole, ½ or ¾ in. coiled stainless steel tubing is banded to the production tubing. On the bottom is a standing valve with a side port injection mandrel above it. Above the injection mandrel is a bottom-hole spring assembly and a plunger.

Makeup gas is injected from the surface down the side string directly into the production tubing. As the gas enters the tubing, it is prevented from entering the wellbore by the standing valve. The gas is forced to U-Tube up the production tubing, driving the plunger ahead of it, which in turn removes liquid from the tubing.

The injection gas is injected for only a short period, just long enough to cause the plunger to surface. Once the plunger surfaces, the well is allowed to bleed down to sales line pressure. As this occurs, liquid enters the production tubing from the wellbore and the plunger drops back to bottom on its own weight.

As the plunger continues to remove liquid from the wellbore, the liquid level in the casing drops. As the liquid in the casing drops, the perforation zone is relieved of hydrostatic pressure, and formation gas enters the casing. The formation gas is produced out the casing.

PLSI developed this technique for low bottom-hole pressure wells in 1992. Initial installations occurred in the Antrium gas zones of northern Michigan. The technology has been economic in this area and to date over 500 installations of this system are in place.

Of the systems described in Fig. 6.54 the progressive plunger system and the GAPL system seem to be the most used plunger-related systems to be used once conventional plunger begins to fail as the well(s) weaken.

6.10   Summary

Plunger systems work well for gas wells with liquid loading problems as long as the well has sufficient GLR and pressure to lift the plunger and liquid slugs.

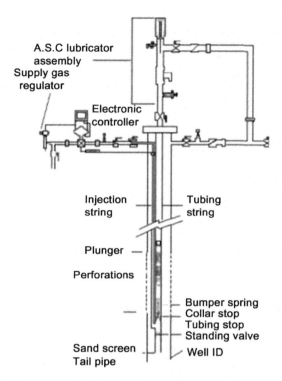

A.S.C lubricator assembly
Supply gas regulator
Electronic controller
Injection string
Tubing string
Plunger
Perforations
Bumper spring
Collar stop
Tubing stop
Standing valve
Sand screen
Tail pipe
Well ID

Figure 6.54 Side string gas supply for plunger lift (PLSI, Midland, TX).

Plunger lift works well with larger tubing so there is no need to downsize the tubing.

Conventional plunger lift works much better if there is no packer, and this can be a problem if the old packer should be removed.

Although the plunger lift can take the well to depletion, the recoverable production may not be quite as much as using a more expensive beam pump system, for example, to pump liquids out of the well in the latter stages of the depletion.

The advancements in recent years include more use of the bypass plungers, the advent of a plunger tracking system, and the ability to predict casing build-up pressures that will surface the plunger at acceptable velocities.

References

1. Lea JF. Dynamic analysis of plunger lift operations. Tech. Paper SPE 10253 (November 1982), p. 2616–2629.
2. Beeson CM, Knox DG, Stoddard JH. Part 1: The plunger lift method of oil production, Part 2: Constructing nomographs to simplify calculations, Part 3: How to user

nomographs to estimate performance, Part 4: Examples demonstrate use of nomographs, and Part 5: Well selection and applications. Petroleum Engineer; 1956.
3. Lea JF. Plunger lift versus velocity strings. *J Energy Resour Technol* 1999;**121**:234−40.
4. Cosby DE, Introduction to Plunger, Gas Well Deliquification Workshop, Denver Colorado, Shale Tec LLC, February 24−26, 2014.
5. Foss DL, Gaul RB. *Plunger lift performance criteria with operating experience- ventura field. Drilling and production practice.* API; 1965. p. 124−40.
6. Phillips D, Listiak S. How to optimize production from plunger lifted systems, Pt. 2. World Oil; 1991.

Further reading

Avery DJ, Evans RD. University of Oklahoma: Design optimization of plunger lift systems. Paper SPE 17585. In: *Presented for presentation at the SPE international meeting on petroleum engineering held in Jinjin*, China, November 1−4, 1988.

Ferguson PL, Beauregard E. How to tell if plunger lift will work in your well. World Oil, August 1, 1985, p. 33−36.

Hacksma JD. Predicting plunger lift performance. In: *Presented at the South West Petroleum Short Course*, Lubbock, TX, 1972, October 5−7, 1981.

Hacksma JD. Users guide to predict plunger lift performance. In: *Presented at Southwestern Petroleum Short Course*, Lubbock, TX, 1972.

McCoy J, Rowlan L, Podio AL. Plunger lift optimization by monitoring and analyzing well high frequency acoustic signals, tubing pressure and casing pressure, SPE 71083. In: *Presented at the SPE rocky mountain petroleum technology conference*, Keystone, CO, May 21−32, 2001.

Rosina L. *A study of plunger lift dynamics* [Master thesis]. University of Tulsa, Petroleum Engineering; 1983.

White GW. Combining the technologies of plunger lift and intermittent gas lift. In: *Presented at the annual american institute pacific coast joint chapter meeting*, Costa Mesa, California, October 22, 1981.

Wiggins M, Gasbarri S. A dynamic plunger lift model for gas wells, SPE 37422. In: *Presented at the Oklahoma city production operations symposium*; 1996.

Hydraulic pumping

Toby Pugh is now retired. Formerly he was a director of Hydraulic Technical Support\Training for Weatherford Completion & Production Systems, a division of Weatherford International Ltd. Before joining Weatherford in 2001, he served as an International Sales Manager at Halliburton Production Services. Most of his 47-year career was spent with the Guiberson Division of Dresser Industries, where he held a variety of positions ranging from Manager of Research and Engineering to Regional Manager. He holds a BS degree in Mechanical Engineering and an MS degree in Aerospace Engineering from the University of Texas at Arlington. He is a registered Professional Engineer; has been awarded six US patents and one European patent; has authored five SPE papers; and has given technical presentations on the use of jet pumps for high-volume production (at the European Artificial Lift Forum), the use of jet pumps to produce heavy oil (at the SPE/IBP Artificial Lift Heavy Oil in Brazil), an economic study comparing the life-cycle costs for ESPs, rod pumps, and jet pumps (SPE Applied Technology Workshop in Colombia), and the use of jet pumps to dewater gas wells and for frac fluid recovery (at the Middle East Artificial Lift Forum). He was the coauthor of a magazine article that was published in the August 2006 issue of World Oil on how using jet pumps increased production and reduced expenses for Nexen-Canada. He also coauthored a magazine article, contributed the chapter, that was published in the November 2015 issue of World Oil on using the first ever subsea jet pump offshore Tunisia. In addition, he wrote the chapter on hydraulic pumps for Dr. William Lyons at New Mexico State University.

7.1 Introduction

Suffice it to say that hydraulic pumps are a form of artificial lift, that is, the well no longer has sufficient energy to produce itself the required volume, and an additional energy must be supplied to compensate the energy deficiency. More specifically, hydraulic piston pumps are used for low-production volumes (approximately 0−200 bpd), while jet pumps can be used for volumes that range from very low to very high (12−35,000 + bpd).

A thorough discussion of hydraulic piston pumps and jet pumps was included in the previous edition of this book. The reader is referred to that edition if such

Gas Well Deliquification. DOI: https://doi.org/10.1016/B978-0-12-815897-5.00007-X

Hydraulic pumping system schematic flow

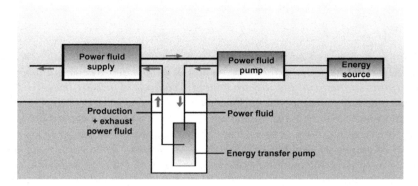

Figure 7.1 Flow schematic for a typical hydraulic system.

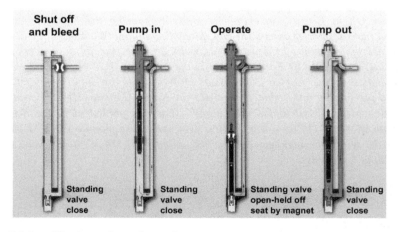

Figure 7.2 Installing/removing a free style pump.

details are needed. However, Figs. 7.1−7.3 are shown for information specific to this discussion.

There are just four essential components of a hydraulic system and from which a large number of variations have been derived.

The preferred style of hydraulic pump installations is known as a free style pump which means the pump is "free" to move down with the power fluid, that is, down to be installed and up to be retrieved. This eliminates the need to use a rig to install/recover the pump.

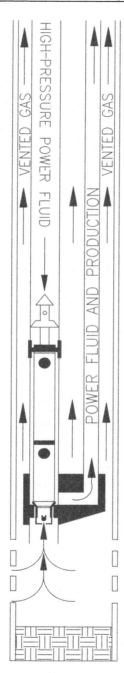

Figure 7.3 Parallel free, open power fluid installation.

The downhole arrangement used for hydraulic pumps is determined by the production requirements of the well. For gas wells, the requirement is to remove the liquids in the wellbore that are creating a back pressure on the reservoir thereby inhibiting the inflow of gas. The liquid volumes are typically low, so an installation with parallel tubing stings is often used.

7.2 Application to well deliquification—gas, coal bed methane, and frac fluid removal

Due to the large volume of wells in the multiple shale formations, it has become increasingly important to be able to remove the liquids from the wellbores in those formations, which are creating a back pressure on them and inhibiting the inflow of gas. The primary method of liquid removal, historically, has been the sucker rod pump, which is a positive displacement pump, due to the relatively low volumes of liquid that needs to be removed (produced). The common problems encountered with the use of such pumps are extreme depths, wellbore deviations (causing rod/tubing wear), orientation of setting, their propensity to gas lock, cost to retrieve, and difficulty in adapting to changing reservoir conditions.

Hydraulic pumps overcome all of the above issues. Jet pumps have been used at depths of up to 20,000 ft., the power fluid column will not cause tubing wear, and they will not gas lock. In addition, they are also highly tolerant to solids as they do not have any moving parts. The ability to control the operation of the jet by varying the injection rate of the power fluid, in conjunction with replacing it by merely changing the direction of the flow, represents a significant cost-saving to the operators. These capabilities have resulted in jet pumps being widely used in areas as diverse as South Texas, West Texas, Mississippi, and Alberta, Canada.

Jet pumps are frequently used for both a wellbore cleanup after a frac job and then to produce the well. They are typically operated until they are no longer applicable (more on this below), beyond the condition where a hydraulic piston pump can be used to complete pumping off the well. Just like a jet pump, a piston pump also overcomes the above-listed problems with the exception of the solids, and has the same operational flexibility as jet pumps. While their use is yet to find as much acceptance as jets, they have, however, been used for this purpose in West Texas, South Texas, and Romania.

The significant differences between a CBM well and the typical gas well/FFR (frac fluid recovery) applications are that CBM wells have even lower production volumes and bottom-hole pressures.

Another difference that has been reported but never demonstrated is that coal fines create more severe downhole problems than sand. Typically, the coal fines in a CBM well are no more damaging than what is experienced with solids in other wells, gas or oil. While some CBM wells have fines that are sharp and angular, the same is also true for other types of wells, such as those which are in the bromide formation in Oklahoma, United States. It has also been reported that the fines in

CBM wells tend to clump together inside the jet pumps and inhibit the flow through the pump. However, absolutely no evidence has ever been provided by those making that claim. A subsequent investigation found an evidence that this problem occurs only on the surface, that is, never downhole.

7.3 Jet pumps

A jet pump is an excellent choice for this application because it is highly tolerant to sand and other particles and can typically be used in applications where the GLR through the pump is less than 1000. Hence, they have been operated at GLRs as high as 6000, *but* as much free gas as possible should be sent around the jet pump and not through it. This is accomplished by using a parallel tubing or concentric tubing installation, setting the intake of the pump below the perforations (if possible) and using a gas anchor attached to the bottom of the jet pump. It will also permit the tubing—casing annulus to function as an additional gas separator thereby allowing the gas to flow uninhibited to the surface. This will also reduce the energy needed to operate the jet as it would not be "producing" the gas along with the liquid.

It should be noted that the jet pumps currently available on the market cannot pump off a well as a low-flowing bottom-hole pressure (i.e., low inflow) will result in power fluid cavitation damage to the throat before the pressure approaches zero within the pump. (*Note*: Throat is the only part of a jet pump that will undergo cavitation damage. Also, it is sometimes assumed that free gas will cause cavitation problems which is impossible.) This type of cavitation is known as power fluid cavitation. But, this nomenclature is actually a misnomer as power fluid will *never* create cavitation damage. The name is used because there was essentially only power fluid present at the time. The conditions for this type of cavitation can be identified by installing a gage below a jet pump that will sense and record the flowing bottom-hole pressure. With this information, those specific operating conditions can then be avoided. For deep wells depending on the tubulars, friction in the lines can also limit performance.

A new style of jet pump is being developed that will not suffer either power fluid cavitation (not enough inflow) or production cavitation (too much inflow for the installed nozzle/throat combination). It is a radically new concept and will take quite some time to develop into a commercial product. While it has already been claimed that a noncavitating pump has been developed, it cannot be scientifically verified in a test lab setting, whereas the performance of the above pump has been verified.

The 1-1/4 in. coiled tubing jet pump was the first hydraulic jet pump used for dewatering these wells. This small jet pump was designed so that it could be used either as a free style pump (free to move with power fluid to install and retrieve) inside a 1-1/4 in. tubing or as a fixed style pump attached to it. In both cases the 1-1/4 in. tubing, which can be either coiled or standard, is run inside another string of tubing (Fig. 7.4), and the produced water and spent power fluid return to the surface through the tubing—tubing annulus. The gas is free to flow to the surface through the casing—tubing annulus. This size of jet pump was initially used due to its small size and the anticipated low-production rates, less than 500 bpd of liquids.

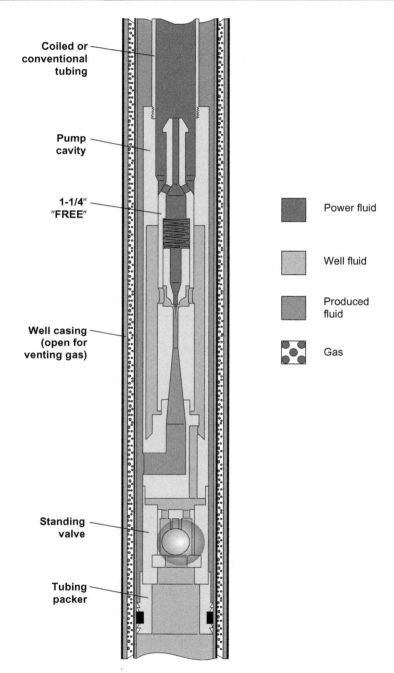

Figure 7.4 1-1/4 in. coiled tubing jet pump installation.

With time, it was decided that using a full size 2 in. jet was preferred to remove the maximum amount of fluid as quickly as possible, especially frac fluid. This change was driven by the operators as they were concerned about the formation damage due to the length of time the frac fluid was in contact with the reservoir. The procedure that evolved was to inject sufficient power fluid so that the jet pump would produce at rates up to 2000 bpd. That injection pressure setting would be maintained until it was determined that the physical presence of the jet pump was inhibiting the inflow of gas from the reservoir. It was then removed (sometimes just unseated) to allow a freer flow of gas. If the well was loaded-up again, the pump would be reinstalled and the procedure should be repeated. This procedure frequently resulted in the well being able to complete the unloading process by "gas lifting" itself. The typical installations are parallel tubing, Fig. 7.3, or a single string of tubing with a casing packer. *Note*: It is normal to recover no more than about 50% of the injected frac fluid.

The power fluid cavitation problem discussed earlier can be easily overcome as demonstrated by an operator in England. A pressure gage was installed below the inlet of the jet pump with real-time measurements sent to the surface power fluid pump. When the inlet pressure was too low, the pumps were turned off. When the pressure recovered, the pumps were restarted. An equation was developed to obtain the pressures for stopping and restarting the power fluid pump. The parameters used for calculating the limits were based on the operating conditions observed when the cavitation problems previously occurred.

While intermittent operation can result in problems for some forms of artificial lift, jet pumps can be easily stopped and restarted without any problems.

Another, and even more significant development, was establishing the methodology for sizing the nozzle and throat areas when very small sizes are involved. It is based solely on the difference of the two areas whereas the standard method uses just the ratio of the two areas. Using only the standard methodology resulted in continually having power fluid cavitation problems.

By including all of the above, power fluid cavitation issues were completely resolved.

Below is the installation graph of the well that had a daily production rate of 12 bpd. Important information that can be gleaned from this graph is that such wells are typically overpressured at the beginning of operations and care must be taken not to start the pump operating in the area identified as production cavitation. It is always necessary to select an initial pressure that keeps the pump out of that region. As the overpressure is reduced, while maintaining a constant injection pressure, the production rate automatically reduces. The injection pressure can then be slowly increased to a higher value, as long as that higher pressure is not in the cavitation zone. This process is continued until the desired injection pressure is achieved. *Note*: The production rate will stabilize when the well's inflow productiviey index/ inflow performance relationship (PI/IPR) matches the outflow for the desired injection pressure line (Fig. 7.5 and Table 7.1).

An important point to recognize when doing an analysis for deliquifying any well is that it is a transient situation and all installation analysis computer programs

Graph for demonstrating unloading

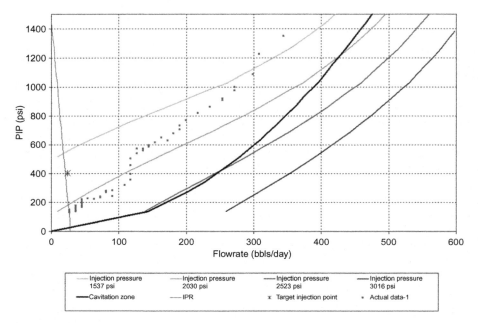

Figure 7.5 Graph demonstrating unloading.

Table 7.1 Nozzle/throat combinations for unloading gas wells

Casing	Depth	Guiberson combination	Oilmaster combination
4-1/2 in.	Any	D nozzle, 6−8 throat	9 nozzle, 9−11 throat
5-1/2 in.	0−5500 ft.	E nozzle, 9−11 throat	10/11 nozzle, 11−13 throat
5-1/2 in.	5500 + ft.	D + nozzle, 9−10 throat	10 nozzle, 11−12 throat

are based on steady-state conditions. The obvious conclusion is that it is not possible to use these programs to do an analysis for unloading a well. To perform deliquification processes using jet pumps, sufficient empirical field data has been collected as to which nozzle/throat combinations can be used to get the best results as per the following information (Table 7.2).

Note: In wells with damaging fines, the part that is typically the most damaged is the throat. It has been found that using a throat made of silicon carbide, instead of the standard tungsten carbide, greatly improves the life of the pump. The outside of the nozzle can occasionally have this issue, so new nozzle materials are being tested. The user should contact the manufacturer for the latest information.

For steady-state operations, the operator typically can select from several different nozzle/throat combinations. There may be constraints imposed by the surface

Table 7.2 Data for steady-state operating point in Fig. 7.5

```
****PERFORMANCE CALCULATION AT SPECIFIED INJECTION PRESSURE OF 2163.
   psig****
PUMP PARAMETERS: PRODUCTION RATE AT WHICH CAVITATION BEGINS: 267. Bpd
                CONSUMED HORSEPOWER: 12. hp
                PUMP INTAKE PRESSURE: 397. psig
POWER FLUID: POWER FLUID PRESSURE AT NOZZLE: 3707. psig
                POWER FLUID INJECTION RATE: 288. bpd
                POWER FLUID FRICTION LOSS TO PUMP: 19. psig
                SPECIFIC GRAVITY: 0.9991
                FORMATION VOLUME FACTOR: 1.0109
                VISCOSITY: 0.6298 cp
PRODUCED NON-WATER LIQUID: SPECIFIC GRAVITY: 0.7403
                FORMATION VOLUME FACTOR: 1.0429
                VISCOSITY: 1.1375 cp
PRODUCED WATER:          SPECIFIC GRAVITY: 0.9991
                FORMATION VOLUME FACTOR: 1.0109
                VISCOSITY: 0.6298 cp
PRODUCED GAS: SPECIFIC GRAVITY: 0.5900
                DENSITY: 1.1945. lbs/ft3
                FORMATION GAS OIL RATIO: 50. ft3/bbl
                DISSOLVED GAS OIL RATIO: 50. ft3/bbl
                FREE GAS OIL RATIO: 0. ft3/bbl
Returns:            Produced Liquid Rate (Surface)      :    12. bpd
                Produced Liquid Rate (Subsurface)    :    12. bpd
                Produced Gas Rate (Subsurface)       :    0. bpd
                Total Return Rate (Subsurface)       :    300. bpd
                Pump Discharge Pressure              :    1605. psi
Minimum Pump Intake Pressure to Avoid Power Fluid Cavitation:
                Pressure Ratio: 0.5746
                Mass Ratio: 0.042
                Area Ratio: 0.232
                %Submergence: 0.00233
                Minimum Pump Intake Pressure: 3 psi
```

equipment, such as injection rate, injection pressure, and horsepower, which would dictate the actual combination to be used. If not, then any of the combinations would be acceptable.

Table 7.3 shows what can be expected when operating a jet pump under varying conditions. Its information should not be considered as exact since the use of different combinations and different well parameters will modify the results.

7.4 Piston pumps

Around 2005, a major oil company stated that it needed a piston pump capable of producing 25 bpd from 10,000 using just 5 bhp in order to deliquify gas wells. Weatherford spent a few years on that effort and came out with its PL-III, which was able to meet these requirements. It has the additional advantage that it *cannot*

Table 7.3 Typical steady-state jet pump performance data

	(ALL CASES ARE FOR A WATER CUT OF 50%)			
DEPTH (FEET)	6000	12,000	6000	12,000
PRODUCTION RATE (BPD)	100	100	400	400
FORMATION GLR (SCF/SB)	1500	1500	100	100
INJECTION PRESSURE (PSI)	3750	3675	2550	4125
INJECTION RATE (BPD)	450	700	850	1450
HORSEPOWER	32	49	40	113

Note: Methane obeys the natural gas laws. So, only the actual volume of free gas is calculated and that is used in the calculations.

gas lock (unlike a sucker rod pump). It accomplishes this by making use of the spent power fluid and does not require recirculating/reproducing previously produced liquids (such as rod pumps).

The PL-III is a slow stoking, hydraulic piston pump, which is ideal for producing low volumes. However, it is a positive displacement pump so the ingestion of gas will always effect its volumetric efficiency.

In addition, just like all positive displacement pumps, solids in the produced fluids have the potential of causing erosional problems due to the close tolerances of the parts in the pump. This style of pump is used when it has been determined that the bottom-hole conditions have dropped low enough that a jet pump is no longer a viable option. Again, as the piston pump is a positive displacement pump, it can be used to complete the process of pumping off the well. In order to minimize capital expenditures, the bottom-hole assembly (BHA) used for the jet pump is designed such that it can also be used with the piston pump so no workover is needed.

While the size of the smallest free style piston pump that has ever been designed was for 1-1/4 in. tubing, the smallest slow stroking free style pump that is currently available is for 2-3/8 in. tubing. In order to use it in a dewatering application, the typical installation has two strings of parallel tubing. One string must be large enough for allowing the pump to go from the surface to its housing in the well (a BHA) and also be a conduit through which the power fluid goes to the pump. The other string is a conduit for returning the spent power fluid and production fluids to the surface. This type of installation allows the gas to flow to the surface through the casing–tubing annulus and not through the pump, which could easily result in a zero volumetric efficiency. Very low volumetric efficiencies mean that energy is being used to produce gas. The absolute minimum amount of gas should go through a positive displacement pump, with zero being the preferred amount. This arrangement limits the casing size to 6-5/8 in.-20 pound/ft. or larger.

Due to the design of the reversing valve in the PL-III pump, the speed of the pump is limited and it should not be considered for production rates above approximately 125 bpd. For rates above this, any suitable pump can be used. For wells with production rates greater than approximately 200 bpd, a jet pump is the

Table 7.4 Analysis for piston pump installation

PUMP VERTICAL DEPTH	7000. FT	TUBING LENGTH TO PUMP	7000. FT
OIL GRAVITY	10.00 API	WATER SPECIFIC GRAVITY	1.03
GAS SPECIFIC GRAVITY	0.80	GAS / OIL RATIO	25. CFPB
WATER FRACTION	0.00	BOTTOMHOLE TEMPERATURE	250. DEG F
SURFACE TEMPERATURE	200. DEG F	CASING I. D.	4.653 IN
POWER FLUID USED	WATER	SURFACE LINE I. D.	3.000 IN
SURFACE LINE LENGTH	50. FT	SEPARATOR PRESSURE	200. PSI
THE POWER FLUID PATH IS THE TUBING DESCRIBED BELOW:			
POWER FLUID TBG O. D.	2.375 IN	POWER FLUID TBG I.D.	1.995 IN
THE PRODUCTION RETURN PATH IS THE TUBING DESCRIBED BELOW:			
PROD. RTN. TBG. O. D.	1.050 IN	PROD. RTN. TBG. I.D.	0.824 IN
PERFORMANCE OF SPECIFIED PISTON PUMP AT SPECIFIED INFLOW CONDITIONS			
THE REQUIRED PUMP DISPLACEMENT FOR THE GIVEN CONDITIONS IS:			361. BPD
THE APPARENT VOLUMETRIC EFFICIENCY DUE TO GAS IS:			**58. PERCENT**
PUMP SPECIFICATIONS:		INFLOW PERFORMANCE:	
KOBE A 2 X 1-5/16 X 1-5/16		PRODUCTION RATE = 200. BPD	
		PRODUCING BHP = 100. PSI	
ENGINE END DISP.	= 4.03 BPD/ SPM		
PUMP END DISP.	= 3.97 BPD/ SPM	**PUMP PERFORMANCE :**	
MAXIMUM SPM	= 121. SPM	OPERATING PRESSURE	= 7760. PSI
DISP. @ MAX. SPM	= 480. BPD		
PRESSURE RATIO	= 1.000	POWER FLUID RATE	= 386. BPD
		STROKES PER MINUTE	= 91.0 SPM
GAS THRU PUMP	GOR = 25. : 1	BRAKE HORSEPOWER	= 57. HP

preferred pump to use. This is considered to be a typical or standard jet pump application.

Producing low volumes is normal, or standard, for piston pumps based on how they have been historically used. Below is the installation analysis for a typical piston pump parallel tubing installation. When using a hydraulic piston pump, it should be remembered that there is a depth limit of approximately ± 15,000 ft. (depending on the particular pump). The minimum production rate will be on the order of 25 bpd, again depending on the particular pump. Also, as with jet pumps, setting the intake below the perforations is always preferable (tubing−casing annulus used as a gas separator) and a gas anchor should be attached to the intake of the pump (Table 7.4).

7.5 Summary

Before using a hydraulic jet pump, ensure that

* it is on a cost level of beam pumping.
* it is suitable for gas wells to remove larger amounts of liquids.
* it will not bring the well to very low flowing pressure.
* the method eliminates costly well servicing.
* all pumping systems work better if a minimum of gas enters the pump.

Similar to Sucker Rod pumping systems, use of hydraulic piston pumps is efficient but solids are not tolerated well.

Further reading

Anderson J, Freeman R, Pugh T. Hydraulic jet pumps prove ideally suited for remote Canadian Oilfield, SPE 94263.

Christ FC, Petrie HL. Obtaining low bottomhole pressures in deep wells with hydraulic jet pumps, SPE 15177.

Clark KM. Hydraulic lift systems for low pressure wells. *Petrol Eng Int* 1980.

Hrachovy MJ, McConnell ML, Damm MW, Wiebe CL. Case history of successful coiled tubing conveyed jet pump recompletions through existing completions, SPE 35586.

Peavy MA, Fahel RA. Artificial lift with coiled tubing for flow testing the monterey formation offshore california, SPE 20024.

Pugh T. Hydraulic pumping, iBook; 2014.

Pugh T, Robinson C. Hydraulic piston pump for dewatering gas wells, SPE 145749.

Liquid unloading using chemicals for wells and pipelines

8

Dr. Sunder Ramachandran is a Technical Advisor for the Oil Field Services — Chemicals division of Baker Hughes, a GE company. He is located in Sugar Land, TX. He has also been involved in the development and understanding of how surfactants change flow regime and can be used to unload liquids from wells and pipelines.

He has been with Baker Hughes from 1996 to 2014. He was briefly with Aramco Services Company from March 2014 to December 2015. He rejoined Baker Hughes in January 2016. He was a member of the Research Faculty at the California Institute of Technology from 1994 to 1996. He has doctorates in Chemical Engineering from Colorado State University and from the Indian Institute of Technology in Madras. He obtained his bachelor's degree in Chemical Engineering from the Indian Institute of Technology, Kanpur. He has over 90 publications on a variety of subjects related to his research, and over 40 additional industry presentations and 19 US Patents. He is a member of AIChE, SPE, ACS, and NACE. He served as the Technology Coordinator of the S Committee (Corrosion Science and Technology Committee) of NACE from 2011 to 2013. He currently is a member of the Technical Advisory Group (TAG) for the United States on the International Standard Organization (ISO) Technical Committee (TC) 156 on Corrosion of Metals and Alloys.

8.1 Introduction

Liquids in gas wells and pipelines in annular flow are removed by movement of the liquid film with the gas along the pipe wall and liquid droplets entrained along the high-velocity core. Physical and chemical conditions can increase or impede the movement of liquid. Chemicals modify fluid properties such as surface tension. In contact with gas, they can create a foam which has a lower density. Chemically modified properties of a fluid can change the relationship between pressure drop and flow. In special instances, the changes can dramatically decrease pressure drop for a given flow rate. For economical treatment with chemicals, it is important to identify opportunities when chemical treatment can result in liquid unloading. Chemicals have been used to unload liquids from gas wells, gas storage wells, plunger-lifted gas wells, gas-lifted oil wells, and pipelines. Surfactants can be used continuously or intermittently. Capillary injected application have been a popular means to introduce chemicals in wells. In offshore and European wells, special

Gas Well Deliquification. DOI: https://doi.org/10.1016/B978-0-12-815897-5.00008-1

designs were needed to satisfy the challenge of safe pressure containment with continuous injection of chemical. For each application special challenges exist with each application. Testing for fluid and materials compatibility is important for each application. Productive use of chemicals is associated with the intrinsic decline rate of reservoirs. Typically, chemical use is most relevant when the reservoir becomes mature. Companies that recognize the shift in maturity of their asset and introduce an appropriate chemical program will increase the ultimate recovery of gas from the reservoir.

8.2   Chemical effects aiding foam formation

8.2.1   Surface tension

Cohesive forces of molecules within a liquid create an elastic tendency to minimize surface area. This elastic tendency is known as surface tension. Surface tension is an important property to determine the interface between different phases. Chemicals that are surface-active preferentially migrate to the surface and decrease the surface tension of a liquid. During the process of interface creation, the elastic tendency to minimize the surface area is related to the concentration of chemicals at the surface. When interfaces are suddenly created, the elastic tendency is related to diffusion of the chemical to the surface as well as its ability to reduce cohesive forces. Dynamic surface tension includes the ability of chemicals to quickly migrate to the surface. One method of dynamic surface tension measurement is the maximum bubble pressure method. The effects of different surfactants on dynamic surface tension as a function of concentration and bubble frequency using the maximum bubble pressure method have been measured.[1] Brines should be kept under a CO_2 sparge during measurement to maintain pH, prevent solid precipitation, and oxygen ingress. When the surface tension of a solution is measured as a function of surfactant concentration, a concentration will be reached when no further reduction in surface tension will be obtained. This concentration is known as the critical micelle concentration (CMC). Typically, surfactants reduce the surface tension of aqueous brines to values between 40 and 60 mN/m.[1] CMC values of typical surfactants in aqueous brines range from 850 to 2100 ppm.[1] This variation is a function of brine salinity, pH, surfactant type, and chemistry. Typical surfactant chemistries vary from nonionic, anionic, cationic, and amphoteric.[2] Surface tension measurements have been presented on some silicon-based chemistries.[3] Studies comparing a commercial surfactant and sodium dodecyl sulfate have revealed that CMC, the equilibrium surface tension, the dynamic surface tension, and the surface elasticity are not good predictors of the dose rate of a surfactant needed to decrease pressure gradient.[4] The surface tension reduction of surfactants in aqueous systems can be as high as 50%. This reduction is not seen with chemicals that change the surface tension of oils.

8.2.2 Foam formation and foam density measurement

Foam is formed by the agitation of surfactant containing brine in the presence of a gas phase. Foams are created when gas gets trapped in a liquid. Foams consist of at least three phases: liquid, gas, and interface. Three stages characterize the lifetime of a foam. These stages are lamella formation, film thinning, and film rupture.[1] Lamella formation occurs in liquid with dissolved surfactant.[5] The adsorption of surface active material to bubble surfaces provide foam a longer term stability.[5] It is important that the surface active material has a low-vapor pressure so that it does not evaporate to form a lamella.[1] Foam acts like a liquid phase as it has a constant volume for a period of time. Foam degrades with time as gas and liquid separate from the foam. Film thinning occurs due to drainage of liquid. The mechanism of film thinning is different above and below the CMC. When the foam is formed above the CMC, the structural disjoining pressure is positive due to the presence of micelles within the film. Surfactants are normally introduced into field systems at concentrations above the CMC. The drainage of liquid decreases when the viscosity of liquid increases. Surface viscosity effects play a critical role in foam stabilization.[6] Foams break when they directly contact liquids that are soluble in the bulk liquid and have a lower surface tension than the bulk phase.[6]

One method for determining foam density is to use a blender test.[1] In this test, a volume of 100 mL of either synthetic or produced fluid at a given condensate/water ratio is agitated at a low speed for 60 s.[1] The total volume of fluid and foam is measured and immediately recorded. The time for 50 mL of fluids to separate from the foam is also recorded. It is important in the test to choose an appropriate condensate/water ratio similar to the condensate/water ratio in the portion of the well where liquid loading is occurring. The effect of salinity and different condensate/water ratio can affect the results of these tests.[7] The effects of surfactants on reducing liquid density in aqueous systems are higher than the reductions obtained in systems with larger amounts of oil.

Another method to use is a foam column test.[8] The test procedure for this test is described in the literature.[8] In this test the percentage of mass of fluid recovered is recorded with time.[8] The foam column test can be water jacketed to allow measurements at appropriate temperatures.[9] Measurements of foam density are best done with fresh fluids or appropriate sparging of CO_2 to maintain pH, prevent solid precipitation, and oxygen ingress. Fresh fluids are preferred especially if the fluids contain hydrocarbons as there are volatile constituents that with age can evaporate from fluids. The conditions of the test are carefully selected to minimize evaporation.

Methods have been developed that evaluate foam performance at high temperatures and pressures using a foam test column.[10]

8.3 Flow regime modification and candidate identification

The annular flow regime occurs for wells and pipelines with high-gas and low-liquid superficial velocities. In the annular flow regime, there is a liquid film and a central core of gas. The annular flow regime has a low overall friction factor. As gas velocities decrease, the flow regime changes to either slug flow or the slug/churn flow regime. The overall friction factor for these flow regimes is significantly higher than the annular flow regime. The transition between the regimes has been described by theories of liquid droplet flow reversal or liquid film reversal.

The equation determining the minimum gas velocity where liquid droplets flow with gas has been a popular method to determine the gas flow rate at which liquid loading occurs.[11] Liquid loading equations were modified to represent liquid unloading behavior of South Texas gas wells.[12] The effect of well inclination has recently incorporated into a new model for liquid loading.[13] The proper use of droplet flow models on wells in different fields has been evaluated.[14] Liquid loading evaluations need to be done using conditions at the bottom of the well.[14]

Experimental studies on the occurrence of flooding in annular dispersed gas liquid pipe flows and the role of the dispersed phase showed that liquid loading is not directly caused by dispersed phase flow reversal.[15] A review of droplet and film models shows that the film model provides a better agreement with experimental data on liquid loading.[16] Droplet theories have been successful due to the fact that the physics behind droplet and film flow reversal are both based on the balance between drag and gravitational forces.[17] A new method for predicting liquid loading uses the assumption that liquid loading initiates when the liquid film starts falling backward.[18] The method accounts for the effect of inclination and diameter.[18]

There are a few flow loop studies of foamed liquids in vertical gas liquid flow.[4] Methods that model the effect of foam on liquid film reversal contain equations that determine the fraction of gas trapped in foam, foam holdup, gas void fraction, and interfacial friction factors.[19] This is used to calculate pressure drop for foamed systems.[19]

Wells can operate at subcritical rates. They do so at "metastable" flow rates with intermittent gas production that has been modeled numerically.[20]

The effect of surfactants that reduce surface tension and create a low density to increase droplet entrainment has been well described in earlier work.[1] Reduction in liquid surface tension results in smaller droplet sizes.[1] Creation of foam results in a lighter fluid that gas can carry away at lower velocities.[1] The effect of chemical on droplet entrainment has been used to model gas wells and gas storage wells.[7] The parameters needed for well identification using this model are provided in Table 8.1.

The models used to identify gas wells calculate the minimum velocity needed to prevent liquid loading in the absence of chemical and the effect of chemical at different dosages to lower the minimum velocity needed to prevent liquid velocity. Successful candidates for foamer application have a gas velocity below the liquid loading velocity, but at a velocity where a sufficient amount of chemical can lower the minimum gas velocity to prevent liquid loading with chemical. One example of the results of such a calculation is in Fig. 3 of Ref. [7] which is shown in Fig. 8.1.[7]

Table 8.1 Parameters needed to identify possible wells for application with surfactant

Well head temperature
Bottomhole temperature
Well head pressure
Bottomhole pressure
Tubing inner diameter
Water production rate
Condensate/oil production rate
Gas production rate
Gas specific gravity
Chloride level in brine
Oil/condensate specific gravity (API)

In Fig. 8.1, the actual gas velocity is shown by the dotted lines. This value is 16.6 ft./s and independent of concentration. The velocity needed to unload the well without chemical is 29 ft./s. Four products at varying concentrations lower this unloading velocity to velocities below 16.6 ft./s. These products would be recommended for this application. The figure here shows the effect of chemical on the critical rate. In this instance, chemicals lowered the critical rate to below 15 ft./s for some products at 10,000 ppm. This is a 41% decrease in critical rate. This reduction is typically obtained in high water cut systems. Smaller decreases are seen in liquids with large amounts of hydrocarbons. The model allows for identification of relevant surfactant packages in different conditions. Some operators avoid chemicals when they have large amounts of hydrocarbons in the liquid phase. This strategy though may not allow them to avail some of the new developments of chemicals that foam hydrocarbons.[8,9]

Pipeline systems have been identified where chemicals have been used to dramatically increase gas production.[21] Parameters used to select the pipeline are collected in Table 8.2.

The model calculates the minimum velocity for the pipeline to be in the annular flow regime.[21]

Models have also been used to examine pipelines where the use of chemicals has been used to decrease liquid holdup and increase gas production.[22]

8.4 Application of surfactants in field systems

One of the applications of surfactants to alleviate liquid loading in gas wells is to inject a surfactant solution down the casing annulus.[23] As surfactant volumes can be low in these applications, often the surfactant solution is diluted. These allow for better pump regulation—assured drainage of the surfactant in these applications.[24]

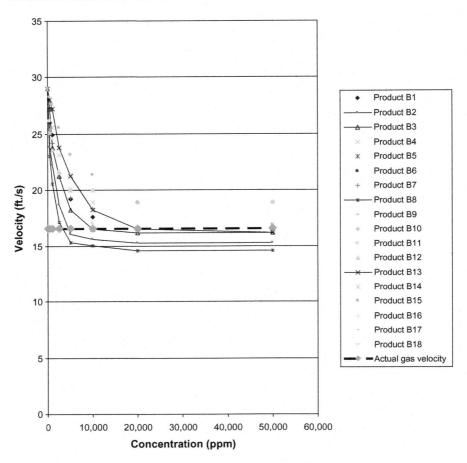

Figure 8.1 Effect of chemical concentration on the velocity needed to unload the gas well. Source: From Fig. 3 of Ramachandran S, Bigler J, Orta D. Surfactant dewatering of production and gas storage wells SPE-84823. In: *Presented at the SPE Eastern Regional/ AAPG Eastern section joint meeting.* Pittsburgh, Pennsylvania; September 6–10, 2003 [7].

Table 8.2 Parameters needed to identify possible pipelines for application with surfactant

Temperature at spot of interest in pipeline
Highest temperature in pipeline
Pressure at spot of interest in pipeline
Pipeline inner diameter
Water production rate
Condensate/oil production rate
Gas production rate
Gas specific gravity
Chloride level in brine
Oil/condensate specific gravity (API)

Another application is to pump a large amount of surfactant with enough chase fluid down the annulus of well to ensure it reaches the tubing intake depth intermittently.[24] One example of this is a case where a rise in water level of a gas storage well interrupted the gas deliverability of the well.[7] The rise in water level occurred at the time when gas deliverability is most important. In one instance, the water level in the gas storage field rose 51 ft. The well was modeled and a batch treatment using 3 gallons of surfactant was designed and implemented. The batch treatment unloaded a large amount of water and the well liquid level was found to be zero after treatment.[7] The well was tested and found to have identical well deliverability it had 40 years earlier.[7] Batch treatments are frequently used to kick off wells operating in a condition of metastable flow.[20] Another means of introducing surfactant in a well in an intermittent form is to use surfactants in a solid stick.[25]

Injection of surfactant down the tubing—casing annulus was used in the Barnett shale area to significantly reduce the need for lift gas.[26] Increases in production of over 22% were seen in application of surfactants in gas-lifted wells in Malaysia.[27]

The use of continuous stainless steel tubing (capillary string) below 1 inch outer diameter to deliver surfactants to the bottom of well tubing is one of the most effective methods. For unloading, backside injection is also very effective. Although soap sticks are not as effective, the use of soap sticks may be the most commonly followed by backside injection. Capillary injected surfactants can substantially increase the gas production from wells experiencing liquid loading.[7,26] The use of capillary injected surfactants to increase gas production has been used in many places. In the United Arab Emirates, they have been used to increase gas production in one well by 73%.[28] Use of capillary injected surfactant increased the duration of production 11 times and increased daily production of a gas condensate well with over 50% condensate in the liquid phase by 60%.[6,29] Capillary injected surfactants have been used in high-temperature wells to increase production.[30]

The predominant metallurgies used with capillary tubing are 2205 duplex, 2507 super duplex, Inconel 825, and Inconel 625.[31] It is important to test the chemical for its thermal stability and compatibility with metals in a capillary application.[8] One method to assess suitability of a chemical for capillary injection is to test the chemical in a capillary coil placed inside an oven.[10] Chemicals are pumped at a set rate and the differential pressure is recorded. The chemical is then kept in the tube for an extended period of time at a given temperature. After exposure to heat, the chemical is pumped again at the same rate and if the differential pressure is low and similar to the differential pressure seen before heating, the chemical is approved for capillary use. New developments have allowed the use of capillary injection in wells with surface controlled subsurface safety valves (SCSSSV).[31] This is done utilizing a wireline retrievable surface controlled subsurface safety valve.[31] A system has been developed to allow capillary tubing to be hung below the tree so that capillary tubing can be installed in wells with surface safety valves.[31] In one offshore system, the lower master gate valve was replaced with a modification that allowed capillary tubing to deliver surfactant through a SCSSSV.[31] In this system the foamer injection point is separate from the hydraulic fluid.[31] Use of surfactant with this system allowed an increase in total recovery from the well to a value as high as 10% in 2 years of operation.[32]

Combination corrosion inhibitor surfactant products have been an important means for preventing liquid loading and corrosion due to CO_2 in wells with declining gas production.[1] Surfactants have been used to increase production and eliminate slugging from subsea oil and gas-producing wells connected by two mile long flow lines to the production platform.[33]

8.5 Surfactant application for increased ultimate recovery

Optimization of production from mature gas wells involves production monitoring, well automation, installation of additional compression, velocity string installation, increased surfactant use, capillary installation, and plunger lift installation.[34] Systematic efforts such as these resulted in a 10% increase in production from the Lobo field in South Texas without any production decline.[34] Liquid loading calculation including the effect of surfactant indicated that a typical gas well in the Canyon Sand area would increase cumulative production of each well between 4 and 80 mmscf/day per well if surfactant was used.[7] The use of surfactant at an early stage is key to a successful operation of a shale gas reservoir. It has been estimated that application of combination foamer, scale corrosion inhibitor package increased the revenue of a Barnett Shale gas well by $170,000 per year in 2004.[35] It is important to produce as much gas as possible from each completed well in a tight gas formation. Use of surfactant in wells that see steep declines in well productivity is an important method in keeping operations in these reservoirs economic.

8.6 Summary and conclusion

The use of surfactants to increase gas production is one of the most important elements in increasing gas production from a mature gas field. Surfactants reduce the cohesive force of molecules to create an elastic tendency to minimize liquid area known as surface tension. Some surfactants create a foam that has a reduced density. Introduction of surfactants in a liquid reduces the gas velocity necessary to keep a central core of gas flowing in the system. When systems in two-phase flow have a central gas core, they operate at lower pressure drops for a given gas liquid flow rate. Surfactants increase entrainment of liquid in gas. Promising candidates for the use of surfactants in gas wells and pipelines experiencing slugging can be identified using relevant computer models. The surfactants can be applied by pumping the fluid down the casing annulus. Surfactants can be used intermittently. In some cases, they can stabilize the flow of wells operating in a metastable condition. Surfactants are found in both liquid and solid forms. They can be used to reduce the need of gas in gas-lifted oil wells. They can be applied using a capillary string.

In such instance, care should be taken in the selection of the appropriate chemical. Chemicals have been used in offshore systems and have helped increase gas production and prevent slugging. Combination corrosion inhibitor and foamer products have been important in maintaining gas production and preventing corrosion due to carbon dioxide. Use of surfactants is an important tool in increasing gas production in mature gas fields without production decline.

References

1. Campbell S, Ramachandran S, Bartrip K. Corrosion inhibition/foamer combination treatment to enhance gas production, SPE-67325. In: *Paper presented at the SPE Production and Operations Symposium*, Oklahoma City, OK; March 24–27, 2001.
2. Willis MJ, Horsup DI, Ngyuyen DT. Chemical foamers for gas well deliqufication, SPE-115633. In: *Paper presented at the 2008 Asia Pacific oil & gas conference and exhibition*, Perth, Australia; October 20–22, 2008.
3. Koczo K, Tselnick O, Falk B. Silicon-based foamants for foam assisted lift of aqueous-hydrocarbon mixtures, SPE-141471. In: *Paper presented at the SPE international symposium on oilfield chemistry*, The Woodlands, TX; April 11–13, 2011.
4. van Nimwegan AT, Portella LM, Henkes RAWM. The effect of surfactants on vertical air/water flow for prevention of liquid loading, SPE-164095. *SPE J* 2016;**21**:488–500.
5. Edwards DA, Wasan DT. Foam rheology: the theory and role of interfacial rheological properties. In: Prudhomme RK, Khan SA, editors. *Foam theory, measurements and applications*. New York: Marcel Dekker Inc.; 1996.
6. Schmidt DL. Nonaqueous foams. In: Prudhomme RK, Khan SA, editors. *Foam theory, measurements and applications*. New York: Marcel Dekker Inc.; 1996.
7. Ramachandran S, Bigler J, Orta D. Surfactant dewatering of production and gas storage wells, SPE-84823. In: *Presented at the SPE Eastern Regional/AAPG Eastern section joint meeting*, Pittsburgh, PA, September 6–10, 2003.
8. Orta D, Ramachandran S, Yang J, Fosdick M, Salma T, Long J, et al. A novel foamer for deliquification of condensate-loaded wells, SPE-107980. In: *Paper presented at the SPE rocky mountain oil & gas technology symposium*, Denver, Colorado, April 16–18, 2007.
9. Debord SB, Lehrer S, Means N, Crosby S. Development and application of foamers to enhance crude oil production, SPE-141025. In: *Paper prepared for presentation at the SPE international symposium on oilfield chemistry*, Woodlands, TX, April 11–13, 2011.
10. Shoeibi Omrani P, Shukla RK, Vercautern F, Nennie E. Improving foamer selection procedure for gas well deliquification, SPE-181592. In: *Paper presented at SPE annual technical conference*, Dubai, UAE, September 26–28, 2016.
11. Tucker RG, Hubbard MG, Dukler AE. Analysis and prediction of minimum flow rate for the continuous removal of liquids from gas wells, SPE-2198. *JPT* 1969;1475–82.
12. Coleman SB, Clay HB, McCurdy DG, Norris III HL. A new look at predicting gas well load-up, SPE-20280. *JPT* 1991;329–33.
13. Belfroid SPC, Schiferli W, Alberts GJN, Veeken CAM, Biezen E. Prediction onset and dynamic behaviour of liquid loading gas wells, SPE-115567. In: *Paper presented in 2008 SPE annual technical conference*, Denver, Colorado, September 21–24, 2008.

14. Sutton RP, Cox SA, Lea JF, Rowlan OL. Guidelines for the proper application of critical velocity calculations, SPE-120625. In: *Paper presented at 2009 SPE production and operations symposium*, Oklahoma City, OK, April 4–8, 2009.
15. van't Westende JMC. Droplets in annular-dispersed gas-liquid pipe flows. PhD Thesis, Delft, Technical University, Delft, Netherlands, 2008.
16. Luo S. A new comprehensive equation to predict liquid loading, SPE-167636. In: *SPE annual technical conferences*, September 30–October 2, 2013.
17. Veeken CAM, Belfroid SPC. A new perspective on gas-well liquid loading and unloading, SPE-134483. *SPE Prod Oper* 2011;343–56.
18. Shekar S, Kelkar M, Hearn WJ, Hain LL. Improved prediction of liquid loading in gas wells, SPE-186088. *SPE Prod Oper* 2017;539–50.
19. Ajani A, Kelkar M. Pressure drop prediction in vertical well under foam flow condition, SPE-181237. In: *Paper presented at the SPE North American artificial lift conference*, The Woodlands, October 25–27, 2016.
20. Dousi N, Veeken CA, Currie PK. Modelling the gas liquid loading process, SPE 95282. In: *Paper presented at offshore Europe 2005*, Aberdeen, Scotland, UK, September 6–9, 2005.
21. Ramachandran S, Orta D, Bartrip KA. Use of specialty chemicals to unload gas wells and deliquify pipelines. In: *Paper presented at Northern Area Western NACE conference*, Calgary, Alberta, February 4, 2003.
22. Ramachandran S, Eyrlander JGR, Wittfeld C, Baker ER, Schorling P. Alleviation of liquid loading in wells and pipelines in mature european gas wells using chemicals. In: *Conference paper at Deutsche Wissenschaftliche Gelleschaft für Erdöl*, Erdgas und Kohle e.V. April 28, 29, 2005.
23. Libson TN, Henry JR. Case histories: identification of and remedial action for liquid loading in gas wells: intermediate shelf gas play, SPE-7467. In: *Presented at 53rd annual fall meeting of SPE of AIME*, Houston, TX, October 1–3, 1978.
24. Lea JF, Tighe RE. Gas well operation with liquid production, SPE-11583. In: *Paper presented at 1983 production operations forum*, Oklahoma City, OK, February 27–March 1, 1983.
25. Hearn W. Gas well deliquification, SPE-138672. In: *Paper presented at the Abu Dhabi international petroleum exhibition & conference*, November 1–4, 2010.
26. Farina L, Passuci C, Di Lullo, Negri E, Pascolini O, Anderson S, et al. Artificial lift optimization with foamer technology in the alliance shale gas field, SPE-160282. In: *Paper presented at SPE annual technical conference and exhibition*, San Antonio, Texas, 8–10 October, 2012.
27. Ahmad SA, Mc Gregor S, Sen YM, Davoren S. Increasing production via foam assisted gas lift in a mature oil well, SPE-189201. In: *Paper presented at SPE symposium on production enhancement and cost optimization*, Kuala Lumpur, Malaysia, November 7–8, 2017.
28. Polouse B, Al Hamadi M. Foamer application for sajaa asset gas wells, SPE-164376. In: *Paper presented at the SPE middle east oil and gas show*, 10–13 March, 2013.
29. Debord SB, Lehrer S, Means N, Crosby S. Novel foamer application for enhanced oil production, SPE-141593. In: *Paper presented at the SPE middle east oil and gas show*, September 25–28, 2011.
30. Kalwar SA, Awan AQ, Rehman AU, Abassi HS. Production optimization of high temperature liquid hold up gas well using capillary surfactant injection, SPE-183676. In: *Paper presented at the SPE middle east oil & gas show*, March 6–9, 2017.

31. Embrey M, Bolding J. Latest advancements in capillary intervention on and offshore, SPE-121677. In: *Paper prepared for presentation at the 2009 SPE/ICoTA coiled tubing and well intervention conference and exhibition*, Woodlands, Texas, USA, 31 March—1 April 2009.
32. Vogelij NA, Veeken CA, Islamov RI, Lugtmeir B, Ros O, De Vries G, et al. Four years of continuous foamer application in southern north sea offshore gas wells, SPE-180031. In: *Paper presented at the SPE bergen one day seminar*, April 20, 2016.
33. Chakraborty S, Lehrer SE. Increasing oil and gas production by the application of black oil foamers, SPE-183904. In: *Paper presented at the SPE middle east oil and gas show*, March 6—9, 2017.
34. Harms L, Urlaub J, Carrier B, Cremar B. Optimizing mature gas wells in South Texas — a team approach, SPE-124911. In: *Paper presented at the 2009 SPE annual technical conference and exhibition*, October 4—7, 2009.
35. Ramachandran S, Duckworth M, Johnson K, Sharp B, Hausam M. Use of a combination foam, corrosion and scale inhibitor to improve productivity in tight gas formations. In: *Presentation at 2005 Denver de-liquification conference*, February 28—March 2, 2005.

Progressing cavity pumps

9

Ken Saveth, Engineering Manager, Reliability PCP/GL/HPU Reliability Oilfield
Services, Baker Hughes, a GE Company.
  Ken Saveth is a graduate in Petroleum Engineering Technology, Oklahoma State
University and over 34 years of experience in the oil and gas industry, mainly in
the training, design, installation, and optimization of artificial lift systems. Most of
the time, Ken has worked globally with Progressing Cavity Pumping (PCP) and
Electric Submersible Pumping (ESP) systems. Although PCPs are used in a plethora of applications, a large percentage of their applications are the deliquification
of natural gas and coalbed methane (CBM) or coal seam gas (CSG) wells. Ken currently is the Engineering Manager in Reliability involved in the improvement of
Progressing Cavity Pumping Run Times and determining Root Cause Failure. His
global travel involves PCP failure investigations as well as leading workshops on
the design, installation, optimization, and operation of Progressing Cavity Pumping
Systems.

9.1 Introduction

Progressing cavity pumping systems have long been used as an artificial lift method
in challenging artificial lift applications and when the overall system operating efficiencies need to be optimized. Although progressing cavity pumps (PCPs) have
been used in artificial lift since the early to mid-1980s, their application to deliquefy coalbed methane/coal seam gas (CBM/CSG) wells was brought about due to
their ability to handle higher percentages of free gas at the pump intake as well as
the coal fines that are often produced along with the water. Today, there are more
than 8000 PCPs producing water from CBM/CSG wells throughout the United
States, Canada, the United Kingdom, Kazakhstan, Russia, China, New Zealand,
Australia, and India.
  One of the main advantages of PCPs is their ability to handle produced solids,
liquids, and gases while maintaining a high overall system efficiencies when compared to any other artificial lift method in the same application. When you include
the comparatively lower capital cost, PCPs are the most often preferred artificial lift
system for many operators producing CBM/CSG wells. There is an extremely large
install base of PCP systems operating in the United States alone.

Gas Well Deliquification. DOI: https://doi.org/10.1016/B978-0-12-815897-5.00009-3

Since CBM/CSG are not the only gas wells that experience liquid loading problems requiring lift or experience the need to handle small amounts of frac sand or possibly other solids, the information provided here is also applicable to other natural gas wells that produce water in addition to CBM/CSG wells.

9.2 The progressing cavity pumping system

The PCP is a positive displacement pump that comprises two components: the rotating component or rotor and the stationary component or stator. The rotor is usually manufactured from high strength steel and coated with a chrome layer ranging from 0.010 to 0.020 in. thick and has the configuration of an external single helix (Fig. 9.1). (*Note: In some more abrasive applications, double chrome plating can be used.* The rotor is the only moving component within the pump.)

The stator consists of a steel tube with an elastomer, or rubber, permanently bonded inside the tube and has the configuration of an internal double helix. When the rotor is inserted into the stator, it creates a continuous seal line (compression/interference fit between the rotor and stator elastomer) that extends from the pump

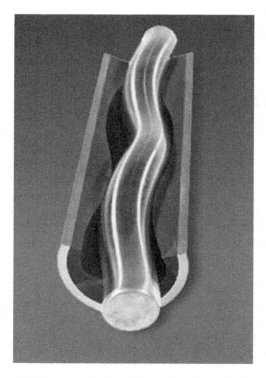

Figure 9.1 Cut-away of PCP pump showing chrome plated rotor (inside) and rubber lined stator (outside) and housing for stator.

suction to discharge. This creates a series of identical but separate cavities that progress from the pump suction to the discharge as the rotor turns. One cavity opens as the other closes, creating a nonpulsing pumping action.

In a conventional installation, the stator is installed on the bottom of the tubing string. The tubing string must be removed from the well to retrieve the pump. Alternatively, there are select models of PCPs (Insert PCP) that can be inserted into 2-7/8″ and larger tubing. An Insert PCP is installed and retrieved with the rod string. In both the cases, a torque anchor or No-Turn tool is used below the stator. The torque anchor is to prevent the tubing string from unscrewing during the normal pumping operations.

A conventional PCP system will typically have the following components: prime mover, drivehead, stuffing box, rod string, PCP, and torque anchor (Fig. 9.2).

In an Insert PCP system, the same components will be present; however, the jewelry associated with the insertable aspect of the system will be part of the system as well (Fig. 9.3).

Fig. 9.4 highlights the general application envelope for PC pumping systems.

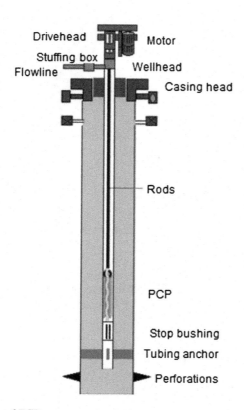

Figure 9.2 Schematic of PCP system.
Source: Image courtesy of Baker Hughes, a GE Company.

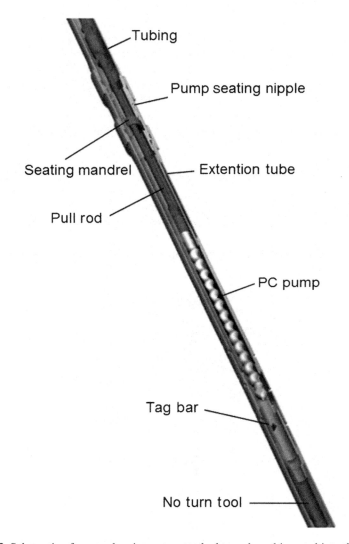

Figure 9.3 Schematic of pump showing rotor attached to rods and inserted into the stator. The tag bar is for spacing and the no-turn tool prevents twisting of the tubing when starting. *Source*: Image courtesy of Baker Hughes, a GE Company.

9.3 Water production

In gas well dewatering applications, PCP systems can produce very high rates at first. US applications range from several thousand barrels of water per day to less than 10 BWPD on the extreme low end. At the other end of the globe in Australia,

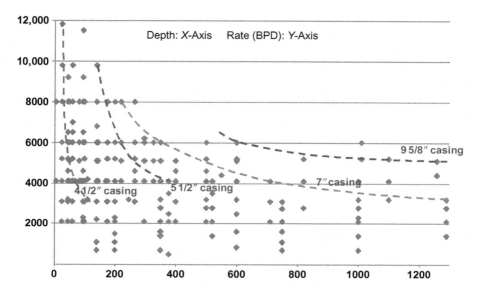

Figure 9.4 General application envelope for PC pumping systems.

CSG or coal seam gas wells can produce close to 5000 BWPD from depths of 2200 ft. and 2500 BWPD from depths of 3000 ft. Bottom-hole temperatures of these applications are around 175°F. Indonesia's light oil applications represent some of the deepest and hottest applications in which PC pumping systems are deployed where pump setting depths are nearing 9000 ft. true vertical depth (TVD) and bottom-hole temperatures around 220°F.

With a wide range of belt/sheave ratios available for a PCP's surface driveheads combined with a variable frequency drive. The production rate of a PCP could range from 500 to 5000 BWPD; however, the total system efficiency would be significantly less at the lower production end. Since initial production rates can decline rapidly over months or even days, the wide production rate range is particularly important in gas well dewatering applications.

9.4 Gas production

Due to the typically low-pressure reservoirs in which gas coalbed or CSG is found, it does not take much liquid head built up in the annulus area to reduce or even stop gas production. For this reason, the reservoir pressure must be drawn down to a very low level to achieve high gas recovery. In many cases, even fluid level must be pumped down to the coal face. In order to accomplish this, the pump setting depth is often landed below the lowest set of perforations to draw the liquid level down into the coal interval. In the case of an openhole completion, the pump can only be set to the bottom of the cased section.

Free gas can create major problems in artificial lift systems. In beam pumping or reciprocating rod lift (RLS) pumps, excess gas can build up below the balls in the traveling valve and does not allow enough pressure to build up under it on the downstroke to lift it off its seat thereby keeping fluid from being produced. It essentially gas locks.

In electric submersible pumps or ESPs, free gas tends to build up around the eye of the impeller where the pressure is lowest. In a similar but different fashion, this buildup of free gas does not allow fluid to continue through the impeller/diffuser geometry adequately thereby gas locking. In ESPs, it can be somewhere around 12% free gas.

It is important to point out that there are methods of gas separation in both RLS and ESP, which change these percentages; however, it is a topic not covered in this chapter.

With PCPs, this does not occur. There are no values to lock or impellers/diffusers around which gas bubbles can accumulate. Being a positive displacement pump, there is none of these components or their associated problems present. For a given pump geometry at a given pump speed, a certain amount of production can be expected. This total volume might consist of both liquid and gas. The higher the percentage of free gas present at the pump intake, the larger the volume of gas within the pump cavity. This then, affects the pump's volumetric efficiency, pump performance and in many cases, the pump's life expectancy. Conventional percentages of free gas at the pump's intake are around 40%.

Higher percentages of free gas also come with the potential for damaging the PCP, in particular, its elastomeric component. Free gas is not a good lubricant between the rotor and the stator and excessively high percentages of free gas at the pump intake and through the pump can cause the pump to overheat. With the reduction of adequate lubrication, the friction between the rotor and the stator will increase, which in turn will cause thermal expansion in the elastomer. This thermal expansion will increase the interference fit between the rotor and the stator further increasing the friction and frictional heat. Unless it is corrected, this cycle will continue to the point where the increased frictional heat will eventually harden the elastomer beyond its temperature limit and there is no longer any interference fit between the rotor and the stator. This is typically called "run dry." Depending on the rotational speed of the PCP and the downhole temperature, this "run dry" can occur relatively fast; often in less than 30 min; and can result in a nonrepairable catastrophic failure.

As the interference fit between the pump's rotor and the stator allows for some slippage between the two components from adjacent cavities and from a cavity of higher pressure to one of lower pressure, the presence of free gas leads to complication. As gas is compressible, the pressure distribution along the pump from the suction to the discharge end is nonlinear thereby causing varying stresses within the stator's elastomer. It often results in the upper cavities, or stages, having to generate higher pressure/stage than the lower ones where the gas is being compressed. This could result in an overpressuring of the upper portion of the pump and an ultimate hysteretic failure.

9.5 Handling of sand/solids/fines

PCPs are well known for their superior ability of handling produced sand, solids, or fines when compared to other methods of artificial lift. This ability, however, is due to a number of different reasons. The elastomer properties, the pump geometry, the angularity of the solids being pumped, and the pump speed all affect the ability to handle solids. When dealing with the handling of solids, the elastomer's resilience and viscoelastic properties allow the pump to continuously or intermittently handle them with minimal or reduced damage to the pump. Depending on how the well was completed, these solids can range from proppant used in the fracturing process of the well to produced sands or fines from unconsolidated reservoirs or coal seams. The pump's geometry is referring to the swept rotor angle (SWA—swept angle), or the angle that the rotor helix makes with the perpendicular to centerline of the rotor itself (Fig. 9.5).

In wells where higher percentages are expected, a lower SWA is needed to allow for more vertical forces imparted to the particle than horizontal forces pushing the particle into the elastomer.

With increased production of solids of any sort, come many issues that one needs to be aware of. Wear of both the rotor and the stator is accelerated. The hardness of the solids, the speed of the pump, and the geometry of the pump all play an important role in how much wear may occur and how fast it will occur. The additional load caused by the solids ingested into the pump will increase the torsional load of the pump as well as the horsepower associated with it. If solids production is expected, it is important to consider the additional torque and power requirements when sizing the rod string and prime mover in a rod driven progressing cavity pump (RDPCP) and the submersible motor and system shafting in an electric submersible progressing cavity pump (ESPCP).

If extreme conditions exist, solids can plug the intake or form bridges in the production tubing. The former can create a "run dry" condition while the latter can

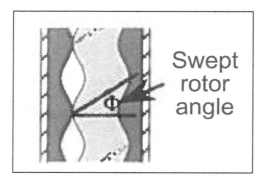

Figure 9.5 Showing the angle that the rotor helix makes with the perpendicular to the centerline of the rotor.
Source: Graphic courtesy of C-FER Technologies.

create additional pressure against which the pump will have to overcome. A high percentage of solids ingested into the pump can also result in severe damage to the pump as well as to the other parts of the artificial lift system.

Oftentimes, if the well is drawn down at a steadier rate rather than rapidly, the influx of sand/fines can be minimized. If higher production rates are needed than are seen at the lower speeds, a gradual increase in pump speed over time will allow the bottom-hole pressures change gradually as well.

Often, solids suspended in the fluid in the production tubing settle out when the well is shutdown for any reason. In such cases, the solids tend to settle out on top of the pump causing difficulties in starting the pump and in many cases locking the pump up. In an RDPCP system, small workover rigs or "Flush-bys," as they are called, would be needed to raise the rods, string and rotor, flush the tubing and pump with fluid, reseating the rotor and then restarting the pump. It is important to note, however, that in order to do this, a polished rod that is longer than the rotor would be needed to ensure that the rotor can be completely pulled out of the stator when raising the rod string without having to remove the surface drivehead from the wellhead. It is important to note that an adequate volume of fluid needs to be pumped back through the pump to ensure that the pump and solids are completely flushed from the system. A rule of thumb is to use one complete tubing volume.

In the case of an ESPCP, the system can be run in reverse for a short duration while pumping fluid down the production tubing. If the tubing is full, this may be adequate. *The duration of running in reverse should not exceed 5 minutes at a time.*

While the system is running, ensuring that the fluid velocity in the production tubing is greater than the sand settling velocity at its lowest production rate is critical in keeping the sand/solids/fines to settle out on top of the pump. The velocity at a given production rate would be affected by the flow area (cross-sectional area of the sucker rod/production tubing configuration) and the fluid viscosity. Reducing the production tubing size will increase the flow velocity thereby allowing better transportation of solids. Since most CBM/CSG wells produce water, the fluid viscosity would not play an important role as it does in more viscous applications like the Canadian or Venezuelan heavy oil plays.

As many of the unconventional wells can be directional or horizontal completions, sand settling and sand bridging are more prevalent. A reduction in the diameter of the production tubing while keeping the same side sucker rods may exacerbate rod/tubing wear in both the rod coupling locations or the rod body in areas that have high inclinations, deviations, or doglegs. This could result in premature rod or tuning failures.

Avoiding solids accumulation in or around the pump intake, adequate rat hole or sump is needed to keep from plugging the pump intake. Ensuring that the proper tag bar/intake configuration is important and as there are multiple configurations available, evaluating what is best for your application is critical. In applications where solids production is expected, extended tag bars with multiple slots and an extended and/or paddle rotor could be used. These configurations are shown in Figs. 9.6 and 9.7.

The particle velocity of the sand particles can affect the pump's elastomer while its presence will create additional friction and subsequently torsional requirements.

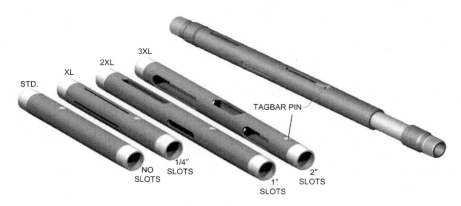

Figure 9.6 Extended tag bars with multiple slots.
Source: Graphic courtesy of Weatherford.

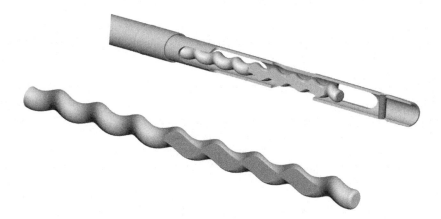

Figure 9.7 Paddle Rotor.
Source: Graphic courtesy of Weatherford.

For this reason, it is important when designing the entire system from the selection of the elastomer to the sucker rod size and the horsepower required. Maximum pump speeds should also be limited to minimize the additional wear associated with the abrasives. 300 RPM is a rule of thumb for this maximum speed; however, in some cases, it can be even lower. Each case needs to be evaluated individually.

Since torque equates to current in an electric prime move or hydraulic pressure in a hydraulic system, means to limit or control this variable via current limits set on a variable speed drive or back-pressure valves can be used to protect the system components from premature failures due to solids that may plug the pump or be pulled into the pump's intake. The control can either shut the system completely down, or change the pump speeds incrementally until the high limit either passes, or it is shutdown because it continues to increase above and beyond a predetermined period of time at this elevated torque.

9.6   Critical flow velocity

As mentioned in the previous section, keeping solids or fines in suspension is critical in preventing these solids from falling back on top of the pump. In heavy viscous fluids, the fluid viscosity generally is enough to carry the particles; however, in CBM/CSG wells where the produced fluids are lighter oils or even water, this fluid velocity plays a crucial role (Fig. 9.8).

9.7   Design and operational considerations

As one begins to consider the design and operational aspects of a progressing cavity pumping system as it applies to CBM and CSG applications, there are several reservoir details that should be kept in mind.

CSG is not generally produced virginally (i.e., without artificial lift) initially because the seam full of water and the hydrostatic pressure at the seam face are such that the gas cannot desorb and be produced. The hydrostatic pressure in the production tubing and acting on the coal face needs to be lowered to the point where the CSG can become free gas and be produced.

Although the initial thought would be to lower the fluid level as rapidly as possible to maximize the gas production equally as rapidly, doing so may pose additional problems. Having too rapid of a delta-P across the coal seam face could potentially break away solids from the formation, which would enter the pump and could create excess gas at the intake.

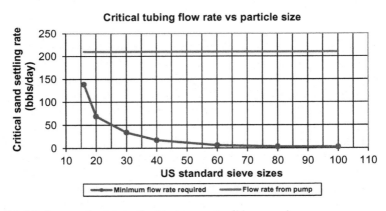

Figure 9.8 Minimum velocity required to transport solids upward.

Dropping the pressure across the face of the coal seam too quickly often results in creating instability within the seam matrix thereby allowing a higher percentage of fines to be produced. Handling these solids was covered in an earlier section.

To minimize this, operators should gradually drawdown the wells to reduce this rapid delta-P. Once the desired minimum hydrostatic pressure across the formation is reached, maintaining this pressure should be the goal. This gradual drawdown takes time and can often take weeks.

9.8 Implications of pump setting depth

How wells are completed and where the pump is set relative to the perforations or coal seam is critical to the overall operation of the PCP as well as the ability to maximize the drawdown of the wells resulting in maximizing the gas production. Typically, setting the pump intake below the producing zone is the ultimate setting depth as it allows for this maximum drawdown to be achieved. Generally, there are either open- or cased-hole completions. Certain completions, however, pose challenges that one needs to be aware of.

9.8.1 Open-hole completion

In an open-hole completion, the PCP can be set below the coal seam, typically in the open-hole section and below the cased-hole portion of the completion if an adequate amount of rat hole is present. In this scenario, the fluid level can often be drawn down to at or below the coal seam thereby maximizing the drawdown and gas production. The challenge here is that drawdowns of such magnitude may create an environment where the uncovered zone produces higher volumes of solids, therefore, uncovering the producing zone needs to be evaluated on a case-by-case basis. The additional challenge in open-hole completions is the fact that the torque anchor used to protect the pump from backing off must be set inside the cased-hole portion of the completion. This forces one to either set the torque anchor above the pump in the production tubing portion that is in the cased-hole area thereby leaving the pump unprotected and open for torsional back-off issues should the ingestion of solids cause torque spikes or not being able to set the pump below the producing zone. Running a gas separator or tail pipe below the pump in this latter case may be a possible solution; however, it should be evaluated on a case-by-case basis.

9.8.2 Cased-hole completion

In cased-hole completions, it is much easier to set the pump below the producing formation and still be able to run a torque anchor provided that there is an adequate amount of rat hole unlike in open-hole completions, cased-hole completions allow for the installation of a torque anchor below the pump (as it should be).

9.8.3 Presence of CO₂ and its effects

Carbon dioxide (CO_2) is often produced along with the methane gas in many CBM/CSG wells. The presence of CO_2 may cause several operating problems with progressing cavity pumping system. In particular, the elastomeric component of the stator.

Contrary to the common thought, the elastomer component of a PCP is permeable. What is meant here is that it has porosity and permeability, and given the right conditions and fluid/gas compositions, permeation of the elastomer can occur.

On a molecular level, CO_2 has a very small molecular structure and under the right conductions downhole can permeate into the elastomer matrix. Some of the elastomer's ingredients can be susceptible to this. It should be noted that it is not necessarily the dissolving of the CO_2 into the elastomer matrix that is often the problem, but is when the dissolved CO_2 comes out of solution that the elastomer can be damaged.

The failure mechanism that is generally associated with CO_2 permeation is referred to explosive decompression or ED. When there is a rapid delta-P across the elastomer/rotor interface, or even a rapid delta-P across the pump itself, the gas that permeated into the elastomer matrix wants to come out of solution. Unfortunately, it does not come out easily or in the same manner as it got in. The rapid decompression caused the CO_2 that is in solution within the elastomer matrix to rapidly expand. In doing so, it creates blisters and/or internal fracturing within the matrix. To mitigate the effects of ED, the following steps can be taken:

- Elastomer selection
  - Select an elastomer that has high mechanical properties.
  - Select an elastomer that allows the gas molecules to pass into and out of the elastomer matrix. (*Note: This is a discussion to be held with the PCP manufacturer.*)
  - It is highly recommended that for a first application into a new field that an elastomer compatibility test be run with the elastomers in question with the well fluids (water and oil) and at the down-hole temperature of the application.
  - If any sort of chemical treatment program is going to be used for corrosion, scale, paraffin, etc., it is essential to check the compatibility of the desired chemical with the elastomers being considered.
  - The pump setting depth (PSD) and producing fluid level (PFL) both address the free gas present at the pump intake. By minimizing the percentage of free gas entering the pump, the CO_2 that is in the gas phase will also be minimized.
  - It is important to point out, however, that as CO_2 is soluble in water, the gas permeating into the elastomer may not completely be eliminated.

9.9 Selection of progressing cavity pumps

Determining the conditions to design a PCP has long been a challenge. Often, well performance data is either field-centric, that is, it is from general field-wide field and not well specific, it is based on theoretical predictions gained from reservoir studies or even a very wide range of operational parameters. Altogether, applications engineers designing and selecting the required PC pumping system are often unsure which operational conditions to use to select the pump. In gas well dewatering (CBM or CSG) wells, it is good practice to design for the *worst-case scenario*. This worst case is most often with the well drawn down to its maximum level. This equates to the highest possible delta-P across the pump as well as achieving an optimum hydrostatic pressure against the producing zone.

Maximum delta-P across the pump is one aspect of proper pump selection. When selecting a PCP for a given application, ensure that the maximum delta-P could be seen is considered in selecting the necessary lift for the pump selection. It has long been a rule of thumb that 80% pressure loading is desired; however, there are many cases where one deviates from this rule of thumb (i.e., consistent inventory across a field, anticipated change in the dynamic fluid level, etc.).

Maximum recommended pump speed is another aspect of proper pump selection. Although many PCP manufacturers show a maximum of 500 RPMs on their literature, it is not the case. Many operators have set maximum pump speeds to 300−350 RPM and is often based on maximum cycles/lifetime and historic mean time between failures (MTBF) data.

Under this scenario, our design objective is to select a pump that will operate at a maximum rotational speed of approximately 400 RPM and around 80%−90% of its maximum rated pressure differential. As there are no adverse effects to the elastomer, other than some water swell, the mechanical properties do not significantly deteriorate. This allows the pump to run 90% loaded without much risk of elastomer degradation due to hysteresis. Designing the pump to initially operate at higher speeds allows a much greater turn-down ratio of the PC pumping system as fluid rates decline.

Once the well has stabilized under pumped-off conditions, the operator can optimize the PC pumping system to increase the run time and reduce the work-over frequency. This usually results in selecting a PCP that will be operating at speeds and loading of approximately 300 RPM and 80% of the maximum rated pressure differential, or less.

As mentioned in previous sections, fluid viscosity and production of solids also play a critical role in pump selection. This would affect the pump's geometry (i.e., long or short pitch).

9.10   Elastomer selection

When considering the elastomer portion of the PCP, there are always several options offered by all PCP manufacturers all of which are variations of ACN (acrylonitrile) and durometer or hardness. By varying the ACN content of an elastomer, the elastomer's ability to run in higher gravity oils, higher temperatures (with some processing variations), and varying water cuts are affected. For CBM and CSG wells, the elastomer with the lowest water swell (as determined in elastomer

Figure 9.9 Flyer for the C-Fer PC-PUMP program which is advanced program for sizing and selection for PCP systems.
Source: Image courtesy of C-FER Technologies.

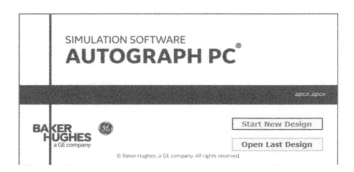

Figure 9.10 Baker Hughes offers the ESPCP so the Autograph program can be used to design this system.
Source: Image courtesy of Baker Hughes.

compatibility tests in a laboratory) and the best mechanical properties is generally the best choice. It is often considered a medium nitrile elastomer and as it happens, is often the most economical.

To accurately size and select PCPs for any application, companies use computer sizing programs that have been developed internally or by third party companies (Figs. 9.9 and 9.10).

Such programs take the data discussed in previous sections and allow the engineer to work several scenarios and able to take into account the plethora of design considerations and able to run numerous pump geometries and case histories to optimize the selected progressing cavity pump for your gas dewatering application.

Further reading

Progressing Cavity Pump Catalogue. Baker Hughes GE.

Saveth K. SPE-25448-MS. Field study of efficiencies between progressing cavity, reciprocating and electric submersible pumps. In: *SPE production operations symposium.* Oklahoma City, OK, 21–23 March 1993.

Saveth K, Klein S. SPE-18873-MS. The progressing cavity pump: principle & capabilities. In: *SPE production operations symposium.* Oklahoma City, OK, 13–14 March 1983.

Saveth K, Klein S, Fischer K. SPE-16194-MS. A comparative analysis of efficiency and horsepower between progressing cavity pumps and plunger pumps. In: *SPE production operations symposium.* Oklahoma City, OK, 8–10 March 1987.

Saveth K, Samuel GR. SPE-39786-MS. Progressing cavity pumps (PCP): new performance equations for optimal design. In: *SPE permian basin oil & gas conference.* Midland, TX, 23–26 March 1998.

Use of beam pumps to deliquefy gas wells

10

James F. Lea's experience includes about 20 years with Amoco Production Research, Tulsa OK, 7 years Head PE at Texas Tech and for last 10 years or so teaching for Petroskills and working for PLTech LLC consulting company. Lea help start the ALRDC Gas Dewatering Forum, has been coauthor of two previous edition of this book, author of several technical papers, and recipient of the SPE Production Award, the SWPSC Slonneger Award, and the SPE award Legends of Artificial Lift.

Lynn Rowlan, BSCE, 1975, Oklahoma State University, was the recipient of the 2000 J.C. Slonneger Award bestowed by the Southwestern Petroleum Short Course Association, Inc. He has authored numerous papers for the Southwestern Petroleum Short Course, Canadian Petroleum Society, and Society of Petroleum Engineers. Rowlan works as an Engineer for Echometer Company in Wichita Falls, Texas. His primary interest is to advance the technology used in the Echometer Portable Well Analyzer to analyze and optimize the real-time operation of all artificial lift production systems. He also provides training and consultation for performing well analysis to increase oil and gas production, reduce failures, and reduce power consumption. He presents many seminars and gives numerous talks on the efficient operation of oil and gas wells.

10.1 Introduction

Beam pumping systems are in the top four methods of deliquification. Among plunger lift, gas lift, surfactants, and SRP systems; probably rank second in the frequency of usage to deliquefy gas wells. They are used to pump liquids up the tubing and allow gas production to flow up the casing. Gas lift, plunger lift, and surfactant lift all produce gas and liquids up the tubing but pumping systems take mostly liquid up the tubing and gas is separated to casing production. One of the technical feats that must be accomplished is to separate most of the gas to the casing gas production because the down hole sucker rod pump functions poorly if at all, if much gas has to be pumped.

Beam pumps are commonly used, available, and are considered to be one of the most economical oil well production methods. However, with the deliquification of

Gas Well Deliquification. DOI: https://doi.org/10.1016/B978-0-12-815897-5.00010-X

gas wells, plunger lift systems are less expensive; so SRP systems are used to deliquefy, then the well/s in question do not have enough gas/liquid ratio, or do not have remaining pressure build up to operate plunger lift systems or conversely they may not be rated for high enough production as electrical submersible pumps (ESPs) and/or Gas Lift could be. Also SRP systems are excluded from wells that may produce excessive solids production or wells where gas separation can be difficult (high gas/liquid ratios and surging and heading production). Although most SRP systems have the pump landed before the KO to avoid wear, for horizontal wells, wells that have high dog leg severity (DLS) may exclude SRP systems. Gas lift, plunger, ESP, and land hydraulic pumps all are less restricted by the presence of high DLS.

Beam pump installations typically carry high costs relative to other deliquefying methods. The initial cost of a beam pump unit can be high if a surplus unit is not available. In addition, electric costs can be high when electric motors are used to power the prime movers (although the motor HP for deliquefying gas wells can be low) and high maintenance costs often are associated with beam pumping operations. Due to the expense, alternative methods to deliquefy gas wells should be evaluated before installing beam pumps. SRP systems can theoretically draw down the well to lower pressure/higher production than for instance plunger, but plunger lift can be so much less costly that plunger still may be preference.

If beam pumps are to be used for gas well liquid production, the beam system often will produce smaller volumes of liquids. Rates are low later in the life of unconventional wells as well. Because of the usually low volumes required to deliquefy gas wells, and the fact that beam pumps do not have a "lower limit" for production and efficiency as do other pumping systems such as ESPs, they are often used for gas well liquid production. However, there are production problems for low rates using SRPs.

The next section, as an introductory section, highlights the system components. Once the components and nomenclature are identified, their designs are discussed and some typical designs for deliquification of gas wells are shown. Pump-off control (POC) is widely used with SRPs, so it is discussed as part of the design process. Design of the rod string is discussed as well as special rod design procedures needed for wells with DLS. Since deliquefying gas wells and pumping gassy wells, gas separation has a section unto itself followed by gas handling pumps for the case where some gas passes through the separator.

10.1.1   The surface unit

Units that most likely to be used on a system to deliquefy a gas well include:

- Conventional units
- Mark units
- Hydraulically powered surface units (many types)

There are many other types of units used as well including rack and pinion drive units, air balance units, etc. Fig. 10.1 shows a small conventional unit lifting water

Figure 10.1 A small conventional unit.

off of a coal seam. Larger units are required when the well is still producing with high production on an unconventional decline curve.

Units (and gear boxes) are specified per the following nomenclature:

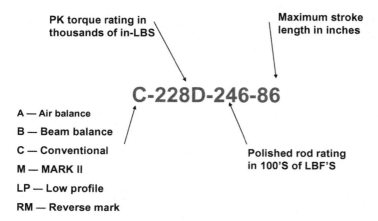

10.1.2 Wellhead

The wellhead is configured so that polished rods can be sealed off in the wellhead packing rubbers. Also the casing is vented for gas production and mostly the liquids exit the well at the top of the tubing. As shown in Fig. 10.2, many times the gas and liquids are combined and transported to the separator.

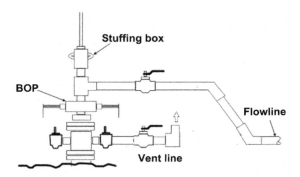

Figure 10.2 Beam system wellhead.

10.1.3   Polish rods

At the top of the rod string, there are a group of polish rods (PRs) that can smoothly slide though the wellhead packing rubbers and a seal is made. PRs carry the weight of the rod string and the fluid load on the upstroke. Special attention should be made to use a coupling such that one end has a PR thread and the other end has a rod thread when fastening the PRs to the underlying rods (Fig. 10.3).

10.1.4   Sucker rods and sinker rods

Sucker rods translate the up/down motion of the unit to the down-hole pump to create a pumping action. Sizing, care and handling, and proper make-up or tightening of the rod threads with the couplings are important. Different rod grades are more/less resistant to corrosion and are stronger/weaker. Note: Rods are referred to as the number of 1/8's in. in the body of the rod so a 7/8's rod is referred to as a size "7" rod (Fig. 10.4).

10.1.5   Sinker bars

Normally the operator will put in some sinker bars over the pump. If the pump has resistance to stroking down (bent barrel, solids), compression will be generated in the rods over the pump. Sinker bars are there to take the compression, and the larger diameter will not create as much rod/tubing damage as regular rods. Sinker bars are larger than the regular rods in the body but the threads and couplings are in such way that they will fit in the normally used tubing. For the most part 2 3/8's and 2 7/8's, tubing is used with beam pumping systems (Fig. 10.5).

Rod guides are commonly installed where deviation creates side loading and rod/tubing wear (Fig. 10.6). They can be installed at locations of noticeable wear when rods are pulled or spaced in advance by analyzing a deviation survey.

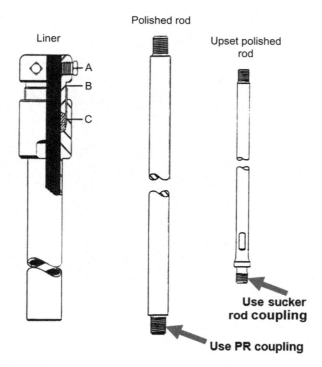

Figure 10.3 Polish rods.

API rod No.	Rod size	Metal area Sq. in.	rod weight In air, lb/ft., Wr
5	5/8	0.307	1.13
6	¾	0.442	1.63
7	7/8	0.601	2.22
8	1	0.785	2.90
9	1 1/8	0.994	3.67
10	1 ¼	1.227	4.53

Sucker rods nominally consist of 25 foot joints (30 ft. in California) with a threaded pin (male) connection on both ends. Manufacturers furnish a threaded coupling (female on one end of each rod. The sizes increase in 1/8 in. increments.

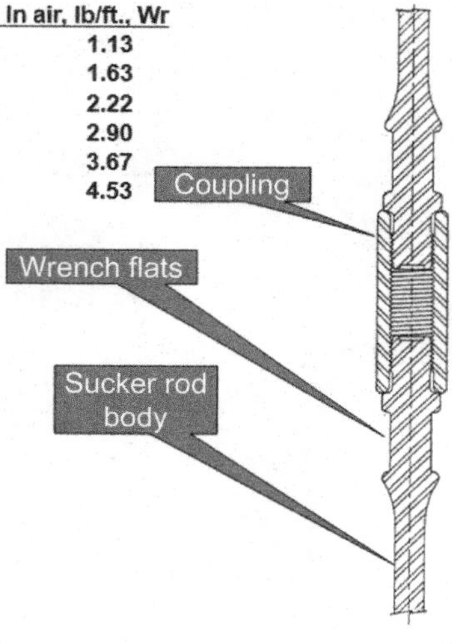

Figure 10.4 Sucker rods.

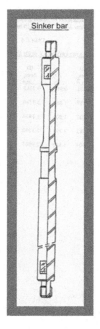

Figure 10.5 Sinker bars.

Use guides at wear locations

At dog-leg locations and/or at above pump to reduce wear due to fluid pound

Molded guides tend to slip less than hand installed

Figure 10.6 Rod guides.

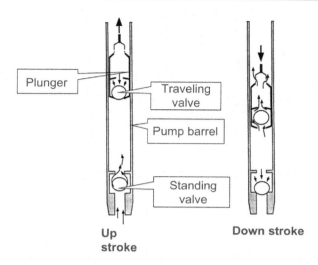

Figure 10.7 Down-hole pump schematic.

10.1.6 Pumps

A pump is tied at the bottom of the rod string. There are a number of types of pumps but one type (insert pump) can be installed and pulled using rods without a tubing. The tubing pump has the barrel fixed to the bottom of the rod string, and the plunger is fixed to the end of the rods. To pull the entire tubing pump the tubing would have to be pulled. For insert pumps a so-called "Top Holddown Pump" is less likely to stick, and a "Bottom Holddown Pump" will pump from a deeper depth without splitting. There are many types of pump designed to handle sand, including tight fit pumps and very loose fit pumps. It is better to keep gas out of pumps using separators but if gas still gets into pumps, special pumps for gas handling are available. Pumps for gassy fluids are designed to specifically handle gas using a high compression ratio (CR) or to make sure fluids are in the barrel on the downstroke to eliminate the so-called gas lock (Fig. 10.7).

The American Petroleum Institute (API) designation for bottom hole sucker rod pumps is illustrated in Fig. 10.8.

For instance a 1-1/4 in. bore rod–type pump with 10 ft. heavy wall barrel and 1 ft. lower and upper extensions, a 4-ft. plunger, and a bottom cup–type seating assembly for operations in 2-3/8 in. tubing would be designated as follows:

$$20 - 125 - RHBC - 10 - 4 - 1 - 1$$

10.1.7 Pump-off controls

To use and on/off pump-off controller, the SRP system has to be sized to produce a rate more than the well will produce. Then when the pump pulls the liquid from the well to a point at or near the pump intake, the pump will start taking gas. When this occurs, the POC is designed to shut down for a given time to allow the liquid

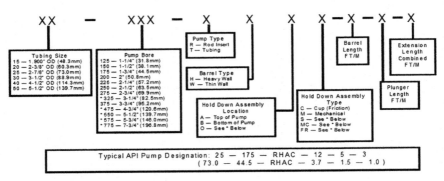

Figure 10.8 API pump designation.

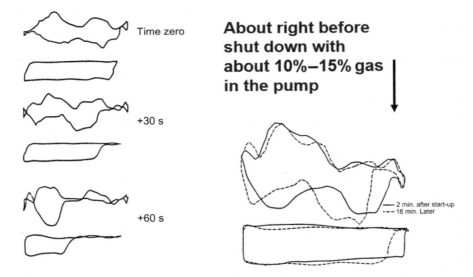

Figure 10.9 Surface and bottom hole dynamometer cards: W/WO gas in the pump.

to rise above the pump intake, and when the system is started again, it will have a near-complete liquid fillage if things are set up correctly.

The surface dynamometer card is a plot of measured rod load (load cell on the carrier bar under the clamp) on the y-axis and measured position (usually a "Hall Effect" transducer that sees when the crank travels by and the rest of the position is mathematically generated) on the x-axis. From the measured surface dyno card a so-called bottom hole card can be generated by a diagnostic computer card. The shape of the surface dyno card changes as gas enters the down-hole pump and especially the shape of the generated down-hole card also change as gas enters the pump. In the left side of Fig. 10.9, there are three sets of surface/bottomhole dyno cards with increased gas in the pump from top to bottom.

From the shape of the bottom-hole dyno card, many POCs can be set such that when a certain allowable amount of gas (reduced liquid fillage) occurs, the POC will shutdown the unit and let liquids recover over the pump.

On the higher volume portion of the decline curve for unconventional or shale wells, the production can be erratic and surging. Operators have found that a variable speed drive (VSD) POC controller seems to work better under these conditions.

The VSD POC uses a deadband on the right where incomplete fillage occurs and when the fillage gets low (too much gas in the pump), the VSD slows the speed and allows more liquid to enter the well for the pump to pick up. When the fillage becomes high (less gas in the pump), the VSD allows the unit to speedup and to pump the liquid level lower. This keeps the pump mostly full and keeps a low average level of fluid over the pump such that production is maintained at a high level.

POCs are important when deliquefying gas wells as the rates of liquids to be pumped can be low, and the SRP can easily pump faster than the well produces allowing gas into the pump, the POC and either slow the unit speed or shut it down for a period and allow the liquid to build back up in the annulus over the pump intake.

10.2   Beam system components and basics of operations

Fig. 10.10 shows an approximate lift-volume range for the application of beam pump systems. This depends on many factors including the presence of solids, deviation, other harsh conditions, and as such is only an approximate overall figure. Any point of operation under the curve is said to be feasible using this figure. Fig. 10.11 shows a typical beam pumping system.

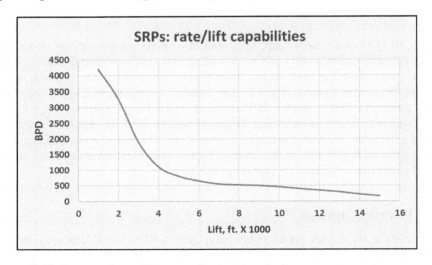

Figure 10.10 An approximate lift-rate application chart for beam pumping systems.

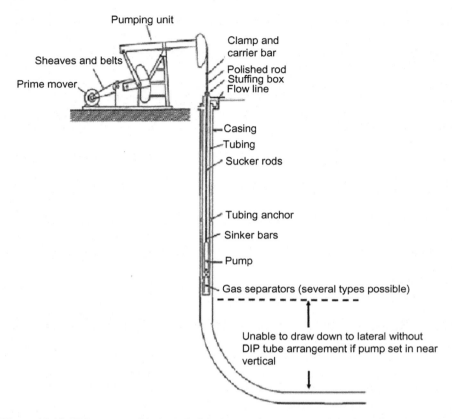

Figure 10.11 SRP system with down-hole pump set in near vertical portion of horizontal well.

Fig. 10.11 shows a beam unit, rods, and down-hole pump with the down-hole pump set above the kick-off point for a horizontal or unconventional well. If set in vertical well, then the perforations would have been shown in relation to the location of the down-hole pump. For horizontal or unconventional wells the pump is set above the well kick-off (KO) with sucker rod pumps (SRPs) in the United States since the KO is so sharp that damage would occur if the pump operated in a curved section of the well. Some horizontal wells, for instance, in Canada, have longer radius, smoother kick-offs, and the bottom-hole pump in those cases can be landed horizontally in the early portion of the lateral.

10.2.1 Prime movers

Starting at the surface the prime mover is typically a NEMA D electrical motor or a single or multicylinder gas engine. Some operators have installed NEMA B motors with variable speed drives in an effort to improve overall power efficiency.

From the below selection of oil field motors, most installations use a NEMA D motor to take advantage of the motor slowing (slip) under load (which cushions load), it has high starting torque and the fact it is still relatively efficient compared to other classes.

Type	~ Efficiency full load	Slip	Starting torque	Application
NEMA B	~92 +	2%–3%	100%–175%	Transfer pumps
NEMA C	~90 +	4%	200%–250%	Positive displacement injection pumps
NEMA D	~88%	8%–13%	275% +	Beam pumps
Ultra hi-slip	Lower	15%–30%	275% +	Special application beam pumps

10.2.2 Belts and sheaves

The belts and sheaves serve to carry power from the motor to the gearbox (GB). Alignment and the right number of belts are important (Fig. 10.12). Each belt is rated to transmit a certain amount of horsepower (HP). For a low HP application only one or two of the sheave grooves would be belted. Fig. 10.12 then shows what is a higher HP system.

10.2.3 The gearbox

The GB reduces rotation speed from the input shaft to the output shaft by a factor of about 30:1 and increases available torque by about the same amount. It is designed to not to be overloaded. The unit most be maintained fairly well balanced (weights adjusted to keep motor and GB loads about the same on up/downstroke) to make sure the GB is not overloaded in opeation. Also maintenance includes checking the condition of the lubricant in the GB (Fig. 10.13). Older GB's may need the addition of a wiper if they are to turn less than about 5 SPM (strokes per minute). Newer GB's are sold with the wiper already installed.

10.3 Design basics for SRP pumping

Most modern SRP design programs will take input concerning the rods, unit, pumping speed of strokes per minute (SPM), estimated rod/fluid-tubing friction, pump size, etc. and generate a surface and bottom hole dynamometer card, rod loads top to bottom, pump stroke length, production and loads on the motor, loads on the gearbox, and loads on the pumping unit. With a little iteration, you generate a design such that selected components are not overloaded and the desired production can be achieved. Fig. 10.14 shows factors considered in the rod string and the top/bottom dyno cards that are generated on the left side of the figure. Rules of thumb, rules from experience, and guidance from the programs can allow a workable design.

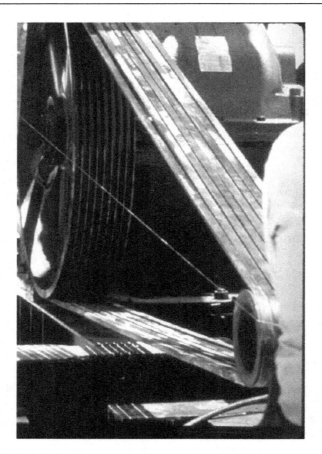

Figure 10.12 Belts and sheaves.

Fig. 10.15 shows input operating parameters (depth, desired production, tubulars, WHP, WHT, BHT, fluid propterties, dampening factors, etc.) and see loading on GB input, rods, and uni input. Also stroke length at the pump is determined and the resultant production. With few iterations, we can determine a design with no overloads that will achieve the desired production.

Fig. 10.16 reminds us to design with a GB with no overloads (but loaded to at least 50% or it is too large), a motor that is not overloaded but loaded to at least 50%, keep minimum rod load at the surface above zero, and if negative loading over the pump, we add sinker bars to absorb the compression, the maximum load on rods at surface should be below the structural rating of the unit and the rod loading anywhere in the rod string should fit the old (or possible newer revised) modified Goodman diagram with no overloads.

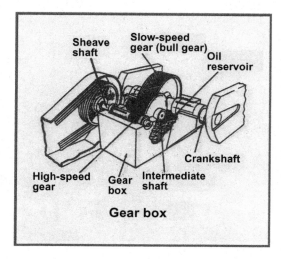

Figure 10.13 Gearbox internals.

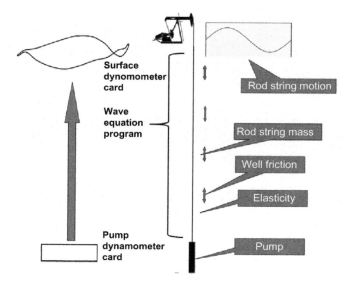

Figure 10.14 Schematic of some of the functionality of a design or predictive computer model (Lufkin).

10.3.1 Example designs

A few examples of design (that could be gas well designs) are shown as follows:
Case #1: Deep, low volume:

Unit: CW Conv.
Depth: 9000 ft.

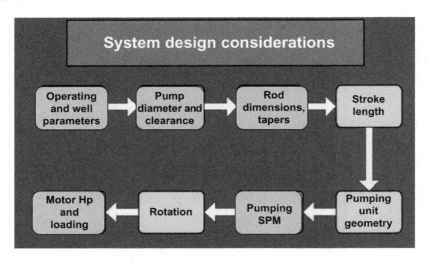

Figure 10.15 Design considerations (Lufkin).

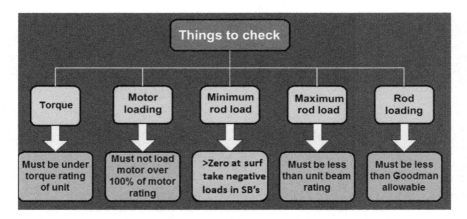

Figure 10.16 Items to check for SRP design (Lufkin).

Stk length: 64 in.
Pmp diameter: 1.25 in.
Tubing: 2 3/8's
Anchored tubing:
76 grade D rods
Surface tubing pressure (WHP), casing head pressure (CHP): 100 psi
PIP: 100
Desired rate: 30 bpd
Dampening factor: 0.1
Unit, pump efficiency: 95%

Results

Rate (100% pump volumetric eff.)	31.9 bbl/D
Rate (95% pump volumetric eff.)	30.3 bbl/D
Rod taper	31.1%, 68.9%
Top steel rod loading	82.2%
Min API unit rating	114-213-64
Min NEMA D motor size	4.32 HP
Polished rod power	2.35 HP
Traveling valve (TV) Load	19,226 lb
Standing valve (SV) Load	14,443 lb

Comments: With the low volume and small pump size the motor HP required is low. Probably would choose a 10 HP motor if available. A small unit is required even though deep.

Case #1: More shallow low volume:

Unit: clock wise (CW) Conv.
Depth: 4000 ft.
Stk length: 48 in.
Pmp diameter: 1.25 in.
Tubing: 2 3/8's
Anchored tubing:
76 grade D rods
WHP, CHP: 100 psi
Pump intake pressure (PIP): 100
Desired rate: 30 bpd
Dampening factor: .1
Unit, pump efficiency: 95%

Results

Rate (100% pump volumetric eff.)	31.6 bbl/D
Rate (95% pump volumetric eff.)	30.0 bbl/D
Rod taper	31.1%, 68.9%
Top steel rod loading	44.6%
Min API unit rating	80-95-48
Min NEMA D motor size	1.85 HP
Polished rod power	1.03 HP
TVLoad	8545 lb
SVLoad	6419 lb

SPM: 4.19

Comments: HP required very small. Very small unit required. Could possibly be plunger but only if well meets buildup pressure requirement and 400 scf/(bbl−1000') or more.

Case #2: More shallow low volume:

Unit: CW Conv.
Depth: 4000 ft.
Stk length: 48 in.
Pmp diameter: 1.25 in.
Tubing: 2 3/8's
Anchored tubing:
76 grade D rods
WHP, CHP: 100 psi
PIP: 100
Desired rate: 30 bpd
Dampening factor: .1
Unit, pump efficiency: 95%

Results

Rate (100% pump volumetric eff.)	31.8 bbl/D
Rate (95% pump volumetric eff.)	30.2 bbl/D
Rod taper	31.1%, 68.9%
Top steel rod loading	44.6%
Min API unit rating	80-95-48
Min NEMA D motor size	1.87 HP
Polished rod power	1.04 HP
TVLoad	8545 lb
SVLoad	6419 lb

SPM: 4.20
Comments: HP required very small. Unit required small.
Case #3: Deep, higher volume

Unit: CW Conv.
Depth: 9000 ft.
Stk length: 120 in.
Pmp diameter: 1.5 in.
Tubing: 2 3/8's
Anchored tubing:
86 grade D rods
WHP, CHP: 100 psi
PIP: 100
Desired rate: 250 bpd
Dampening factor: .1
Unit, pump efficiency: 95%

Results

Rate (100% pump volumetric eff.)	263.0 bbl/D
Rate (95% pump volumetric eff.)	249.8 bbl/D
Rod taper	27.0%, 26.7%, 46.3%
Top steel rod loading	97.4%
Min API unit rating	640-305-120

Min NEMA D motor size	49.19 HP
Polished rod power	22.47 HP
TVLoad	23,848 lb
SVLoad	16,962 lb

SPM: 8.9

Probably 50 HP motor, much larger unit needed.

Case #4: Shallower, higher volume

Unit: CW Conv.
Depth: 4000 ft.
Stk length: 120 in.
Pmp diameter: 1.5 in.
Tubing: 2 3/8's
Anchored tubing:
86 grade D rods
WHP, CHP: 100 psi
PIP: 100
Desired rate: 250 bpd
Dampening factor: .1
Unit, pump efficiency: 95%

Results

Rate (100% pump volumetric eff.)	262.9 bbl/D
Rate (95% pump volumetric eff.)	249.7 bbl/D
Rod taper	34.0%, 66.0%
Top steel rod loading	66.8%
Min API unit rating	456-133-120
Min NEMA D motor size	19.09 HP
Polished rod power	10.74 HP
TVLoad	9539 lb
SVLoad	6479 lb

SPM: 8.84

Comments: Probably use 25 HP motor. Unit needed is 456 but can go to 320 if rotate opposite direction.

This completes showing a few example SRP designs.

10.3.2 Rod designs with dog leg severity present

We know that for horizontal wells the pump is sent right above the KO to avoid pump and rod wear. However, there can still be DLS along the rod string so below are some recommendations for rod design when DLS is present.

First of all a survey is needed to see the well profile. A gyro survey is more accurate than an measured while drilling (MWD) survey. The survey should have points every 10 ft. but 2 per 100' is minimum. The build rate for rod installations should be <20°/100' but since pump landed before KO, this is usually not a problem.

DLS is the change in degrees of inclination per 100'. For sucker rod lift the DLS should be <5°/100'. Even better would be <3.0°/100'. If greater than 10°/100' then maybe well not suited for sucker rod lift.

The survey can then be entered into one of the few wave equation models, and the programs have utilities to then calculate the side loading along the rod string. Roughly, if side loading less then ∼70 lbs may be 0.3−1 failures/(well-year). Side loading from 70 to 150 lbs could expect failures of 1−2 failures/(well-year) and if greater than 150 lbs, then failures could exceed 2 failures/(well-year). Although a lot of the industry follows these recommendations as standard operating procedure (SOP), many of the recommendations originated from Russel Stevens, Rod Consulting.

Fig. 10.17 shows a very high side loading calculation from a survey and wave equation program.

Rod guides to protect against rod/tubing wear for side loading carry about 40 lbs/guide. So side loading determines how many guides per rod should be used. Some operators never put on less than three guides per rod. However, if less than 100 lbs then you may need no guides. For Fig. 10.16, one rod would need at least three guides to absorb the highest side loading.

Some operators add a stabilizer bar just above the pump and some add between sections of sinker bars (Fig. 10.18).

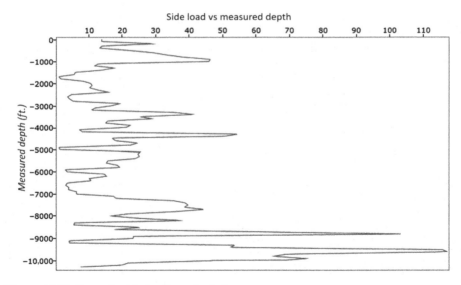

Figure 10.17 Example side loading calculations.

Figure 10.18 Stabilizer bar.

Table 10.1 Sinker bar specifications (Flexbar)

Effective weight sinker bars (C or K grade)					
Description	Minimum tubing ID	lbs/ft.	lbs/bar	API pin size	Decimal OD equiv.
*1 ¼ in.	2 in.	4	100	5/8 in.	1.250 in.
*1 3/8 in.	2 in.	5	125	5/8 in.	1.375 in.
1 3/8 in.	2 in.	5	125	3/4 in.	1.375 in.
1 1/2 in.	2 in.	6	150	3/4 in.	1.500 in.
*1 1/2 in.	2 in.	6	150	7/8 in.	1.500 in.
1 5/8 in.	2 1/2 in.	7	175	7/8 in.	1.625 in.
1 3/4 in.	2 1/2 in.	8.2	205	7/8 in.	1.750 in.
*2 in.	2 1/2 in.	10.7	267.5	1 in.	2.000 in.

10.3.3 Sinker bars

Operators add sinker bars above the pump to absorb the possible compression in the rods due to pump resistance on the downstroke. The user is allowed to enter a "pump resistance" that is defaulted to 200 lbs in most programs. If from field and failure symptoms or from loads inferred from calculated down-hole cards from a diagnostic, the compression loads appear to be higher or if the user wants additional safety factor against buckling in the rods, more sinker bars are specified. The program is then run, and rod compression is then found to extend into the rods over the pump. The technique is to add sinker bars until the rod load just above sinker bars is positive by a few hundred pounds. This may take some iterations as and when the sinker bars are added, the distance that compression extends upward can/ will change. Regardless, this method is widely used to design sinker bars using a wave equation program.

Large diameter sinker bars can fit inside tubing even though the rod diameter is large, because the pins and couplings (which determine fit) are smaller (see Table 10.1).

10.3.4 Design with pump-off control

Often if a beam pump is used to dewater a gas well, then relatively small amounts of liquid must also be produced to allow the gas to flow. The usual procedure is to pump liquids up the tubing and allow the gas to flow up the casing. Because small rates of liquids may be produced, it is not unusual for the beam system to pump at a rate higher than the well can deliver liquids over time. When a beam pump is operated at a rate beyond the capacity of the reservoir to produce liquids, the liquid level in the well is pumped near or below the pump intake and the well is said to be "pumped-off." This condition allows gas to enter the pump barrel and inefficient and damaging conditions may exist.

There is considerable literature[1,2] concerning beam pump systems on "pump-off" control. With gas in the pump barrel, the pump plunger initially compresses the gas on the downstroke of the pump before contacting the liquid. If sufficient gas is allowed into the barrel, the plunger can contact the fluid causing "fluid pound" with sufficient force to ultimately damage the pump and rod string. This is of primary concern in gas wells due to the relatively high volumes of gas produced with typical low volumes of liquid.

The pump-off controller enables the beam pump to operate with sufficient liquid levels to prevent damage while operating the pump at a high efficiency. The controller essentially stops the pump when the well has been pumped off or nears being pumped off. However, some pumping systems are often allowed to operate in the pumped-off condition with continual gas interference at the pump. This results in poor efficiency and can result in "fluid pound" as the plunger contacts the fluid in gas/liquid filled barrel on the downstroke. Fluid pound can lead to mechanical damage to the system.

The beam pump system should be designed to be able to pump the fluid level in the annulus down to the minimum value consistent with efficient pump operation and prevention of fluid pound.

To achieve this design objective the pump should be designed to pump at a rate given by

$$\text{Design rate} = \frac{\text{maximum inflow capacity} \times 24\,\text{h/day}}{\text{pump volumetric efficiency} \times \text{h pumped/day}} \tag{10.1}$$

The pump volumetric efficiency is essentially the percent fillage of liquids in the pump barrel. For effective pump-off control, 20 h/day pumping time is a good rule-of-thumb. The maximum reservoir inflow capacity should be used for the desired daily rate. Example 10.1 illustrates this equation.

Example 10.1 Design system pumping rate for POC

A gas well with a maximum liquid flow capacity of 300 bfpd is to be put on beam lift to pump off the liquids. For what rate should the pump be designed, assuming a pump volumetric efficiency of 80%?

$$\text{Design rate} = \frac{300 \times 24}{0.80 \times 20}\,\text{bfpd} = 450\,\text{bfpd} \tag{10.2}$$

Using this technique the pump is designed to operate about 20 h/day with an 80% volumetric efficiency. The pump-off controller will turn the well off when it reaches fluid pound conditions. The operator usually sets the downtime based on production considerations or some controllers will make adjustments in downtime automatically for optimum downtime.

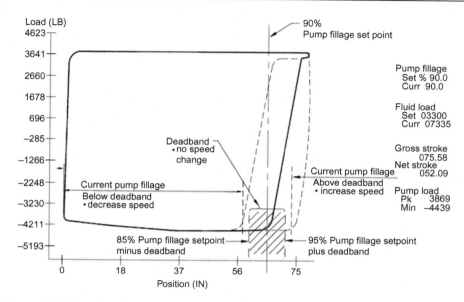

Figure 10.19 Deadband shown on a bottom hold dynamometer card for VSD POC control.

Using a typical volumetric efficiency of 80% and 20 h/day pumping time, a simpler rule-of-thumb is to simply design the beam pump system to deliver a rate equal to 1.5 times the reservoir maximum inflow capacity.

Design rate = 1.5 × maximum inflow capacity

Variable speed drive pump-off control

For the heading and surging that pumps experience in horizontal wells, some operators prefer the VSD POC to allow better control of the production and fluid levels. The VSD POC has a deadband on the right of the calculated down-hole card and when incomplete fillage strays to the left of the deadband (more gas in the pump) then the VSD will slow down and allow liquid to fill over the pump intake. When the fillage reaches values to the right of the deadband, then the unit will speedup and lower the fluid level. This is pictured in Fig. 10.19.

10.4 Handling gas through the pump

If separators are not successful in eliminating gas from the pump, then special pumps or pump construction will assist in handling gas through the pump as a second resort.

Fig. 10.20A shows a schematic of the stationary barrel sucker rod pump with a cycle showing no gas in the pump. At the start of the upstroke the plunger begins moving upward. This movement increases the volume of the compression chamber

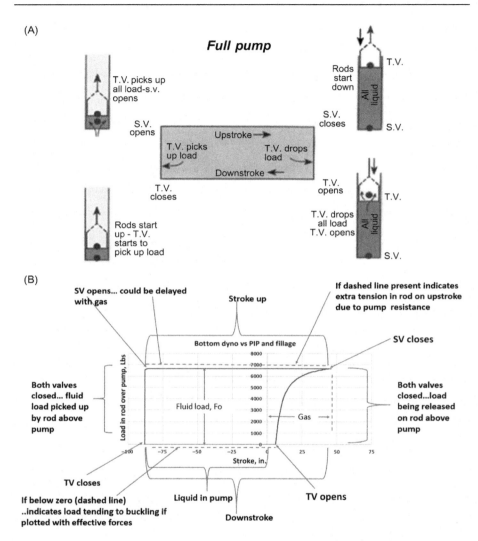

Figure 10.20 (A) Normal pump cycle with little gas in the pump. (B) Pump cycle with significant gas in the pump barrel.

between the traveling valve located in the plunger and the standing valve in the barrel. As the compression chamber volume increases the pressure decreases until it is lower than the hydrostatic pressure in the casing/tubing annulus. At this point the standing valve opens and admits fluid into the compression chamber. For solid fluid with no gas, this happens almost immediately when the plunger starts upward.

Fluid continues to fill the compression chamber until the plunger reaches the top of the stroke. As the plunger begins to travel downward at the start of the downstroke, fluid tries to escape from the compression chamber through the standing

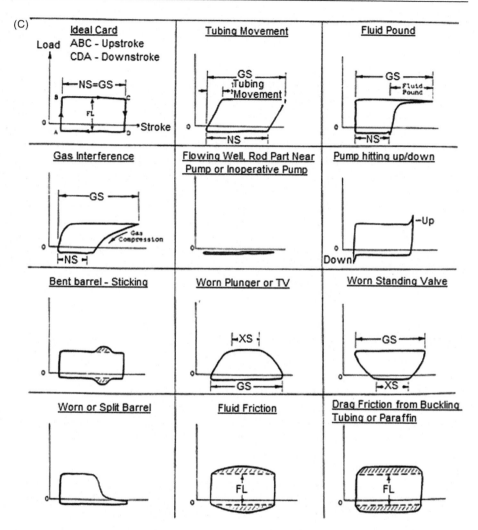

Figure 10.20 (C) Various pump dynamometer cards showing example problems.

valve. The flow of fluid past the standing valve ball pulls the ball back onto the seat, thus closing the standing valve.

As the plunger continues downward it decreases the volume of the compression chamber and raises the pressure. The pressure increases until it is higher than the fluid pressure in the tubing above the plunger. At this point the traveling valve opens and the plunger continues downward, and falls through the fluid, until it reaches the bottom of the stroke.

When the plunger reaches the bottom of the stroke and starts upward, fluid tries to flow back through the traveling valve into the compression chamber. This fluid flow past the traveling valve ball causes it to move back onto the seat, sealing the fluid pressure and closing the traveling valve. As the plunger moves upward in the barrel, it lifts the fluid column toward the surface and begins the fluid production cycle again.

And before solution to the problem of gas in the pump, let us define gas locking as problems to be avoided in the pump along with just the fact that gas in the pump takes space and reduced the liquid volume per cycle produced.

Fig. 10.20B shows a pump cycle with significant gas to be handled by the pump. One of the main differences with gas in that when the plunger starts its downward motion, the gas under the plunger is immediately compressed to some degree but does not immediately reach a pressure sufficient to open the TV. So the plunger continues downward through a load release distance until the gas under the plunger reaches a pressure greater than that above the plunger and the TV opens. The shape of the card indicates that fillage is affected by the presence of gas in the pump barrel. Fluid is still produced but the rate is reduced because part of the barrel is filled with gas and not liquid. Fig. 10.20B is a card showing gas interference. It could be thought of a "fluid pound" card if the load release slope is more near vertical and the load is released from the rods above the pump quickly over a short plunger travel. Fig. 10.20C shows some various pump dynamometer cards whose shapes correspond to different pumping situations and problems.

10.4.1   Gas lock or loss of valve action: summary

The pressure inside the pump chamber is a function of plunger travel and compressibility of fluid in the pump. Sucker rod pumps intake well fluids through the SV and discharge fluids into the tubing through the TV. Gas pumped into the tubing can be detrimental to the operation of the sucker rod pumping system. No pump action can occur when too much gas is pumped into the tubing. Use of back pressure and/or gas separation may be required in order to maintain pump action. A recommended practice is to keep the gas out of the tubing by setting the pump intake below the perforations. This is not possible in horizontal wells, so when a rat hole is not available, the recommended practice is to use a properly sized down-hole gas separator. Using a specialty pump such as a VSP pump will discharge gas into tubing, but prevent down-hole rod-on-tubing wear problems created by compressing gas on the down stroke. Using a long stroke length typically increases the pump's compression ratio, but gas in the pump is discharged into the tubing. Proper spacing of the pump to minimize dead space at the bottom of the stroke increases the compression ratio, but gas in the pump is discharged into the tubing. Slippage through the pump clearances partially fills liquid into a pump chamber full of gas increasing the compression ratio, but the gas in the pump is discharged into the tubing. If the pump has clearance between the plunger and the barrel, then it is impossible to gas lock the pump. Sufficient back pressure can prevent tubing fluids from unloading. Unloading tubing fluids is usually caused by poor down-hole gas separation because the gas in the pump is discharged into the tubing. Pump action stops when too much gas is pumped into the tubing because excessive gas discharged into the tubing lightens the tubing liquids causing the tubing liquids to unload from the tubing. The real problem is loss of pump action, not gas lock. When operators state their pump is gas locked, then the typical problem is the pump has no differential pressure across the valves and the pump will not pump.

Since leakage occurs by leaking liquid from above the plunger downward through the plunger-barrel clearance (on the upstroke), then there should always be some liquid above the SV on the downstroke, so it is difficult to understand how a cycle could occur without the TV opening when and if the plunger does hit liquid. It is easier to understand how the pumping efficiency could be lowered, perhaps drastically, by gas in the pump barrel.

10.5   Gas separation

10.5.1   Principle of gas separation

Pumps work or work the best when they pump mostly liquids. If gas gets into the pump, less liquid will be produced and too much gas may prevent pumping completely.

The principles of gas separation (below) include the following:

Gas bubbles travel upward in wellbore fluids $\sim \frac{1}{2}$ ft./s

If the downward velocity of fluids (before the fluids enter the pump or separator) less than $\sim \frac{1}{2}$ ft./s then the bubbles will rise up the casing and not go down to enter the pump.

Maximum liquid rate such that gas separation can be possible

For 1 in.^{2} of area where fluids travel down before entering the pump, downward velocity of $\frac{1}{2}$ ft./s the allowable rate for gas separation is:

$$\text{BPD} = (1 \text{ in.}^2/144 \text{ in.}^2/\text{ft.}^2)(\frac{1}{2} \text{ ft.}/s)(1/5.615 \text{ ft.}^3/\text{bbl})(3600 \text{ s/h} \times 24 \text{ h/day})$$
$$= 53.42 \text{ BPD}$$

Or round to the number to 50 BPD for conservatism and to provide easy to remember number.

Example for natural separation with intake set below perforations is shown in Fig. 10.21.

Example: From Fig. 10.21, assume:

Casing ID = 4 in.
Tubing OD = 2.375 in.
Downflow area (shown in Fig. 10.21) = $\pi (4^2 - 2.375^2)/(4) = 8.132$ in.^{2}
Max flow to achieve separation is 8.132 in.^{2} (of downflow area) $\times$ 50 bpd/in.^{2} = 406.6 BPD

Smaller flows and larger downflow areas also allow separation (Table 10.2).

In this table the rate is 410 BLPD for the case in the example earlier. The earlier example calculated a BLPD rate of 406.6. The slight difference related to what grade of casing/tubing was used for calculations.

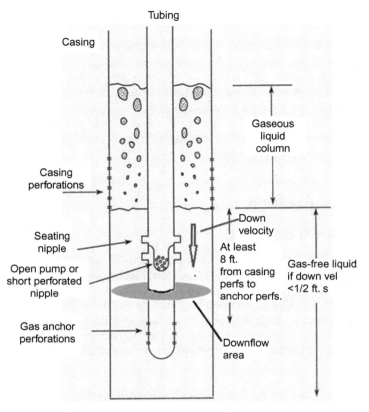

Figure 10.21 Natural separation with downflow area shown.

Table 10.2 Liquid capacity of natural separators (setting below perforations)

Gas separator capacity—pump below fluid entry zone				
Casing size (in.)	Dip tube size (in.)	Description	Annulus area[a] (in.^{2})	Liquid capacity (BLPD)
Conventional				
7	3½	Perforated tubing sub	23.1	1150
7	2 7/8	Perforated tubing sub	26.7	1335
7	2 3/8	Perforated tubing sub	28.8	1440
5½	2 7/8	Perforated tubing sub	12.7	635
5½	2 3/8	Perforated tubing sub	14.8	740
4½	2 7/8	Perforated tubing sub	6.1	305
4½	2 3/8	Perforated tubing sub	8.2	410
Higher capacity if needed				
5½	1½	Perforated line pipe	16.4	820
4½	1¼	Perforated line pipe	10.4	520

[a]Annulus area between casing and perforated tubing sub (or line pipe).

Poor boy separator

For some wells you cannot set the intake below the gas influx as you can for using the natural separator above. For instance, there may be no "rat hole" in the well or if in a horizontal well, you cannot set below the later (for most horizontals). Then you can use a so-called "poor boy" separator where you essentially build a down-flow area and passageway into the separator.

Fig. 10.22 is an example of using a "poor boy" separator.

Example calculation for using a poor boy separator

Assume:

Separator housing ID = 1.85 in.
Dip tube = 1 in.
Downflow area = $\pi(1.995^2 - 1^2)/4 = 1.901$ in.^{2}
Max. rate = 1.901 in.$^2 \times 50$ BLPD/in.$^2 = 95$ BLPD

For a larger downflow area the maximum rate or capacity could be much bigger (Table 10.3).

Fig. 10.23 shows Harbison–Fischer's rules for constructing a poor boy separator. These rules are commonly adapted for use by the industry.

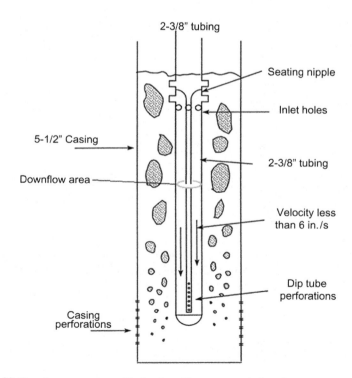

Figure 10.22 Poor boy separator with the downflow area and downflow passageway.

Table 10.3 Capacity of "poor boy" separators (Echometer)

Gas separator capacity table separator above fluid entry zone			
Outer barrel description and size (in.)	Dip tube size (in.)	Annulus area SQ (in.)	Liquid capacity (BPD)
3 1/2 perforated tubing sub	1.250	4.87	260
2 7/8 perforated tubing sub	1.000	3.32	177
2 7/8 perforated tubing sub	1.250	2.52	134
2 3/8 perforated tubing sub	0.750	2.26	121
2 3/8 perforated tubing sub	1.000	1.77	94
2 3/8 perforated tubing sub	1.250	0.96	51
2 3/8 perf tub sub and 1.5 in. pump	1.760	0.69	37

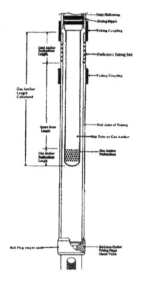

Harbison–Fischer formulas for gas anchor calculations for poor boy and oversize gas separators

$$V = \frac{(P^2)(S)(L)}{(60)(M^2 \text{-} G^2)} \qquad X = \frac{(1.5)(L)(P^2)}{(M^2 \text{-} G^2)}$$

P = Plunger diameter, in.

S = Strokes per minute

L = Stroke length (downhole), in.

M = Mud anchor inside diameter, in.

G = Dip tube outside diameter, in.

V = Calculated velocity of fluid/gas mixture in dip tube / mud anchor annulus, in./s, (should be less than 6 in./s)

X = Calculated length between upper perforation on dip tube to lower perforation on mud anchor. This gives a gas separation volume of 1.5 times the pump volume. A ratio of 2:1 can be used but should not be exceeded, replace the 1.5 with 2.0 in the formula for X.

Figure 10.23 Rules for constructing a poor boy gas separator.

However, more recently Echometer has published a new rule for the dip tube length that gives enough length to keep the downward movement of gassy fluids from reaching the end of the dip tube on the upstroke. The new rule typically recommends a dip tube length that is shorter than the old rule (above) that the dip tube should be long enough such that the separator volume should be 1.5 times the pump volume.

The new proposed dip tube length by Echometer is

Dip tube length = Vb × 60/(SPM × 2) × constant
Vb normally 6 in./s
SPM is strokes per minute
Constant: 1.233 for safety and velocity profile

So for Vb = 6 and SPM = 5
The dip tube length = 6 × 60/(5 × 2) × 1233 = 44.4 in.

This value is typically much smaller than using the 1.5 pump volume rule would give for dip tube length, which dictates separator length.

The following is an innovation to allow a larger downflow area than that one would obtain using typical oil field tubulars:

Collar-sized separator: (Echometer innovation)

- "Decentralized" by allowing separator to rest on LOW side of wellbore. (Tubing anchor if used should be three tubing joints above separator.)
- Annular area, dip tube size, and wall thickness chosen to minimize pressure drop, and provide maximum flow areas.
- Area of ports equal to area between dip tube and outer barrel.
- Use large ports distributed all around outer barrel to facilitate entry of liquid (Fig. 10.24).

Use of the collar-sized separator eliminates some of conventional poor boy separator deficiencies:

- A good separator must balance annular flow area, separator flow area, dip tube diameter, and pressure drop.
- Outer barrel OD same as collar OD.
- Thin short wall outer barrel and short dip tube.
- Large inlet ports distributed around outer barrel facilitate entry of liquid.

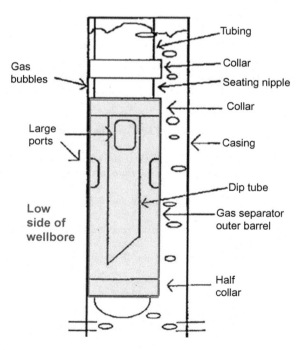

Figure 10.24 Collar-sized separator.

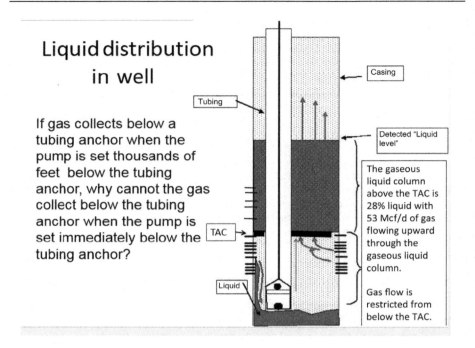

Figure 10.25 Packer a problem with gas separator?

Use of a tubing anchor (tubing anchor) above the intake may trap gas around the intake. This potential problem is illustrated in Fig. 10.25.

Possible problem when using a TAC.

Fluid level depression test to check if packer is problem, considering gas that may collect around the intake:

1. Confirm that the liquid level is above the TAC and the pump chamber is not filled with liquid when pumping.
2. After the well has been shutdown for 10 min, if the pump is full for only a few strokes plus the fluid level is above the TAC, then the fluid level depression test should be run on the well.
3. Close the casing valve to build casing pressure and depress the liquid level. Continue to pump the well.
4. Shoot fluid levels every 15 min as the casing pressure builds-up and the liquid level is being depressed.
5. Acquire four to five additional shots after the fluid level is pushed below the TAC and stabilized.

The fluid level depression tests verified that a gaseous liquid column existed above the TAC with a free gas (no liquid) between the TAC and pump. The high pressure gas below the TAC can restrict liquid production from the well.

Gas rate limitation when using a gas separator

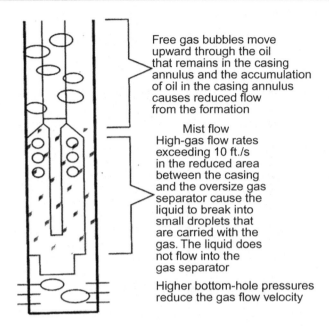

Free gas bubbles move upward through the oil that remains in the casing annulus and the accumulation of oil in the casing annulus causes reduced flow from the formation

Mist flow High-gas flow rates exceeding 10 ft./s in the reduced area between the casing and the oversize gas separator cause the liquid to break into small droplets that are carried with the gas. The liquid does not flow into the gas separator

Higher bottom-hole pressures reduce the gas flow velocity

Figure 10.26 Gas velocity limitation between ID of casing and gas separator.

Gas should not flow too fast in the gas separator in the annulus as mist flow develops and does not allow liquids to easily enter the gas separator. The gas velocity between the casing and the separator should not exceed 10 ft./s (Fig. 10.26).

10.5.2 Casing separator with dip tube: for use in horizontal wells

A packer separator uses the casing to form the outside of the downflow area and as such has a bigger capacity than possible when using a poor boy separator. Because of the higher capacity of the packer separator, it is commonly used for horizontal unconventional wells since the initial rates are very high for these wells. Since pumps are set in the near vertical portion of the well to prevent failures due to rod and pump wear, use of a tailpipe can move the intake down in the well to offset setting the pump higher in the well (Fig. 10.27).

The tailpipe with packer configuration is very effective and will increase production in a well when the pump is set a considerable distance above the formation. The tailpipe reduces the pressure required to push the formation fluids to the pump so a lower pumping bottom-hole pressure (PBHP) exists.

Details of one such packer separator is shown in Fig. 10.28.

When a dip tube is added below the separator, it can be sized considering the design flow rate of liquids and the gas produced. The size of the dip tube can then be determined that will give stable flow from near the lateral up to the packer

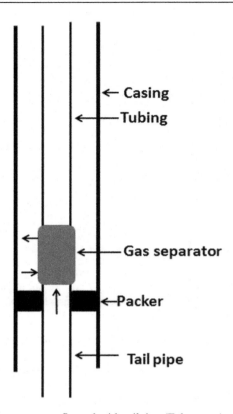

Figure 10.27 Packer separator configured with tailpipe (Echometer).

separator and to maximize the use of produced gas to help lift the liquids to the pump separator and pump intake. If done correctly, this can greatly reduce the pressure drop from the lateral to the separator intake (Figs. 10.29 and 10.30).

There are other ideas in the industry that include the use of a tailpipe with gas separation intended for use in a horizontal well. One such technique is illustrated in the Fig. 10.31.

10.5.3 Compression ratio

In some cases the produced gas volume is so high that most of the gas cannot be separated. In this case the pump must be designed to minimize the effects of the free gas that will enter the pump.

As discussed in Section 10.2 the traveling valve must open on the downstroke in order for the pump to work effectively. When pumping gas through the pump, this means that the pump must compress the gas in the pump on the downstroke to a pressure greater than the pressure above the traveling valve in order to force the traveling valve off its seat. If the traveling valve does not open the pump

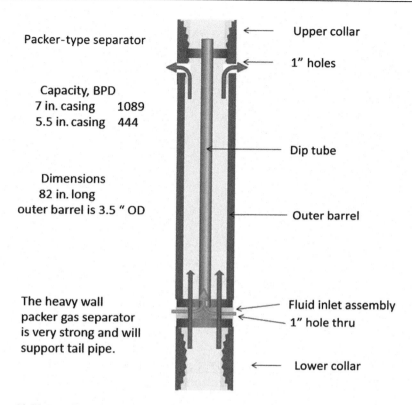

Packer-type separator

Capacity, BPD
7 in. casing   1089
5.5 in. casing   444

Dimensions
82 in. long
outer barrel is 3.5 " OD

The heavy wall
packer gas separator
is very strong and will
support tail pipe.

Upper collar

1" holes

Dip tube

Outer barrel

Fluid inlet assembly
1" hole thru

Lower collar

Figure 10.28 Details of one such packer separator.

action will continue but the pump cannot pump liquid. This condition is called "gas locking."

Beam pump installations can be designed, so that they are not susceptible to gas lock regardless of the amount of gas passing through the pump.[3] If the CR of the down-hole pump is high enough to always open the traveling valve, it will not gas lock even if it contains 100% gas. This certainly will not improve the volumetric efficiency of the pump, but the pump will not gas lock.

The compression of the pump discussed in this section is how much the fluid below the traveling valve is compressed on the downstroke. It is desired that this pressure is sufficient to build up enough pressure to open the traveling valve on the downstroke. If the traveling valve always opens on the downstroke, then the pump will not gas lock.

The definition of CR is given by (see Fig. 10.32):

$$CR = \frac{Down - hole\ stroke + spacing\ clearance + dead\ space}{Spacing\ clearance + dead\ space} \qquad (10.4)$$

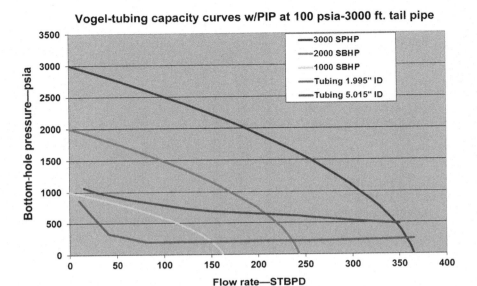

Figure 10.29 Flow path of 5.015 is unstable over the range of flows while the use of 2 3/8's for the dip tube gives a range of stable flow down to about 80 BLPD (in figure, SIBHP is shut in bottom-hole pressure).

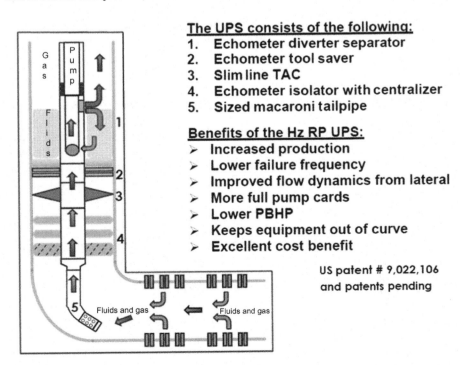

The UPS consists of the following:
1. Echometer diverter separator
2. Echometer tool saver
3. Slim line TAC
4. Echometer isolator with centralizer
5. Sized macaroni tailpipe

Benefits of the Hz RP UPS:
➤ Increased production
➤ Lower failure frequency
➤ Improved flow dynamics from lateral
➤ More full pump cards
➤ Lower PBHP
➤ Keeps equipment out of curve
➤ Excellent cost benefit

US patent # 9,022,106
and patents pending

Figure 10.30 Schematic of one such method of use of packer separator with a tailpipe that can be used in horizontal well.

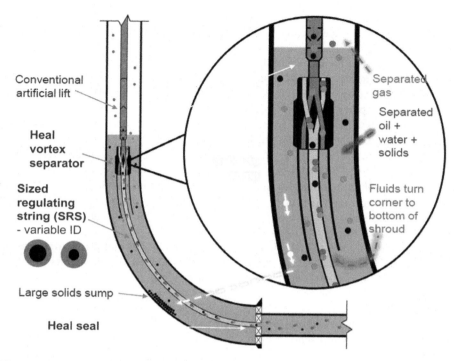

Figure 10.31 Another dip tube method for horizontal well applications (Heal Systems).

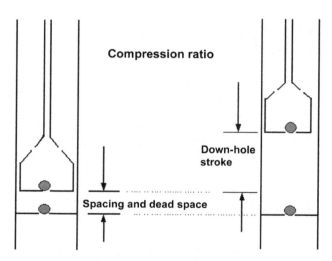

Figure 10.32 Beam pump compression ratio.[3]

The key to attaining a high CR is to maximize the pump stroke while minimizing the spacing clearance and dead space. Typically, little can be done to increase the stroke but careful spacing can drastically increase the CR.

The pull rod must be cut in the shop, so that the clearance between the traveling valve and the standing valve is less than approximately 0.5 in. when the pump is at its downmost position. In the well the pump must be spaced to a bare minimum taking care that the pump does not "tag" or strike bottom on the downstroke, but the standing valve assembly comes close to the traveling assembly on the downstroke to minimize the dead space in the pump.

The traveling valve seat plug can be specified as a "zero clearance" seat plug. This is an "all thread" seat plug that does not extend below the traveling valve cage and thus saves about 1 in. of length, enabling the plunger assembly to be spaced 1 in. lower during pump assembly. It requires a special hex or square wrench to tighten it on the inside of the seat plug since the outside of the seat plug is no longer available. (Source: B Williams, HF Pumps.)

Several standing valve cage designs are available from pump manufacturers, which reduce the wasted space inside the cage. Some of these designs significantly lower the upswept volume, especially when combined with a zero clearance seat plug and properly selected valve rod length. (Source: B. Williams, HF Pumps.)

If this is done, many pump gas handling problems will be solved. It is easily overlooked because you cannot see how long the pull rod is until you disassemble the pump.

10.5.4   Variable slippage pump to prevent gas lock

The Harbison—Fischer (H—F) variable slippage pump shown in Fig. 10.10 is primarily for gas locking conditions. This pump has eliminated gas lock in each field test to date.

Leakage is allowed to occur from over the plunger to under the plunger at the end of the upstroke due to a widened or tapered barrel. This reduces pump efficiency, but the liquid allowed to leak below the plunger insures that the traveling valve opens on the downstroke and that gas lock will not occur (Fig. 10.33).

10.5.5   Pump compression with dual chambers

The pump shown in Fig. 10.34 works by holding back the hydrostatic pressure in the tubing on the downstroke while still allowing fluid and gas to enter the upper chamber. The fluid is compressed once on the downstroke into an upper smaller chamber. Then it is compressed on the upstroke into the tubing. If the upper and lower CRs are 20:1 then the overall CR is 400:1.

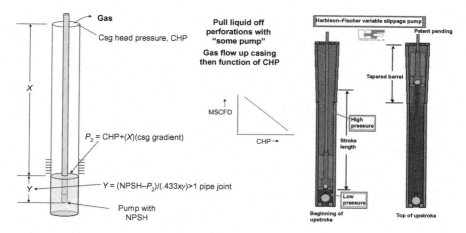

Figure 10.33 Example of a pump that uses designed leakage to prevent gas lock (Harbison–Fischer).

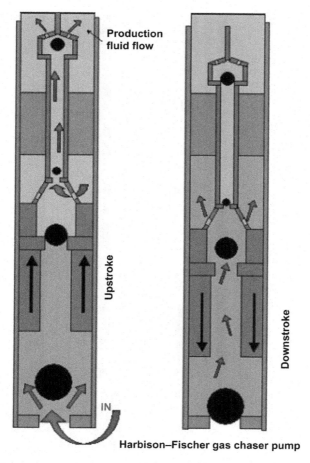

Figure 10.34 Two-stage compression pump (Harbison–Fischer).

10.5.6 Pumps that open the traveling valve mechanically

There are several pumps that involve a mechanism to automatically open the traveling valve on the downstroke, thereby preventing gas lock. Some have sliding mechanisms and others have devices that directly dislodge the traveling valve from its seat if not already dislodged by pressure. The pump assembly in Fig. 10.12 is an example of the latter, using a rod to force the traveling valve ball off the seat on the downstroke (Fig. 10.35).

10.5.7 Pumps to take the fluid load off the traveling valve

Fig. 10.36 shows a slide above the pump that seals the pressure above the pump from being on the top of the traveling valve on the downstroke. There are many other pumps that use this concept.

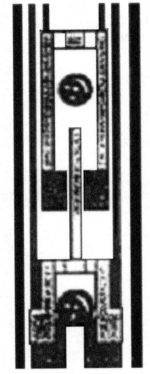

**Hart gas lock breaker
standing valve assembly**

Figure 10.35 Example of a pump that mechanically opens the traveling valve on the downstroke with a rod that lifts the traveling ball off its seat.

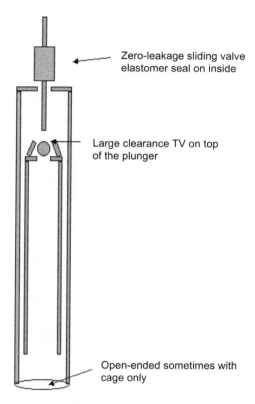

Zero-leakage sliding valve
elastomer seal on inside

Large clearance TV on top
of the plunger

Open-ended sometimes with
cage only

Figure 10.36 Quinn multiphase flow pump: slide above pump closes on downstroke to take fluid
load off of the valve. For usual pump with TV and SV the load off of the TV will allow the TV
to open with gas and liquids below the pump, and reduce fluid pound (Quinn Pumps, Canada).

10.5.8   Gas Vent Pump to separate gas and prevent gas lock (Source: B. Williams, HF Pumps.)

A patent-pending H−F Gas Vent Pump,[2] is shown in Fig. 10.37. This unusual sucker
rod pump was introduced recently and has been able to pump without gas locking by
separating the gas from the fluid before it gets into the pump. A strategically placed
hole in the barrel, or in a coupling joining two barrels allows gas to escape from the
compression chamber into the casing/tubing annulus during the top part of the
upstroke. It also allows fluid to enter the compression chamber with minimum
required pressure, allowing the well to be pumped down further than with other sucker
rod pumps. This positive fill feature gives the operator the option of slowing down the
pumping rate substantially, thus saving energy and wear on the pumping equipment.

There are many other specialty pumps. First, try to separate the gas using
completion techniques with the pump below the perforations. If this fails, try a gas
separator. If this still fails then try the more exotic pumps to handle gas.

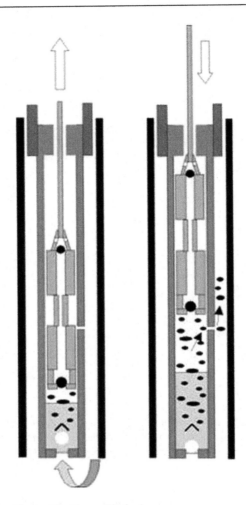

Figure 10.37 Gas Vent Pump (Harbison–Fischer).

10.6 Inject liquids below a packer

In recent years, methods have been developed to separate the liquid and gas phases down hole and then reinject the liquids back into the formation below a packer. This eliminates both the need for disposal of the liquids (water) at the surface and the means required to lift the liquids to the surface. Once the liquids are reinjected the gas can flow freely up the casing/tubing annulus.

There are several commercial devices available[4–6] to do this. The concept shown in Fig. 10.38 uses gravity as both the separation mechanism and the injection mechanism. The water is pumped up the tubing. The bypass seating nipple allows water pressure and flow to bypass the pump. The pressure exerted on the formation below

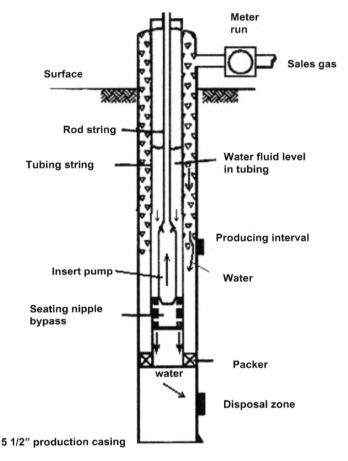

Meter run

Surface

Sales gas

Rod string

Tubing string

Water fluid level in tubing

Producing interval

Insert pump

Water

Seating nipple bypass

Packer

water

Disposal zone

5 1/2" production casing

Figure 10.38 Example of beam pump system to inject liquids below packer so gas can flow unobstructed (Harbison—Fischer bypass seating nipple).

the pump injects the water. The higher the fluid column in the tubing generated by the pump, the greater the pressure on the formation. If a larger pressure is needed than a full column of liquid in the tubing, then a back pressure regulator can be placed on the surface of the tubing. Cases of 300 psi and greater are reported to inject at the desired rate.

10.7 Summary

Beam pumps are often used for dewatering a gas well but special methods may be required to prevent gas interference.

- Gas interference is most often and easily handled by setting the pump below the perforations and flowing the gas up the annulus.
- If the pump does not fit below the perforations for gas separation, then separators or, as a last resort, specialty pumps may be required to combat pump gas interference.
- To handle produced water the beam pump system may be incorporated with a system to inject water below a packer to a water zone. This method eliminates water hauling charges and leaves a free path for gas to flow to the surface.
- Since gas flows up the casing, even if the pump is set below the perforations and all fluid is off the formation, there is still more than the surface CHP pressure on the formation. Therefore for beam pumping and any pumping system that pumps liquid up the tubing and flows gas up the casing, the casing pressure must be low if low formation pressures are to be achieved.

References

1. Lea JF. New pump-off controls improve performance. *Petrol Eng Int* 1986;41−4.
2. Neely AB. Experience with pump-off control in the Permian Basin, SPE 14345. In: *Presented at the annual technical conference and exhibition of the SPE*, Las Vegas, NV, September 22−25, 1985.
3. Parker RM. How to prevent gas-locked sucker rod pumps. *World Oil* 1992;47−50.
4. Enviro-Tech Tools Inc. Brochure on the DHI (Down Hole Injection) tool.
5. Grubb A, Duvall DK. Disposal tool technology extends gas well life and enhances profits, SPE 24796. In: *Presented at the 67th annual SPE conference in Washington DC*, October 4−7, 1992.
6. Williams R, et al. Gas well liquids injection using beam lift systems, southwestern petroleum short course, Lubbock, Texas, April 2−3, 1997.

Further reading

Clegg JD. Another look at gas anchors. In: *Proceedings of the 36th annual meeting of the southwestern petroleum short course*, Lubbock, TX, April; 1989.
Dunham CL. Supervisory control of beam pumping wells, SPE 16216. In: *Presented at the Production Operations Symposium*, Oklahoma City, Oklahoma, March 8−10, 1987.
Elmer W, Gray A. Design considerations when rod pumping gas wells. In: *First conference of gas well de-watering, SWPSC/ALRDC*, March 3−5, 2003, Denver.
McCoy JN, Podio AL. Improved downhole gas separators. Southwestern Petroleum Short Course, Lubbock, Texas, April 7 and 8, 1998.

Gas lift

James F. Lea's experience includes about 20 years with Amoco Production Research, Tulsa, OK; 7 years as Head PE at Texas Tech; and the last 10 years or so teaching at Petroskills and working for PLTech LLC consulting company. Lea helped to start the ALRDC Gas Dewatering Forum, is the coauthor of two previous editions of this book, author of several technical papers, and recipient of the SPE Production Award, the SWPSC Slonneger Award, and the SPE Legends of Artificial Lift Award.

Larry Harms started his consulting company, Optimization Harmsway LLC, after retiring from ConocoPhillips in 2015. He has over 35 years of experience in the application of compression to optimize production. He conducted training courses for hundreds of operations, maintenance, and engineering personnel on compression, production optimization, systems nodal analysis, artificial lift, and gas well deliquification.

11.1 Introduction

Gas lift is an artificial lift (AL) method whereby external gas is injected into the produced flow stream at some depth in the wellbore. The additional gas augments the formation gas and reduces the flowing bottom-hole pressure (BHP), thereby increasing the inflow of produced fluids. For dewatering gas wells the volume of injected gas is designed so that the combined formation and injected gas will be above the critical rate for the wellbore,[1] especially for lower liquid producing gas wells. By returning the gas rate to above the rate at which gas carries liquids up the tubing in a mist flow regime (critical rate of gas) the so-called problem of liquid loading in the tubing is solved and the well can produce at higher rates and lower producing bottom-hole pressures. Care must be taken to flow gas up the tubing above the critical rate but not too much above the critical rate or friction will result and slow production possibly as much or more than liquid loading. The critical rate or critical velocity is discussed in detail in Chapter 3. For higher liquid rates, much of the design procedure may more closely mirror producing oil well gas lift techniques.

Although gas lift may not lower the flowing pressure as much as an optimized pumping system (pump systems pump the liquids up the tubing and gas flow up the casing for most dewatering pump systems), there are several advantages of a gas

Gas Well Deliquification. DOI: https://doi.org/10.1016/B978-0-12-815897-5.00011-1

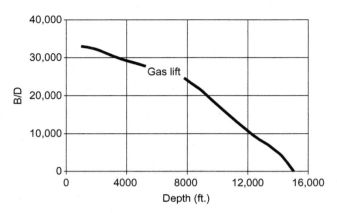

Figure 11.1 An approximate depth-rate feasibility chart for conventional continuous gas lift.

lift system which often make gas lift the AL method of choice. For gas wells in particular, when producing a low amount of liquids, the producing BHP with gas lift may compare well with other methods of dewatering unless pump can be set below perforations which is impossible for horizontal wells. For higher liquid rates, the achievable producing BHP may be higher than pumping techniques.

Of all AL methods, gas lift most closely resembles natural flow and has long been recognized as one of the most versatile AL methods. Because of its versatility, gas lift is a good candidate for removing liquids from gas wells under certain conditions. Fig. 11.1 shows the approximate depth-pressure ranges for application of gas lift, developed primarily for oil wells and generally based on injection gas pressures below 1900 psig.

The most important advantages of gas lift over pumping lift methods are as follows:

- Most pumping systems become inefficient unless pump intake can be set below the producing interval when the gas/liquid ratio (GLR) exceeds some critical value, typically about 500 scf/bbl (90 m³/m³), due to severe gas interference. Setting below perforations is of course impossible for horizontal wells, so for pumping systems in gassy wells, a workable gas separator system is required. Although remedial measures are possible for conventional lift systems, gas lift systems can be directly applied to high GLR wells because the high-formation GLR reduces the need for additional gas to lower the formation flowing pressure.
- Production of solids will reduce the life of any device that is placed within the produced fluid flow stream, such as a rod pump or electric submersible pump. Gas lift systems generally are not susceptible to erosion due to sand production and can handle a higher solids production than conventional pumping systems.
- For some applications, a higher pressure gas zone may be used to auto-gas-lift another zone.
- In highly deviated wells, it is difficult to deploy some pumping systems due to the potential for mechanical damage to deploying electric cables or rod and tubing wear for beam pumps. Gas lift systems can be employed in deviated wells without mechanical problems.
- There are relatively new techniques that allow injection of gas below a packer (for example, Xtra-Lifts and Perf-Lift,).

Gas lift has features to address the earlier production situations.

Another advantage that gas lift has over other types of AL is its adaptability to changes in reservoir conditions. It is a relatively simple matter to alter a gas lift design to account for reservoir decline or an increase in fluid (water) production that generally occurs in the later stages of life of the field. Changes to the gas lift installation can be made from the surface by replacing the gas lift valves via wireline without pulling tubing and reusing the original downhole components. However, many onshore lower volume gas well gas lift operators usually choose to use conventional mandrels where the tubing must be pulled to access gas lift valves and to replace valves.

Techniques can be used which apply gas lift without valves, packers, or any downhole equipment other than tubing is also possible. This involves injecting around the end of the tubing, as one example, which requires sufficient gas injection pressure and techniques to lower the fluid in the casing to allow this to happen.

The two fundamental types of gas lift used in the industry today are "continuous flow" and "intermittent flow." This is the conventional breakdown. However, one could say there is gas lifting gas wells and there is gas lifting oil wells. Gas wells can also be lifted by continuous or intermittent gas lift so the conventional discussion will be presented, although many gas wells are being lifted by continuous flow. Gas lift is used for some unconventional wells when deviation, and/or high gas, and/or solids cause pumping problems. Also unconventional wells have high rates to begin and then the rates decline quickly. Gas lift can be designed for changing rates. Examples are in this chapter which bear on use of gas lift for unconventional production. Gas lift for gas wells usually makes use of conventional IPO (injection pressure operated) valves, if valves are used to unload initial liquids from the well. For lower pressure wells, one may be able to inject down the casing around the end of the tubing using no unloading valves. These might be continuous gas lift. Intermittent gas lift or gas-assisted plunger lift (GAPL) may also be used for lower rates.

11.2 Continuous gas lift

In continuous flow gas lift a stream of relatively high-pressure gas is injected continuously into the produced fluid column through a downhole valve or orifice. The injected gas mixes with the formation gas to lift the fluid to the surface by one or more of the following processes:

- Reduction of the fluid density and the column weight, so that the pressure differential between the reservoir and the wellbore will be increased.
- Expansion of the injected gas, so that it pushes liquid ahead of it which further reduces the column weight, thereby increasing the differential between the reservoir and the wellbore.
- Displacement of liquid slugs by large bubbles of gas acting as pistons.
- For gas wells, produced and injected gas up the tubing are designed to exceed the critical flow rate so loading will not occur. Initially the well has static fluids in it (casing and

tubing) and these liquids must be removed to inject as deeply as possible (an objective for best operation). Use of unloading valves allow the injected gas to reach the bottom of the well or deeply into the well with a minimum of available compressor pressure.

For lower pressure wells that support a smaller initial height of static fluid, the smaller static fluid height might be pushed into the formation with well shut-in and pressure buildup. Then the gas can be injected down the casing and around the end of the tubing with no unloading valves. This is sometimes called gas circulation.

Even if the well pressure and the initial static fluid height are high, a high compressor pressure will allow injection deeply in the well. This could be done with no unloading valves. This can be called high pressure or single point gas lift. These methods are discussed in the chapter.

11.3   Intermittent gas lift

Often in gas wells as the BHP declines, a point is reached where the well can no longer support continuous gas lift, and the well is converted to intermittent gas lift. This conversion can also employ the identical downhole equipment (mainly the gas lift valve mandrels) yet fully adapt the well to intermittent flow. In this case the unloading valves are replaced with dummy valves to block the holes in the mandrels and prevent injection gas from passing into the production stream. The operating valve is then replaced with a production pressure valve with a newly set pressure capacity reflecting the desired fluid level to be reached in the tubing before the well is lifted.

Fitting the operating valve with the largest possible orifice will greatly improve the efficiency in an intermittent gas lift system. The large orifice diameter exerts a minimum restriction to the flow of the injection gas. The injection gas will then quickly fill the tubing below the fluid, ultimately lifting the "slug" of liquid to the surface with the minimum amount of lift gas.

The optimum time to convert a gas lift well from continuous lift to intermittent lift is a function of the reservoir pressure, the tubing size, the GLR, and the flow rate of the well. The individual well conditions will dictate the optimum time for conversion but Table 11.1 lists some good rules of thumb to use to estimate the best time to convert to intermittent lift.

It is becoming common practice to use a plunger (see Chapter 6: Plunger lift) to increase the production from wells on intermittent lift. The lift gas is injected below

Table 11.1 Maximum flow conditions for intermittent lift

Tubing size (in.)	Maximum flow rate for intermittent lift
2-3/8	150 bpd
2-7/8	250 bpd
3-1/2	300 bpd
4-1/2	Not recommended

the plunger, and the plunger acts as a physical barrier between the lift gas and the fluid to reduce the fluid fallback around the gas slug that is characteristic of intermittent lift operations. The plunger extends the life of the well by more effectively removing water from the formation. A plunger with extensions can be used, so that it can pass by gas lift mandrels if needed. When plunger is used over a standing valve, and the gas lift lifts the plunger and liquid slug above a standing valve, then this is more similar to a gas-powered long stroke pump or plunger-assisted chamber lift than conventional intermittent lift.

11.4 Gas lift system components

Fig. 11.2 shows a typical continuous gas lift system that includes:

- a gas source;
- a surface injection system, including all related piping, compressors, control valves, etc.;
- a producing well completed with downhole gas lift equipment (valves and mandrels and packer in this case); and
- a surface processing system, including all related piping, separators, control valves, etc.

The gas source is often reservoir gas produced from adjoining wells that have been separated, compressed, and reinjected. A secondary source of gas may be required to supply any shortfall in the gas from the separator. The gas is compressed to the design pressure and is injected into the well through the gas lift operating valve, where it enters the tubing string at a predetermined depth.

For conventional gas lift, valves or orifices are used to port gas to the tubing, rather than holes or simply the end of the tubing string, so that the gas stream is well dispersed within the liquid column and the flow continues smoothly.

However in this chapter, "gas cycling[2]" is discussed. This is a method to flow additional gas down the annulus and into the bottom of the tubing. This is possible

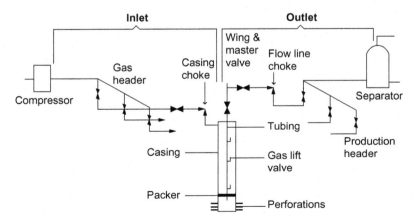

Figure 11.2 Continuous gas lift system.

because the amount of gas is high in the tubing relative to the fluids so severe slugging does not occur as it would be with a lower operational GLR as would be typical for gas lifting oil wells.

It is recognized that there are single well compressors for gas wells where small compressors (reciprocating or screw, for example) are used to lower the surface pressure of a flowing or plunger lifted gas wells to solve liquid-loading issues. In some cases, when there is excess capacity available at high enough pressure, these units can be used to inject gas at a single point and keep the well above critical rate as discussed further in this chapter.

11.5 Continuous gas lift design objectives

The concept of gas lift for an oil well is that it can increase well production by reducing the density of the produced fluid, thereby decreasing the flowing BHP. The concept of gas lift applied to a gas well is that the added gas at depth can return the gas rate to and "above" critical rate and eliminate liquid loading which will decrease the flowing BHP. For both oil and gas wells, over injection of gas can build up friction in the tubing and the flow lines and if excessive, can increase the flowing BHP and reduce production. Gas lift is accomplished by introducing the injection gas, at an optimum (usually maximum) depth, pressure, and rate, into the produced fluid stream. The valve through which the gas is injected into the wellbore fluid stream under normal operating conditions is called the "operating valve."

Nodal analysis (see Chapter 4: Nodal Analysis and Appendix B) can be used to evaluate several tubing sizes and GLRs and to determine possible production increases for the different tubing sizes and GLRs. The well becomes a candidate for gas lift when the artificially increased GLR in the tubing significantly increases the well production. Another way of thinking of gas lift for gas wells is to inject a sufficient additional rate of gas to keep the total gas velocity (from produced + injected gas) high enough to produce stable tubing flow without creating excessive friction.

The efficiency of a gas lift system is highly influenced by the depth of the operating valve. As the depth of the operating valve is increased, more and more of the hydrostatic pressure of the heavier fluid (and gas) column is taken off of the formation, reducing the BHP and increasing production. Typically, before a gas lift well is brought on line, it is filled or partially filled with kill fluid from the workover operation. To bring the well into production the well must first be "unloaded" by injecting high-pressure gas into the annulus to displace the kill fluid down to the operating valve. Pushing the liquid level to the depth of the operating valve requires an extremely high surface injection pressure. In most installations, this high injection pressure is not available; however, if this can be made available high-pressure/single-point gas lift (SPGL) (described later in this chapter) can be done. Several gas lift valves are used to allow the available surface pressure to feed gas to the well at increasing depths until the operating valve at maximum depth is reached.

This process is called "unloading the well" and the additional valves are called "unloading valves."

The series of unloading valves are placed at various depths and have different opening/closing pressures to step the injection gas down to the design injection depth. These unloading valves are designed to have a particular port size and set to specific opening pressures to allow the annular fluid level to pass from one valve to the next. The design of the gas lift system includes the size, pressure rating, depth and spacing of the unloading valves, the optimum depth of the operating valve to maximize recovery, the size of the operating valve orifice and the injection rate, and pressure of the lift gas.

The correct spacing of the unloading valves is critical. Valves spaced too far apart for the injection parameters will not allow the well to completely unload. In this case, injection gas will enter the production stream too high in the well, significantly lowering the system efficiency and, more importantly, the well's production.

Determining the best gas lift design requires considerable knowledge of the well conditions, both present and future. These calculations usually are performed by sophisticated commercial software packages or design charts supplied by gas valve manufacturers. The complete fundamentals of gas lift design and optimization are beyond the scope of this text, although field applications[3] of gas lift technology for gas wells are presented in this chapter.

Please note that for installations in gas wells after the flowing period, reservoir pressures have dropped such that the "kill fluid" is mostly water, could probably be diesel or even condensate, and the well may not stay full of liquids with any of these fluids due to low reservoir pressures. Reducing the density of the kill fluid in designs results in less valves and lower injection depth potential for a given gas injection pressure, resulting in more efficient injection and possibly increased production at less cost with a small risk that a swabbing rig might have to be brought in if the well had to be unexpectedly killed or became extremely liquid loaded with a more dense fluid than that used in the design.

11.6 Gas lift valves

The key to a properly designed gas lift system is the proper choice of gas lift valves. Gas lift valves fall into one of the three major categories:

- Orifice valves;
- Injection pressure operated (IPO) valves; and
- Production pressure operated (PPO) valves.

Schematic examples of injection and PPO valves are given in Fig. 11.3. Most gas lift for gas wells use the Type 1 arrangement. However, for some high rate early time unconventional wells, the Type 4 arrangement may be used to accelerate annulus production.

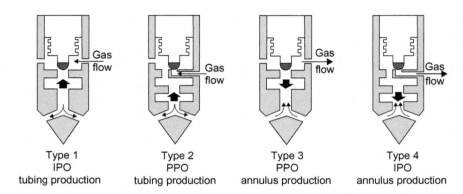

Figure 11.3 Typical gas lift valve types.

11.6.1 Orifice valves

Strictly speaking, orifice valves are not valves because they do not open and close. Orifice valves are simply orifices, or holes, providing a communicating port from the casing to the tubing. Because they do not actually function as valves, orifice valves are used only as operating valves to provide the correct injection flow area as required by the valve design and to properly disperse the injected gas to minimize the formation of slugs. Orifice valves are typically used only for continuous flow applications. The valve includes a check to prevent tubing to casing flow. The orifice valve, if used, is at the bottom of the unloading valves.

11.6.2 Injection pressure operated valves

IPO (sometimes called casing pressure (CP) operated, Type 1 in Fig. 11.3) valves are the most common valve used in the industry to unload gas lift wells. Although somewhat influenced by the pressure of the flowing production fluid, IPO valves are controlled primarily by the pressure of the injection gas.

Fig. 11.4 shows a schematic of an IPO gas lift valve where the injection pressure is applied to the base of the bellows and the produced fluid pressure is applied to the ball (stem tip) through the valve orifice area. Since the bellows area is much larger than the orifice area, the injection pressure dominates control of the valve operation.

Injection pressure valves act as back-pressure regulators and close when the back pressure (casing pressure) reaches a predesignated "minimum" value. Typically, this minimum value is designed to be when the kill fluid in the casing/tubing annulus, being pushed downward by the injection gas during the unloading process, just reaches the next lower valve. This allows the upper valve to close the flow of injection gas, forcing the pressure to continue pushing the fluid level further down the annulus to eventually reach the operating valve.

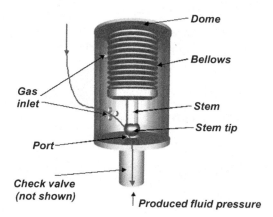

Figure 11.4 Schematic of gas lift valve.

11.6.3 Production pressure operated valves

PPO valves (sometimes called tubing pressure valves) are primarily operated by changes in pressure of the production fluid. Unloading is then controlled primarily by the reduction in hydrostatic pressure in the production stream by injecting lift gas.

PPO valves are used typically:

- where the production fluid is produced through the annulus and injection gas is down the tubing;
- in dual completions where two gas lift systems are installed in the same well to produce two differently pressured zones;
- to reduce requirements for interaction with the well by the operator to "valve down";
- for intermittent lift.

PPO valves are ideal for intermittent lift applications since the valve is designed to remain closed until a sufficient fluid load is present in the tubing, at which time the valve opens producing the liquid.

Once an unloading valve closes during the unload process, it should remain closed. Both IPO and PPO valves use a charged bellows (pressurized with nitrogen), a spring, or sometimes both to obtain the valve closing force. The nitrogen-charged bellows is the most common. The bellows type valves are set to the design pressure in a controlled laboratory environment by the valve manufacturer.

All gas lift valves are equipped with reverse flow check valves to prevent backflow of fluid through the valve. For subsea completions, where minimal intervention is a design objective, the spring-loaded valve may provide the most reliable since in the event of a bellows rupture the spring will keep the stem on seat and the valve will remain closed. The spring-loaded valve is also not sensitive to temperature variations as is the nitrogen-charged bellows. For a well with a packer and check valves, once the well liquids are unloaded then they will not return to the casing if the compressor goes down or is shutoff in the future. Few of gas lift installations for dewatering gas wells use PPO valves in North America.

11.7   Gas lift completions

The heart of a gas lift installation is the gas lift valves. Their placement in the tubing string is fixed during the installation of the tubing by the gas lift mandrels. Gas lift mandrels are placed in the tubing string to position each gas lift valve to the desired depth.

There are two basic "conventional" gas lift systems in use today. They are systems using conventional mandrels with threaded nonretrievable gas lift valves and systems using side pocket mandrels (SPMs) with retrievable gas lift valves.

Conventional mandrels accept threaded gas lift valves mounted on the outside of the mandrel. These valves can only be retrieved and changed by pulling the tubing and are usually not run where workover costs are high.

SPMs allow the gas lift valves to be retrieved using slickline from the surface without the need to pull the tubing. Conventional mandrels are used on nearly all gas wells being gas lifted onshore. SPMs are used offshore where it is expensive to pull tubing.

11.7.1   Conventional gas lift design

A schematic of a gas lift system using conventional mandrels is shown in Fig. 11.5 (left). With this system, gas lift mandrels and valves are installed at the surface when the tubing is run in the well. The valves are threaded into the mandrels and therefore cannot be removed without removing the entire tubing string. Gas lift designs using conventional mandrels are among the lowest cost gas lift designs available.

An added benefit of using conventional mandrels, particularly when removing liquids from gas wells, is that they can readily integrate with plunger lift systems. This is not the case for installations using SPMs. The ID of a conventional mandrel is relatively uniform but the internal pocket of a SPM is eccentric to permit the insertion of gas lift valves via slickline. This presents a problem for plunger operations because as the plunger assembly enters the SPMs eccentric pocket it allows gas to bypass liquid. This typically results in a loss of plunger velocity and in some cases makes it difficult for the plunger to reach the surface. Some operators have successfully adapted extensions to the plunger to effectively straddle the pocket. Most gas lifted gas wells use conventional mandrels/valves in the United States.

SPMs were developed to reduce the cost of changing a gas lift system to maintain a gas lift valve design that optimizes production as well condition change. A

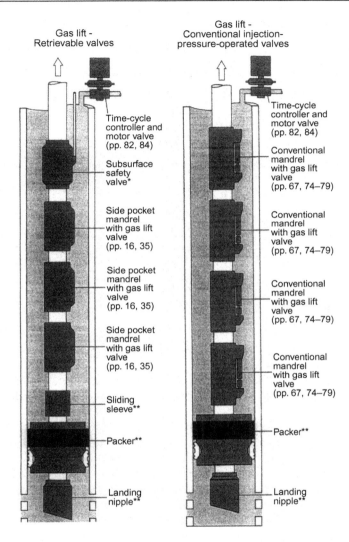

Figure 11.5 Conventional and retrievable completions.
Source: Courtesy Schlumberger-Camco.

schematic of a SPM is shown in Fig. 11.6. The primary feature of SPMs is the internally offset pocket that accepts a slickline retrievable gas lift valve. The pocket is accessible from within the tubing using a positioning or kick over to place and retrieve the valves. The gas lift valves use locking devices that lock into mating recesses in the SPM. Both conventional and SPM mandrels are installed in the well in the same manner, but only the SPM system is serviceable with slickline operations for postcompletion repair or well maintenance.

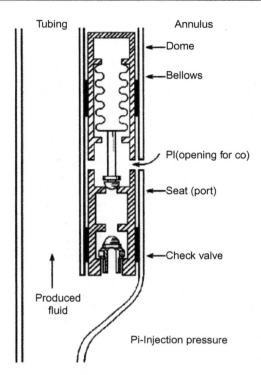

Figure 11.6 Gas lift valve in side pocket mandrel. Conventional mandrels and valves are external to the tubing with only a small hole present through the tubing to introduce gas.

Most/many US onshore gas lift wells use the conventional mandrels that require pulling the tubing to change/repair/adjust valves. SPMs can be used onshore but there is a cost to maintain a wireline system and is usually not done.

The high pressure gas in a gas lift system is usually supplied by a central compressor that compresses the gas produced by the field for reinjection into those wells on gas lift. If the field gas supply is insufficient to meet the needs of the AL system, more gas is generally obtained from the sales line.

Gas lift compression can also be supplied for individual wells when one or two wells in a field are being lifted with gas lift. These small well site compressors are typically skid mounted for easy mobilization when it becomes necessary to move the system from one well to another. Fig. 11.7 shows a typical system for an individually compressed low pressure well on gas lift. This might be a system on a gas well to help lift liquids.

11.7.2   Chamber lift installations

When the complete configuration prevents the point of injection from achieving the desired depth, or when the volume of gas in an intermittent lift installation is less than acceptable, a chamber lift design can be used.

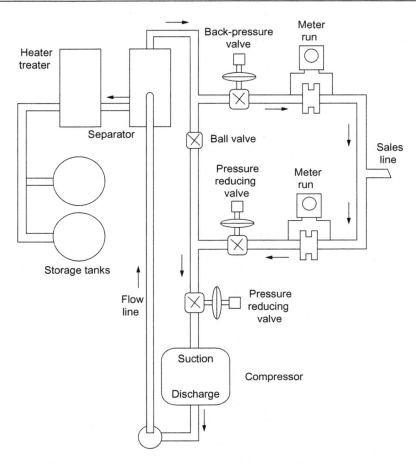

Figure 11.7 Typical compression system for low-pressure gas lift system.

The concept of chamber lift is to create a large diameter volume (chamber) to collect liquids. The larger diameter of the chamber, as opposed to the tubing, allows higher volumes of liquid to accumulate while keeping the liquid column height to a minimum. Lower liquid column heights put less hydrostatic pressure on the formation. Increasing the diameter of the chamber can drastically reduce the hydrostatic head since the BHP is reduced by the square of the chamber diameter. For example, increasing the chamber diameter from 2-3/8 to 3-in. will drop the hydrostatic pressure at the bottom of the hole by almost half for the same volume of liquid.

Typically the chamber consists of a portion of the casing, as shown in Fig. 11.8. Chamber packers isolate the chamber and a dip tube is frequently used in the top packer to allow the gas collected in the chamber to bleed off into the casing above the packer. Chambers can also be built at the surface and installed in the tubing string.

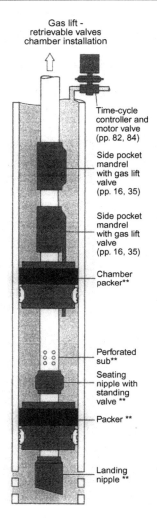

Figure 11.8 Chamber lift design.

Chamber lift is one method of producing a relatively high volume of liquids in a low pressure formation without loss of gas production due to excessive liquid head in the tubing.

In Fig. 11.8 a "chamber" is formed between two packers. Well liquids are allowed to enter the space between the packers at low pressure. After the chamber is filled, gas is injected into the top of the chamber, displacing the liquids into and up the tubing. An additional gas lift effect is added to the liquids as rise due to gas injected from gas lift valves spaced higher in the tubing. A time-cycle controller is provided to control the cycles. Chamber lift should bring a well to depletion but plunger lift does well and does not require as complex as a completion as does chamber lift.

11.7.3 Intermittent lift and/or gas-assisted plunger lift

Intermittent lift is usually done with packer and gas injected down the annulus, through injection and unloading valves and over a SV, under a liquid slug, to carry the liquid to the surface. However, if there is no interface between the gas and liquid then there can be liquid fallback of the liquid slug as it travels to the surface. Therefore intermittent and plunger lift can be combined. If plunger lift is not feasible due to a shortage of gas or if a plunger lift well is declining in rate as the well weakens, the gas assist with a plunger can be used to extend the plunger-related lift (GAPL).

As shown in Fig. 11.9, the gas assist can be sent down the annulus to an open-ended tubing and under the plunger/liquid and up the tubing. In this case, no additional downhole hardware is required. However, if better control of the gas is thought to be needed and if liquid rates are high, then unloading valves located in SPMs or conventional mandrels can be used to unload liquids from the well.

For the GAPL with open-ended tubing the slug size can be determined once the plunger is at bottom under the liquid by the difference between the surface casing pressure (CP) and tubing pressure (TP). For the GAPL the pressure in the casing should build to the same value that was used in conventional plunger lift to lift a slug of a given size. This technique is used to extend the life of the well when conventional plunger lift weakens. The big drawback is to have a source of higher pressure gas, such as a compressor.

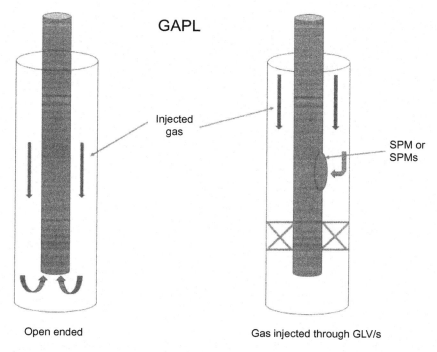

Figure 11.9 Gas-assisted plunger lift or GAPL.

11.7.4 Horizontal or unconventional wells

Over the past decade the number of horizontal wells has ballooned worldwide. Many of these wells are on gas lift to either increase oil production or, in gas wells, to more effectively remove associated liquids so gas can produce more efficiently.

Some operators have attempted to install gas lift in the horizontal section of the hole but found this not to be practical for a variety of reasons.

- Gas lift operates by reducing the hydrostatic head on the formation. In a horizontal or near horizontal section of the hole, there is very little vertical head. Placing gas lift valves in the horizontal lateral gains little benefit over valves located up-hole in the vertical section.
- In the horizontal section of the well the two-phase (gas/liquid) flow tends to become stratified, allowing the gas to pass over the top of the fluid without pushing the fluid to the surface. This greatly reduces the efficiency of the gas lift.
- Servicing gas lift valves with slickline becomes increasingly difficult with increased wellbore inclination.
- Risk of instability/collapses in the horizontal section.

Another problem with the unconventional or horizontal wells is that a very high rate of production comes in initially and then after $1-3$ years the production drops and flattens for a number of years to sometimes a fairly low rate. This decline then is a problem for many methods of AL, including gas lift but the effects on gas lift

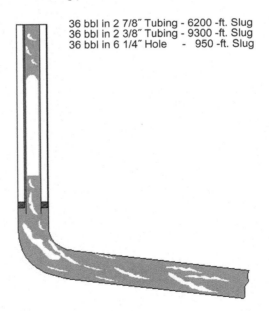

Loading problems with horizontal wells

36 bbl in 2 7/8″ Tubing - 6200 -ft. Slug
36 bbl in 2 3/8″ Tubing - 9300 -ft. Slug
36 bbl in 6 1/4″ Hole   -   950 -ft. Slug

Figure 11.10 Flowing horizontal well. (If gas lift is used to enhance production, install mandrels in sections where there is some significant vertical component.)

may not be as dramatic as on other methods of lift. Fig. 11.10 shows horizontal well before mandrels (and valves) are installed.

Yet another problem is that the production coming from a lateral may be very erratic in slug flow. However, most gas lift produces (at least near the surface) in slug flow inside the tubing. This is perhaps less of a concern for gas lift than for other lift methods.

Initially the well will flow for a while. It is important to flow at a high rate; however, much of the present value profit (PVP) of the well is in the first few years, it can be even better at higher rates. So the operator can flow up the tubing or up the casing/tubing annulus or up both tubing and casing (if allowed), and the size of the tubing can also be changed to maximize the flow in any of the cases. Changing tubulars to increase the flow rate can be costly but this is done for unconventional wells. Once a program of flowing the well perhaps through different tubulars as the well declines is established, the next step is to decide when to use AL and what form of AL to be used. In this chapter, use of gas lift is discussed and it will be presented for the case of initial high rates and then how it looks as the rates decline all the way down to fairly low rates. These events can be shown using Nodal Analysis, and cases will be shown for high/medium and low rates of production using gas lift (shown in Figs. 11.11 and 11.12).

11.7.5 Examples of using gas lift to deliquefy gas wells

Examples will be presented from a low rate, fairly shallow well to a higher rate deeper well and then examples of an unconventional well with a steep decline using gas lift will be shown. The IPR used in the following examples uses the back-pressure equation ($Q = C(Pr^2\text{-}Pwf2)^n$) where C and n are input and are constants that are previously determined from producing data.

Example 11.1 Low rate gas well: gas lift to unload

Data for Example 11.1: gas well

C and n back-pressure equation for inflow performance relationship (IPR)

Well temperature (WHT): 110 well head pressure (WHP): 200 psi; bottom-hole temperature (BHT): 170°F; gas gravity (GG): 0.6; liquids: 44 bbls/MMscf; oil gravity (API): 50; WG: 1

Reservoir: Pr (shut-in pressure): 1111 psi; C (constant in back-pressure equation determined from using producing data): 0.000236 Mscfd/psi^{2n}, n (exponent in back-pressure equation that usually varies from 0.5 to 1 and determined from producing data) : 1.0; water cut (WC) : 0.5

6000 ft of 2 3/8's (1.995 ft. OD) tubing roughness: 0.018 in., gray tubing multiphase flow (MPF) correlation.

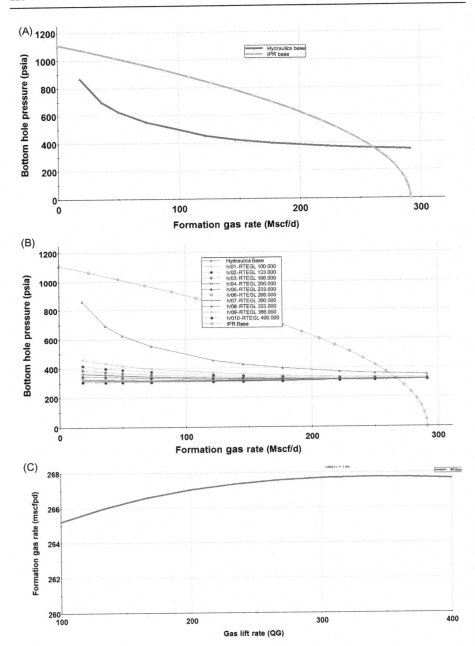

Figure 11.11 (A) Example 11.1, flowing, no lift. (B) Tubing performance with gas added. (C) Gas lift performance curve. (D) Gas lift design. (E) Gas lift valve parameters.

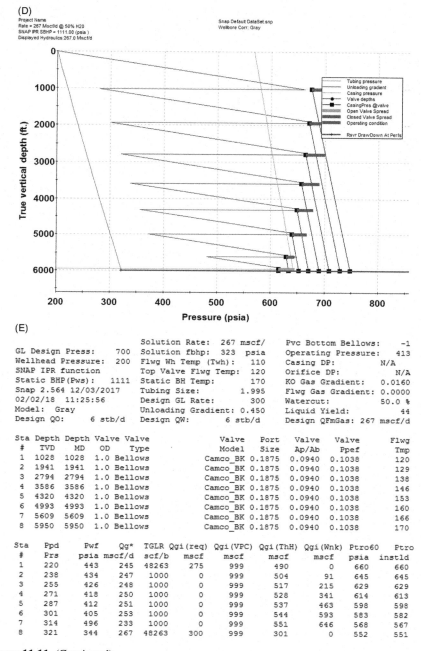

(D)

Project Name
Rate = 267 Mscf/d @ 50% H2O
SNAP IPR SBHP = 1111.00 (psia)
Displayed Hydraulics:267.0 Mscf/d

Snap Default DataSet.snp
Wellbore Corr: Gray

Legend:
- Tubing pressure
- Unloading gradient
- Casing pressure
- Valve depths
- CasingPres @valve
- Open Valve Spread
- Closed Valve Spread
- Operating condition
- Rsvr DrawDown At Perfs

(E)

		Solution Rate:	267	mscf/	Pvc Bottom Bellows:	-1		
GL Design Press:	700	Solution fbhp:	323	psia	Operating Pressure:	413		
Wellhead Pressure:	200	Flwg Wh Temp (Twh):	110		Casing DP:	N/A		
SNAP IPR function		Top Valve Flwg Temp:	120		Orifice DP:	N/A		
Static BHP(Pws):	1111	Static BH Temp:	170		KO Gas Gradient:	0.0160		
Snap 2.564 12/03/2017		Tubing Size:	1.995		Flwg Gas Gradient:	0.0000		
02/02/18 11:25:56		Design GL Rate:	300		Watercut:	50.0 %		
Model: Gray		Unloading Gradient:	0.450		Liquid Yield:	44		
Design QO:	6 stb/d	Design QW:	6 stb/d		Design QFmGas: 267 mscf/d			

Sta #	Depth TVD	Depth MD	Valve OD	Valve Type	Valve Model	Port Size	Valve Ap/Ab	Valve Ppef	Flwg Tmp
1	1028	1028	1.0	Bellows	Camco_BK	0.1875	0.0940	0.1038	120
2	1941	1941	1.0	Bellows	Camco_BK	0.1875	0.0940	0.1038	129
3	2794	2794	1.0	Bellows	Camco_BK	0.1875	0.0940	0.1038	138
4	3586	3586	1.0	Bellows	Camco_BK	0.1875	0.0940	0.1038	146
5	4320	4320	1.0	Bellows	Camco_BK	0.1875	0.0940	0.1038	153
6	4993	4993	1.0	Bellows	Camco_BK	0.1875	0.0940	0.1038	160
7	5609	5609	1.0	Bellows	Camco_BK	0.1875	0.0940	0.1038	166
8	5950	5950	1.0	Bellows	Camco_BK	0.1875	0.0940	0.1038	170

Sta #	Ppd Prs	Pwf psia	Qg* mscf/d	TGLR scf/b	Qgi(req) mscf	Qgi(VPC) mscf	Qgi(ThH) mscf	Qgi(Wnk) mscf	Ptro60 psia	Ptro instld
1	220	443	245	48263	275	999	490	0	660	660
2	238	434	247	1000	0	999	504	91	645	645
3	255	426	248	1000	0	999	517	215	629	629
4	271	418	250	1000	0	999	528	341	614	613
5	287	412	251	1000	0	999	537	463	598	598
6	301	405	253	1000	0	999	544	593	583	582
7	314	496	233	1000	0	999	551	646	568	567
8	321	344	267	48263	300	999	301	0	552	551

Figure 11.11 (Continued)

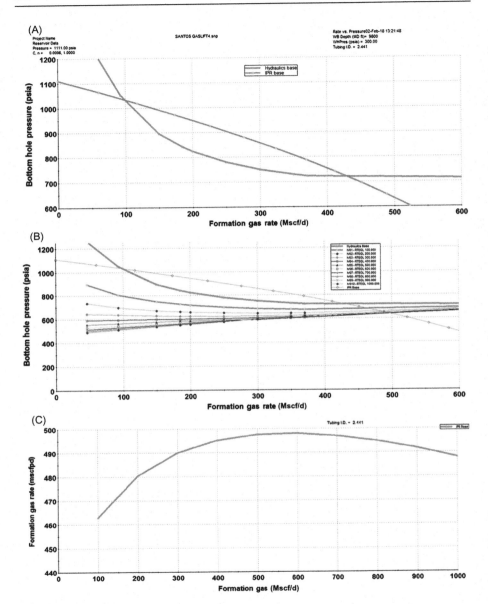

Figure 11.12 (A) With no lift: loaded. (B) See effects of gas injected at bottom. (C) Gas lift performance curve. (D) Mandrel spacing and unloading pressures. (E) Some valve information for the design of earlier.

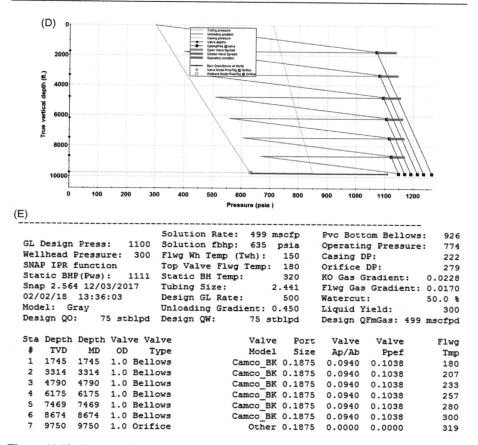

(D)
(E)

```
                                Solution Rate:    499 mscfp    Pvc Bottom Bellows:     926
GL Design Press:      1100    Solution fbhp:    635  psia    Operating Pressure:     774
Wellhead Pressure:     300    Flwg Wh Temp (Twh):  150    Casing DP:              222
SNAP IPR function             Top Valve Flwg Temp:  180    Orifice DP:             279
Static BHP(Pws):      1111    Static BH Temp:      320    KO Gas Gradient:     0.0228
Snap 2.564 12/03/2017         Tubing Size:       2.441    Flwg Gas Gradient:   0.0170
02/02/18  13:36:03            Design GL Rate:      500    Watercut:            50.0 %
Model:  Gray                  Unloading Gradient: 0.450    Liquid Yield:           300
Design QO:     75 stblpd      Design QW:      75 stblpd    Design QFmGas: 499 mscfpd
```

Sta #	Depth TVD	Depth MD	Valve OD	Valve Type	Valve Model	Port Size	Valve Ap/Ab	Valve Ppef	Flwg Tmp
1	1745	1745	1.0	Bellows	Camco_BK	0.1875	0.0940	0.1038	180
2	3314	3314	1.0	Bellows	Camco_BK	0.1875	0.0940	0.1038	207
3	4790	4790	1.0	Bellows	Camco_BK	0.1875	0.0940	0.1038	233
4	6175	6175	1.0	Bellows	Camco_BK	0.1875	0.0940	0.1038	257
5	7469	7469	1.0	Bellows	Camco_BK	0.1875	0.0940	0.1038	280
6	8674	8674	1.0	Bellows	Camco_BK	0.1875	0.0940	0.1038	300
7	9750	9750	1.0	Orifice	Other	0.1875	0.0000	0.0000	319

Figure 11.12 (Continued)

The results shown in Fig. 11.11A show that the flowing well is liquid loaded (unstable according to Nodal because the intersection of the tubing performance curve and the IPR is to the left of the minimum in the tubing performance curve).

See Appendix D for definitions of gas lift terms which will help interpret the results presented here and also other gas lift designs and reports. So now to investigate if gas lift will stabilize the well or in other words solve the loading problem. Here other methods to solve the problem such as smaller tubing, compression, and surfactants could be possibly used. However, the focus is on gas lift so the effects of adding gas to the bottom of the well are investigated later.

The results of adding more and more gas to the bottom of the well give a new tubing performance curve for each injection rate. The tubing plots begin to turn up indicating stable solutions. The original plot of the loaded well with no gas added is still shown as the top tubing performance curve.

Since it is hard to see what plots have a certain amount of gas added, the intersection of the tubing curves (with gas added) is plotted. This is called the gas lift performance curve or, in this case, it is the gas-in versus the gas-produced plot.

From the aforementioned performance plot, the injection rate can be selected. An old rule for oil well is that you should inject about half the gas that gives you the maximum rate. From the aforementioned plot, the peak production occurs when the gas injected is 330 Mscf/d that gives about 267.7 Mscf/d. In this case, assuming costs of high-pressure gas are low, the maximum rate is designed by injecting 300 Mscf/d as the design point for gas lift. Designing gas lift for these conditions (Figs. D and E) gives the gas lift plot showing the locations of the unloading valves with the operating valve at the bottom.

Example 11.2 High rate gas well: gas lift to unload

Data for Example 11.1: gas well
  C and n back-pressure equation for IPR
  WHT: 150; WHP: 300 psi; BHT: 170°F; GG: 0.6; liquids: 300 bbls/MMscf; API: 35; WG: 1
  Reservoir: Pr: 1111 psi; C: 0.0006 Mscfd/psi^{2n}; n: 1.0; WC: 0.5
  9800' of 2 7/8's (1.995 in. OD) roughness: 0.018 in.; gray tubing MPF correlation.

Fig. 11.12A shows loaded well performance with no gas lift. Intersections of tubing performance curves are shown in Fig. 11.12B and the IPR allow calculations to determine the gas lift performance curve are shown in Fig. 11.12C. Fig. 11.11D and E shows results of gas lift design done at the maximum point of the gas-in/gas-out curve (Fig. 11.11C).

Choosing operating point of inject, 500 Mscf/d get the gas lift design, as shown in Fig. 11.12D and E.

So this example is deeper, has higher gas rate and more liquid than Example 11.1. As a result, it takes more CP to operate gas lift and more gas injected, but gas lift does work to unload the well, stabilize the well, and obtain more rate of gas from the reservoir. Further optimization for the design could be considered but this example just shows a workable design.

11.7.6 Horizontal unconventional well

Unconventional wells exhibit a sharp decline in production. Fig. 11.13 is an example of an unconventional oil well showing a sharp decline in the first 2–3 years dropping from about 4500 to 400 bpd in 4 years. If equivalent gas, the rate drops from 26 to 1.7 MMscf/d in the same time. This is just one possible decline curve for an unconventional well.

So based on the aforementioned example decline curve the following examples will study different flow paths through various tubulars and tubular combinations both with and without gas lift over the life of the decline curve. The left column indicates a single tubing or casing or when two shown is annulus flow. The equivalent diameter is shown in the second column.

2 3/8	1.995
2 3/8 4 ½	2.756
3 ½	2.992
2 7/8−5 ½	3.641
4 ½	3.958
2 3/8−5 ½	4.261
3 1/2 7	4.699
5 ½	5.044
2 7/8 7	5.153
7	6.366

Early (Fig. 11.13A) rates are high and the well will flow. Diameters for the possible flow path can be examined to see how they compare with early pressure drop. The aforementioned list of diameters was used and a 44 bbl/MMscf liquid ratio with a 50% WC was input (Fig. 11.13B).

The results are not unexpected with the open 7-in. casing flow giving the lowest pressure drop at the high flow and the smaller casing or tubing or annular flow areas giving higher pressure drops. Also it is seen that the larger flow paths give rise to loading at some high flow rates. For example, the 7-in. casing flow is showing loading at about 10 MMscf/d. The smallest tubings (which give very high pressure drops at high flows) do not load until under 1−2 MMscf/d.

These types of results are to be expected, and the operator can choose the flow path early on that gives the lowest pressure drop and is possible to achieve or is economical to achieve to maximize flow.

Earlier examples show using gas lift to unload wells with loading problems successfully. Here there are very high flow rates early on and then the decline curve shows the rates drop quickly over the first 2 years or so. However, this chapter deals with gas lift so the question becomes could gas lift be used to accelerate the flows earlier on for this example well? Fig. 11.13C takes a diameter that is in the middle of the diameters chosen for analysis and plots one tubing curve (3.958-in. ID). Then some gas lift injection rates at the bottom of the tubing are simulated. The results show that gas lift gives a higher tubing pressure drop for higher rates and only when getting to low rates where liquid loading occurs does gas injection show positive results by preventing liquid loading. So be cautious of using gas lift early on for a gas well since a gas gradient has a lot of friction to start and more gas injected just adds to the friction pressure drop. Only when liquid loading starts does the gas lift help the situation. Of course smaller tubings at the low rates can maintain flow above critical to lower rates and if gas lift is needed with smaller

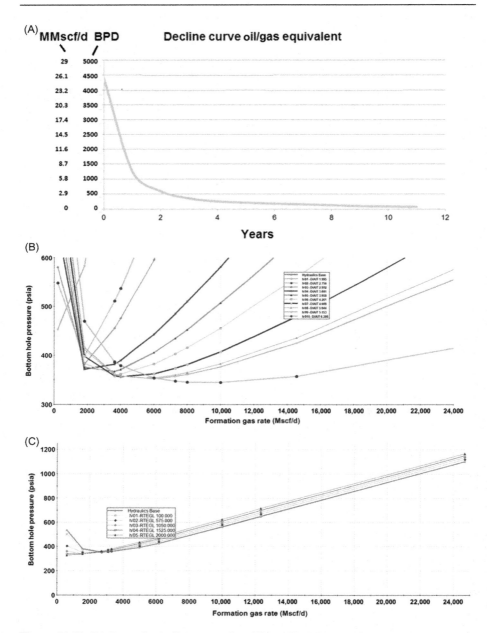

Figure 11.13 (A) Example decline curves for oil/gas. (B) Tubing performance curves by diameter. (C) 3.958-in. IS with gas lift. (D) Oil well tubing curves (see conditions below). (E) Oil well tubing curves with gas lift (3.958-in. ID).

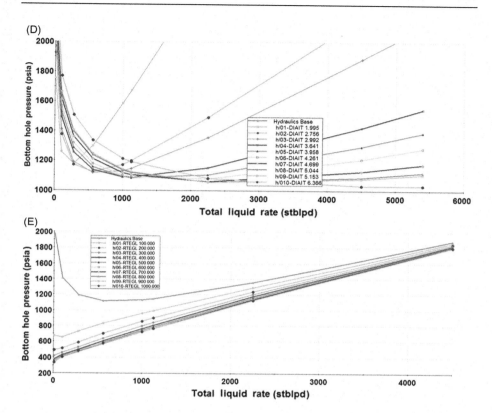

Figure 11.13 (Continued)

tubing, the amount of gas injected to combat loading will be less than for larger tubing.

Although this chapter deals with gas well deliquification, while discussing gas lift, an oil well is examined for contrast. Fig. 11.13D shows tubing performance curves for an oil well over the life of the decline curve shown in Fig. 11.13A. Again the results are to be expected with the larger flow paths giving the lower pressure drop but becoming unstable at higher rates. A WHP of 200 psi is used and a gas oil ratio (GOR) of 555 scf/bbl-oil is input with a 50% WC.

So again a single tubing size (3.958-in. ID) is separately examined as shown in Fig. 11.13E. The top curve is for the tubing performance with no gas added. Then several tubing curves are shown with more and more gas injected near the bottom of the tubing. The results show (unlike for the gas well) that the addition of gas lift lowers the tubing pressure drop only when larger amounts of gas are added, does additional gas lift gas show diminishing results. So unlike a gas well, oil well shows more drawdown with gas lift gas added even at the higher rates. And at the low production rates the benefits are even better by eliminating loading and allowing the use of the tubing to much lower production rates.

In general, for oil or gas wells, injection rates should be such that the tubing performance curve is to the right of the inflection point where the point of application is planned to be. It is up to the user to decide what flow paths to use for early and later much lower production rates but this example shows that the early (and later) use of gas lift for oil wells (well with much more liquid resident in the tubing) can be beneficial even at high production rates and to eliminate loading at low rates. Note that even without the IPR, one can see the range of flows that are stable and to the right of the minimum of the tubing performance curve.

11.8   Single-point/high-pressure gas lift[4]

A technique that is becoming increasingly popular for providing higher rates from liquid-rich wells and lower producing BHP (PBHPs) with lower lift points is SPGL using "high-pressure" injection gas which requires no packer or valves and uses a single point of injection as deep as desired into the well. This type injection is shown in Fig. 11.9, GAPL, in the diagram on the left, "Open-ended."

Actually, this technique has been around for many years and in many ways is even better suited to later life gas wells than conventional gas lift.

The evolution and application of gas lift is historically based on low GLR oil wells with pressure maintenance and limited drawdown requirements, limited compression pressures of no more than 1350 psig, limited casing strengths and packers in every well.

Today, things have changed substantially with low producing BHP eventually being required in almost all wells and higher GLRs being prevalent even in oil wells. High-pressure compressor cylinders are now available for standard units due to compressed natural gas and other applications and high-strength/working pressure casing is being installed in most wells to allow facing down casing and no packers needed.

If lift gas is available at a surface pressure (plus the pressure due to the weight of the gas column) which is above the pressure gradient of the fluid to be lifted from the well, SPGL is always a viable option without the use of gas lift valves. This technique is used regularly with nitrogen being injected down coiled tubing to unload or jet wells.

Therefore applying SPGL with high-pressure injection gas serves to unload and kick off a well early in its life even after stimulation/original completion (6000 psig injection pressure can be used to lift a 10,000′ well with 7000 psig reservoir pressure/0.7 psi/ft. kill fluid).

Later in the life of a gas well when it approaches critical rate flowing up the tubing, much lower pressure gas can be injected down the casing-tubing annulus to supplement the gas rate and keep the well above critical rate.

This means that SPGL can potentially be used as the only lift method needed during a well's life, and that this can be done with no well interventions with a rig or wireline unit. There may be a requirement to change the surface compression to fit the injection pressure and rate needed to lift the well as these change over a well's life. This is also required in conventional gas lift although the range of injection pressures is smaller.

In SPGL, tubing or annular flow can be done without any changes in downhole (as would be needed in conventional gas lift) resulting in rate flexibility from very high to low. Also, depth of injection is not limited due to a packer or requirements to be able to change valves/orifices so the lift point can go as far as the toe of a horizontal well with one simple tubing string if desired. In addition, plunger lift can be easily applied at any time during the life with or without lift gas.

SPGL also has an added advantage in that the injection pressure is essentially equal to the PBHP providing direct feedback on well optimization efforts and information about the well's current position on the IPR curve.

Some concerns have been expressed about well stability with SPGL versus conventional gas lift. Conventional gas lift valves/orifices do disperse the gas in fluid better, possibly reducing slugs, however, for most gas wells the liquid fraction is low, and since the flow is kept out of the slugging region by keeping the well above the critical rate, this should not be a concern. For most horizontal wells, slugging cannot be eliminated in any scenario since it is inherent to these wells based on geometry. Slugging can be handled with good facility design.

Gas wells later in their lives usually will push all of the fluid into the producing interval after a shut-in period. Also if the lift gas injection is started before the onset of liquid loading, no large fluid slugs would be expected. For these wells, there is no liquid initially in the wellbore, and injection gas can be added as the well is opened to sales before liquid slugs form. This means in most cases the required lift gas pressure, using no gas lift valves or packer, is no more than slightly above the inflection point on the J curve (critical rate). This technique is a subset of SPGL called continuous gas circulation (CGC) which was pioneered by J. Hacksma.

In cases where these type of wells are let to load up for a long period of time and/or have some kind of treatment fluid pumped into them, a small percentage may need to be swabbed or a nitrogen tank brought out to unload them.

As an example, the loaded gas well, shown in Example 11.1, could be gas lifted with the same results at the same depth as the conventional gas lift design (eight valves with 1000 psig gas lift supply pressure) using SPGL to 5950' with no gas lift valves, mandrels or packers and a surface injection pressure of 2400 psig, eliminating the downhole equipment. This includes being able to lift a 0.45 psi/ft. kill fluid out of the hole if needed. Alternately, if injection was started on the well in a timely manner (before onset of liquid loading) and no capability to unload kill fluids from the well is needed (as typical for most mature gas wells where the existing tubing string can be used and no well work is required), this well could be lifted with 400 psig injection pressure and no gas lift valves.

As in conventional gas lift, a source of higher pressure gas is always needed for SPGL and in very high-pressure cases a booster wellhead compressor would be required even if there was a gas lift injection system in the field. For wells in fields with no gas lift system, wellhead compressors can be a very useful tool enabling both the lowering of wellhead pressure and injection gas supply if desired.

11.9   Gas lift summary

Gas lift is introduced. Examples show how gas lift can be used to unload a gas well or prevent a gas well from loading. An example for an unconventional gas well is shown with a high rate initially and then quickly declining to lower rates. It is shown that for gas wells, early addition of gas lift (after the well is unloaded of completion/stimulation fluid) to a high-rate gas well adds pressure drop to the tubing and is useful only as the well approaches low rates and critical velocity or rate. This is contrasted to an oil well where it is shown that addition of gas lift can be beneficial even for high production rates and also down near the point of instability or loading. So, for an oil well, it is shown that gas lift can be a high rate method of AL, but for a gas well, it cannot.

Also the concept of SPGL (including CGC) is introduced, which is sometimes even better suited to gas wells than conventional gas lift with valves and a packer. These conclusions are from the few examples shown here and case-by-case situations should be examined. Tubing size is an important consideration. Adding gas lift to both a gas well and an oil well to eliminate loading or instability at low rates can be beneficial.

References

1. Trammel P, Praisnar A. *Continuous removal of liquids from gas wells by use of gas lift.* Lubbock, TX: SWPSC; 1976. p. 139.
2. Boswell JT, Hacksma JD. Controlling liquid load-up with continuous gas circulation, SPE 37426. In: Presented at the production operations symposium, Oklahoma City, OK, March 9−11, 1997.
3. Stephenson GB, Rouen B. Gas-well dewatering: a coordinated approach, SPE 58984. In: Presented at the SPE international petroleum conference and exhibition in Villahemosa, Mexico, February 1−3, 2000.
4. Elmer D, Elmer W, Harms L. High pressure gas-lift: is the industry missing a potentially huge application to horizontal wells? SPE 187443. In: Presented at the SPE annual technical conference and exhibition, San Antonio, TX, October 9−11, 2017.

Further reading

Winkler HW. Gas lift solves special producing problems. *World Oil* 1998;**209**(11):35−9.

Electrical submersible pumps

12

David Divine has a Bachelor of Science degree in electrical engineering from Texas Tech University and is a registered engineer in the State of Texas. He has over 40 years' experience in the oil industry. His experience includes service with Texaco Inc. where he developed the first practical variable speed submersible pumping system and is the author of several papers on this topic. He cofounded Submersible Oil Systems (SOS), a company that designed and manufactured a variable speed controller for submersible pumping. After SOS was acquired by Centrilift-Hughes, he served as vice president over the Systems Division of Centrilift-Hughes. He was a cofounder of Electric Submersible Pumps, Inc. (ESP, Inc.). As vice president of engineering, he helped develop many of the current standards for testing used submersible equipment and has improved many of the standards for the testing of new equipment.

In 1997 he became an independent consultant. He began working for Baker Hughes incorporated as a Technical Representative at the Educational Development Center in Claremore, Oklahoma, for Centrilift in September 2002. In July 2007 he began working with Wood Group ESP, Inc. as a principle engineer. His duties included improving the method and equipment used for varnishing new ESP motors. In 2009 he became the team leader for the development of "Solutions," the ESP sizing software for the Wood Group ESP. Wood Group ESP was acquired by GE Oil and Gas in May 2011 and he continued in the position of solutions development team leader until his retirement from GE Oil and Gas in 2016. Shortly thereafter, he began working for Valiant Artificial Lift Solutions as a principal engineer. He is a member of the Texas Tech Electrical Engineering Academy and has been a member of the Industrial Advisory Board at Texas Tech University. He is the 1995 recipient of the J. C. Slonneger Award presented by the Southwestern Petroleum Short Course to individuals who have advanced the field of artificial lift. He has been presenting seminars and schools on submersible pumping for the past 40 years. He is the coauthor of a Manual and software on the subject of electric submersible pumping.

12.1 Introduction

Historically the electric submersible pump (ESP) was used to dewater mines. Later, it was used to produce oil wells with very high water cut and low gas liquid ratio (GLR). By today the industry has advanced, and the ESP has its found uses in many low water cut and high-GLR wells. The literature abounds with case studies where ESPs are used in wells producing very little water and high volumes of gas.

Gas Well Deliquification. DOI: https://doi.org/10.1016/B978-0-12-815897-5.00012-3

This chapter covers the downhole components and how they may be used to produce wells with very high GLR and high gas void fractions (GVFs) with pump intake conditions such as dewatering gas wells.

The ESP system is depicted in Fig. 12.1A and B. The components will be reviewed from the bottom to the top, starting with the SP motor and ending with the ESP. Producing liquids from a gas well or any high-GLR well using an ESP

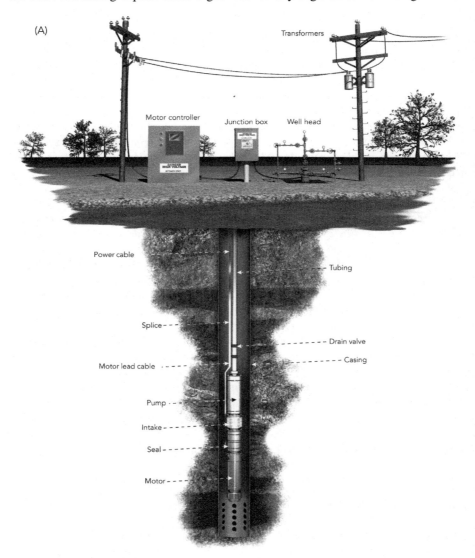

(A)

Transformers

Motor controller   Junction box   Well head

Power cable

Tubing

Splice

Drain valve

Casing

Motor lead cable

Pump

Intake

Seal

Motor

Figure 12.1 (A) An electrical submersible pump (ESP) installation diagram. (B) A detail of the ESP assembly.
Source: Courtesy Valiant ALS.

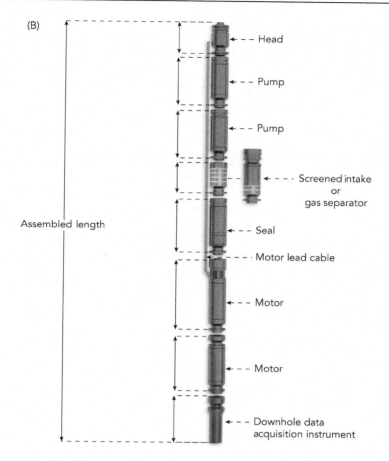

Figure 12.1 Continued

involves mitigating the effects of the free gas that may enter the pump or keeping the gas from entering the pump.

Section 12.2 describes the ESP induction motor (IM), the permanent magnet motor (PMM) designs, and the cautions when designing the system for low-flow gassy wells.

Section 12.3 covers the seal or protector that seals well fluids out of the motor, equalizes the pressure of the oil in the seal and motor with the pressure in the well, provides a bearing to carry the thrust of the pump above the seal, and provides chambers to allow for the thermal expansion and contraction of the oil in the motor and seal. It also discusses about the, cautions when light oils and gas are produced.

Section 12.4 describes about the pump intakes, the methods to avoid gas, and the gas handling ability of the pump. Standard intakes and intake placement are also covered. Estimating and calculating the GVF to determine the probability of successful operations is discussed. Gas separator types and uses are presented.

Section 12.5 covers the details of ESP. Different stage types are defined along with their gas handling ability. Pump performance curves are defined. Stage thrust and pump construction are discussed. The Turpin and Dunbar gas interference correlations are defined and used in sizing examples. The design of tapered pumps with and without gas separators is covered.

Section 12.6 covers the chapter's summary.

12.2 The electric submersible pump motor

A schematic diagram of a submersible pump installation is depicted in Fig. 12.1A. In most installations, the ESP assembly is serviced into the well starting with the motor(s).

The motor may be a two-pole, three-phase IM or a four-pole, synchronous PMM. The IM may be a single motor as shown in Fig. 12.1A or two or more motors bolted together in tandem as shown in Fig. 12.1B.

Unlike surface motors that are filled with air, the ESP motor is filled with a high-dielectric strength oil. This oil provides lubrication, and electrical insulation, assists with cooling, and provides an incompressible liquid, so that the pressure inside the motor may be equalized with the pressure in the well bore.

The bottom of the motor may have a built-in base, a bolt-on base, or a downhole data acquisition instrument, as shown in Fig. 12.1B, bolted onto the bottom.

A single motor will always be an upper tandem motor. This construction contains the motor lead connection, commonly known as the pothead, which allows the three-phase power cable to be connected to the three-phase motor windings. If additional horsepower is required, then a center tandem motor can be connected (or plugged) into the base of the upper tandem motor. The nameplate voltages (NPVs) and the nameplate horsepowers (NPHPs) of the upper and lower tandem motors are added to obtain the total NPV and the total NPHP, respectively. The amperage rating for all the tandem motors will be the same for the total motor.

An upper tandem motor and a center tandem motor are cross-sectioned in Fig. 12.2A and B, respectively. They are depicted with shipping caps bolted to the top and bottom of the motors. There will be a thrust bearing at the head of every motor to support the weight of the shaft, rotors, and rotor bearings of the motor. It is not designed to carry any of the thrust of the pump shaft. The motors contain a single three-phase wound stator that runs almost the length of the motor. The rotors, located on the shaft, are from 13 to 15 in. in length and will be separated by a rotor bearing to keep the them from being magnetically pulled into the stator. As the motor increases in length, so will the stator length and the number of rotors on the shaft. Maximum motor lengths for a single motor vary from 30 to 36 ft. depending on the manufacturer and diameter.

The motors are wound in a "wye" or "star" configuration so that the motor base, when connected, will tie (or short) the three-phase windings together at the neutral

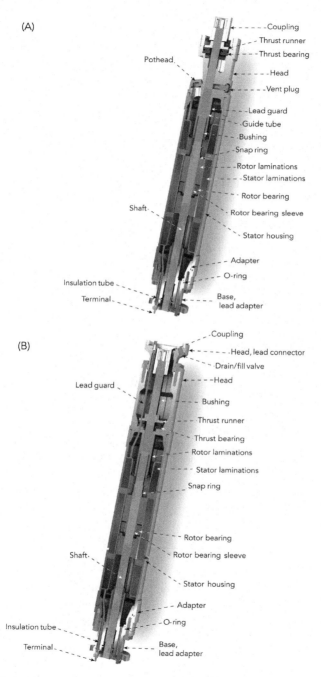

Figure 12.2 (A) An ESP upper tandem induction motor cross section. (B) An ESP lower tandem induction motor cross section.
Source: Courtesy Valiant ALS.

point. In a balanced three-phase system the neutral point will have all three-phase currents flowing through it, but the currents are 120 degrees out of phase, so the vector sum of the three currents is zero.

When a downhole sensor is connected to the base of the motor(s), it provides the wye-point or neutral-point connection for the all three phases and the connection for the signal that is used to provide information to the surface.

12.2.1 Electric submersible pump induction and permanent magnet motor RPM

IM or PMM synchronous RPM is defined by the following equation:

$$RPM_{SYNC} = \frac{Hertz \times 120}{Motor\ poles} \qquad (12.1)$$

The IM, when connected to a 60-Hz power system, or a variable frequency drive (VFD) operating at an output frequency of 60 Hz, has a synchronous RPM of 3600 but the actual RPM will be less. The IM rotor must slip or rotate slower than the rotating magnetic field produced by the stator windings to induce current into the rotor, which develops the rotor's magnetic field. The IM RPM is defined in the following equation:

$$RPM_{IM} = RPM_{SYNC} - RPM_{SLIP} \qquad (12.2)$$

where the slip RPM will vary from 180 to 50 depending on the load imposed by the pump, seal and intake load, and the motor diameter. The IM's full load RPM, when 60 Hz power is applied, will be approximately 3420 RPM.

The full voltage, 60 Hz, composite curve for a 562 IM is shown in Fig. 12.3A. This motor is 5.62 in. in diameter and the length will vary from 6 ft. (40 HP) to 35.5 ft. (450 HP). The average HP per foot is about 10.5. The average HP density is about 62 HP per cubic foot.

The relationship between HP, voltage, amperage, efficiency, and power factor is defined in the following equation:

$$Output\ HP = \frac{voltage \times amperage \times \sqrt{3} \times efficiency \times power\ factor}{746\ W/HP} \qquad (12.3)$$

At 100% load the RPM is 3490, the amperage is at 100% of nameplate, the efficiency is 87% (0.87), and the power factor is 81% (0.81). At 50% load the RPM is 3555, the amperage is at 65% of nameplate, the efficiency is 82.5% (0.825), and the power factor is 66% (0.66).

Reducing the load on the motor to 50% of nameplate will reduce the efficiency from 87% to 82.5% or a 5.2-point reduction in efficiency.

The full voltage, 60 Hz, composite curve for a 456 IM is shown in Fig. 12.3B. This motor is 4.56 in. in diameter and the length will vary from 5 ft. (10 HP) to

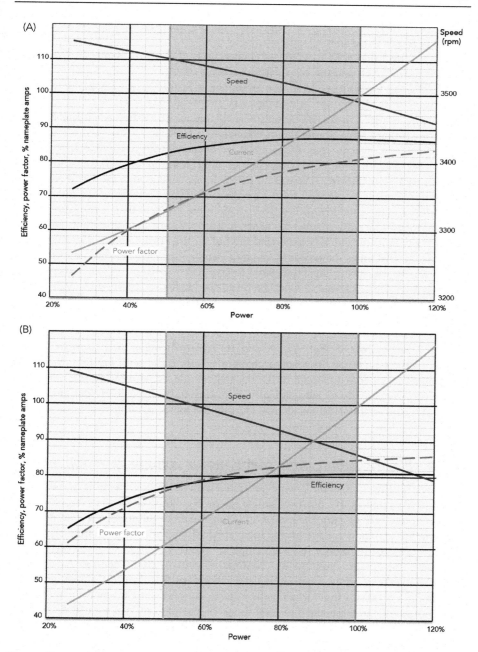

Figure 12.3 (A) A composite performance curve for 5.62 in. diameter induction motors at nameplate voltage and 60 Hz. (B) A composite performance curve for 4.56 in. diameter induction motors at nameplate voltage and 60 Hz.
Source: Courtesy Valiant ALS.

31.2 ft. (120 HP). The average HP per foot is about 3.2. The average HP density is about 33 HP per cubic foot.

At 100% load the RPM is 3430, the amperage is at 100% of nameplate, the efficiency is 81% (0.81), and the power factor is 85% (0.85).

At 50% load the RPM is 3510, the amperage is at 60.5% of nameplate, the efficiency is 77% (0.77), and the power factor is 75.5% (0.755).

Reducing the load from 100% to 50% will reduce the efficiency from 81% to 77% or a 4.9-point reduction in motor efficiency will occur.

The four-pole, synchronous PMM must be operated with a variable speed drive (VSD), also called a VFD. When operated at 120 Hz, the motor will rotate at a synchronous RPM (3600 RPM). There is no RPM slip associated with the PMM since the rotor magnetic field is supplied by the permanent magnets on the rotor. Since there are no rotor losses in the PMM, it is 10%−12% more efficient than a comparable IM. The PMM uses rare earth magnets to generate a strong rotor magnetic field. This field is stronger than the electrically generated field of an IM. As a result, a PMM can generate three to five times the power of an equivalently sized IM. This translates into the horsepower density being increased three to five times. Owing to their higher horsepower density, PMMs may eliminate the need for tandem or triple tandem IMs. The required power may be delivered in one PM motor section.

Fig. 12.4A and B is PM performance curves for a 4.56 in. diameter (456 series) and 5.62 in. diameter (562 series) motor, respectively. Note the higher efficiency as compared to the IMs in Fig. 12.3A and B. Also efficiency is a function of the applied motor voltage as the loading changes just as it did with the IMs. Since the PM motors are operated using a VSD, the motors can be adjusted to the most efficient level by manually or automatically, adjusting the volt per hertz coming out of the VSD as the well conditions change causing the motor loading to change.

12.2.2   Electric submersible pump motor voltage variation effects

The curves in Fig. 12.3A and B were generated with the motor operating at nameplate voltage (NPV) and 60 Hz power. The curves in Fig. 12.5[1] show the effects of high and low voltage on the performance of an IM operating at full load as a percent variation from the optimum voltage for the load, at NPHP and at nameplate frequency.

Notice that the minimum full-load amps are plotting at 102.5% of the NPV. This is because the motors built to the National Electrical Manufactures Association (NEMA) standards allow for rounding the laboratory full-load test amperage up to account for manufacturing variations that can occur and some normal voltage variations that might increase the full-load amps of the motor. The nameplate FLA is used to select the correct wire size, motor starter, and overload protection devices necessary to serve and protect the motor.[2]

The voltage could be adjusted up to 2.5%, so that the peak efficiency is coincided with the minimum amperage. The voltage can be "trimmed" or adjusted to account for the motor loading. A method for automatically adjusting motor voltage based on motor loading was patented in 1984 by NASA engineer, Frank Nola[3] (US Patent 4,439,718).[4]

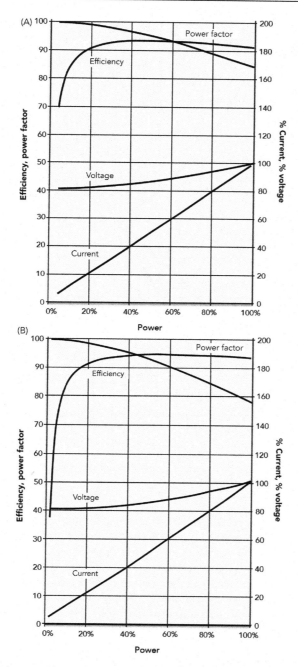

Figure 12.4 (A) A composite performance curve for 4.56 in. diameter permanent magnet motors at 60 Hz. (B) A composite performance curve for 5.62 in. diameter permanent magnet motors at 60 Hz.
Source: Courtesy Valiant ALS.

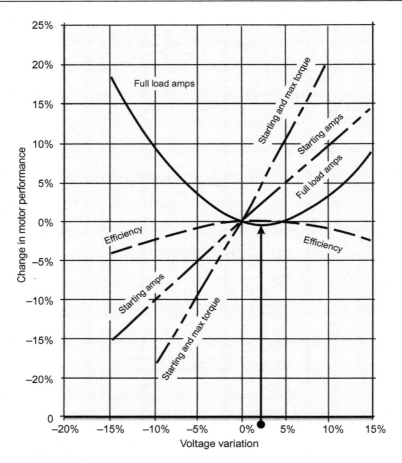

Figure 12.5 Induction motor performance versus voltage variation in percent.
Source: Courtesy Valiant ALS.

Motor load saturation tests can determine the optimum voltage for any IM over a large range of loading. Fig. 12.6 is a plot of motor per unit amps versus per unit motor terminal voltage for constant per unit horsepower levels.[5]

ESP IM manufacturers have tested their motors and developed mathematical models that allow the prediction of the optimum motor voltage based on the motor load. This allows for a maximum motor efficiency and a minimum motor temperature rise (TR) to produce the conditions of the well.

12.2.3 Defining electric submersible pump motor frame sizes

ESP motors are generally rated by HP, as shown in Fig. 12.7A. This is a listing of 456 series motors rated by HP. Allowed motor loading is also listed at 50% to 150%.

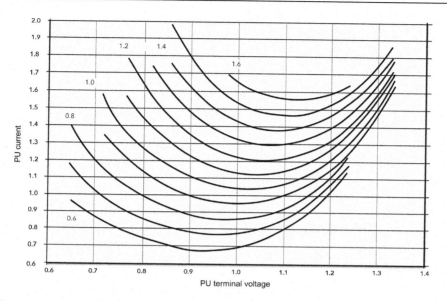

Figure 12.6 Induction motor per unit terminal voltage versus per unit current at per unit loads. Source: Courtesy Valiant ALS.

After running saturation tests for the motor series, as shown in Fig. 12.6, the motors were rated by a frame size and the HP rating for a frame size can vary from 50% to 150%, as shown in Fig. 12.7B.

12.2.4 Electric submersible pump motor, or frame, winding temperature

Motor winding operating temperature is determined by the following factors:

1. Ambient temperature of fluid passing through the motor.
2. TR above the ambient temperature due to the motor operating load, with specific heat = 1.0 and water as the cooling fluid at a velocity of 1 ft./s.
3. TR occurs when the velocity of the cooling fluids is less than 1 ft./s.
4. TR occur when the average specific heat of the cooling fluids is less than one.
5. TR due to any voltage unbalance at the motor terminals.
6. TR due to harmonics in the motor voltage from a VSD.

To determine the motor (or frame), winding TR at the optimum operating voltage, tests with oil as the cooling fluid are run at different velocities and then the same procedure is repeated using water as the cooling fluid. The motor is run long enough at each test velocity, so that the TR above the ambient temperature of the cooling fluid is stable. The motor is shut down and its winding resistance is measured and from its average. The winding temperature is calculated. The temperature of the cooling fluid and the winding TR are also calculated. Heat rise versus fluid velocity for water and oil is plotted in Fig. 12.8.

(A)	456 Series motors		
Standard HP	Length (ft.)	Min load	Max load
40	11.1	0.5	1.5
50	14.5	0.5	1.5
60	16.9	0.5	1.5
70	19.3	0.5	1.5
80	21.7	0.5	1.5
90	24	0.5	1.5
100	26.4	0.5	1.5
120	31.2	0.5	1.5

(B)	456 Series motors		
Frame size	Length (ft.)	Min HP	Max HP
40	11.1	20	60
50	14.5	25	75
60	16.9	30	90
70	19.3	35	105
80	21.7	40	120
90	24	45	135
100	26.4	50	150
120	31.2	60	180

Figure 12.7 (A) 4.56 in. induction motors at standard HP, length, and minimum and maximum per unit loading. (B) 4.56 in. induction motors at standard HP, length, and minimum and maximum HP rating.
Source: Courtesy Valiant ALS.

The TR model assumes the optimum voltage model and it was used to optimize the motor efficiency. The estimated motor temperature can then be calculated based on the ambient temperature of the cooling fluid, motor load, cooling fluid velocity, cooling fluid specific heat, voltage unbalance, and voltage harmonic content. Specific heat of water is one, oil's is approximated at 0.4 relative to water, and free gas is estimated at 0.15 relative to water. These models have names such as application-dependent and temperature-dependent motor ratings.

When deliquefying a gas well, the average specific heat of the cooling fluid may increase the winding temperature to an unacceptable level. A pressure–temperature sensor should be used to monitor the motor winding temperature and provides a shutdown for overtemperature. The motor should be supplied with a thermocouple or a resistive temperature device to connect to the downhole sensor.

When the motor is operated using a VSD, also called a VFD, and the well conditions change, causing the motor loading to change, the motor can be adjusted to the

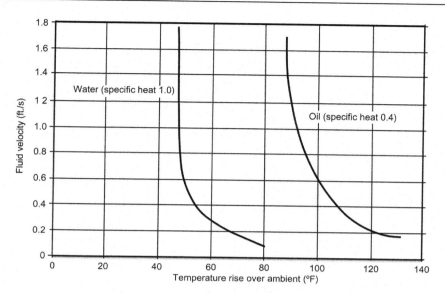

Figure 12.8 Cooling fluid temperature rise above the ambient (°F) versus fluid velocity for water and oil.
Source: Courtesy Valiant ALS.

most efficient level by manually or automatically[6−8] adjusting the volt per hertz coming out of the VSD keeping the motor temperature as low as possible.

Adjusting the motor voltage to optimize the efficiency does come with a caveat. Increased loading at optimum voltage reduces the motor RPM, whereas decreased loading at optimum voltage increases the RPM, as seen in Fig. 12.9. However, when run with a VSD, the Hz can be adjusted to compensate the increase or reduction in RPM.

12.2.5 Electric submersible pump motor insulation life

ESP motors have a basic insulation level of 25 kV. This is the insulation level that is designed to withstand surge voltages rather than only normal operating voltages. NEMA[9] insulation classes, maximum hot spot temperature, and typical materials are listed in Fig. 12.10A. The ESP motor insulation should be class N or class S. Fig. 12.10B plots average insulation life in hours versus the winding hot spot temperature in degrees Celsius for classes A, B, F, H, and N insulations.[10,11] ESP motors are insulated with a minimum of class N insulation. At 160°C (320°F), the average life of the class N insulation would be 300,000 hours or about 34 years. Average insulation life doubles for every 10°C (18°F) decrease in temperature and is halved for every 10°C increase in temperature. Class N insulation is rated for a hot spot temperature of 200°C (392°F). The hot spot temperature is determined by the ambient temperature plus the TR due to the motor loading and operating efficiency plus an adjustment for cooling fluid velocity and specific heat plus a hot spot allowance as determined by each motor manufacturer. As defined by NEMA,

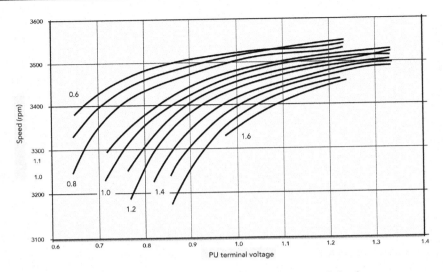

Figure 12.9 Induction motor per unit voltage versus RPM at per unit loads.
Source: Courtesy Valiant ALS.

the 200°C rating equates to an average motor insulation life of 2.25 years. For acceptable life, ESP motors are designed with an average temperature below the rated hot spot temperature. Insulation does not suddenly fail if the hot spot temperature is reached, but useful operating life declines rapidly.

Insulation life versus motor winding hot spot temperature for class N insulation using °F is depicted in Fig. 12.10C. The HP rating of the ESP motor can be calculated based on the allowable TR of the motor insulation and the shaft torque loading limit and the insulation average life can then be calculated based on the hot spot temperature of the loaded motor.

NEMA specifies, for class H insulation, that for a 40°C ambient, a TR of 120°C is allowed with a 15°C hot spot allowance, for an operating winding hot spot temperature of 175°C. This equates to an average insulation life of 20,000 hours.

12.2.6 Applying the National Electrical Manufactures Association method to the electric submersible pump motor's class N insulation

Given a 100 Hp load, an ambient bottom-hole temperature of 180°F (82°C), a 400 series 100 frame size ESP motor, 50% water cut with 20% free gas, fluid velocity of 1 ft./s, cooling the motor and using the motor manufactures test data, the operating temperature at the optimum voltage can be calculated at 311°F (155°C). Fig. 12.10C predicts the average life of class N insulation at this operating temperature to be over 500,000 hours or over 58 years. No one would expect an ESP motor to run that long. There are many other contributing factors that will lead to a motor failure long before the operating temperature causes the motor insulation to fail.

NEMA/UL letter class	Maximum hot spot temperature allowed (°C)	Typical materials
(A)	90	Unimpregnated paper, silk, cotton, vulcanized natural rubber, and thermoplastics that soften above 90°C [5]
A	105	Organic materials such as cotton, silk, paper, and some synthetic fibers [6]
	120	Polyurethane, epoxy resins, polyethylene terephthalate, and other materials that have shown usable lifetime at this temperature
B	130	Inorganic materials such as mica, glass fibers, asbestos, with high-temperature binders, or others with useable lifetime at this temperature
F	155	Class 130 materials with binders stable at the higher temperature, or other materials with usable lifetime at this temperature
H	180	Silicone elastomers, and Class 130 inorganic materials with high-temperature binders, or other materials with usable lifetime at this temperature
N	200	As for class B, and including Teflon™
R	220	As for IEC class 200
S	240	Polyimide enamel (Pyre-ML) or Polyimide films (Kapton® and Alconex Gold®)
	250	As for IEC class 200, further IEC classes designated numerically at 25°C

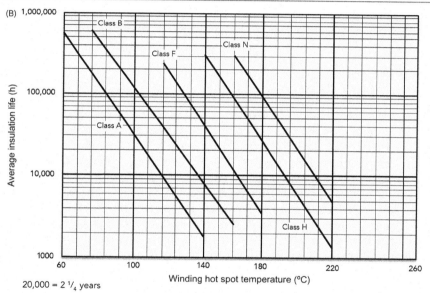

20,000 = 2 ¼ years

Figure 12.10 (A) National Electrical Manufactures Association (NEMA)/Underwriters Laboratory (UL) insulation classes are defined. (B) Average insulation life versus operating temperature in °C. (C) Average insulation life versus operating temperature in °C for class N insulation used in ESP motors.
Source: Courtesy Valiant ALS.

Therefore a smaller frame motor and lower cost motor could be considered that it will run at a higher HP density and an elevated operating temperature. Reducing the frame size shortens the motor and increases the HP density and TR.

Given the same conditions, a 400 series, 80 frame ESP motor is selected. Again, using the test data the operating temperature at the optimum voltage can be calculated at 336°F (169°C). Fig. 12.10C predicts the average life of class N insulation at this operating temperature to be around 200,000 hours or nearly 23 years.

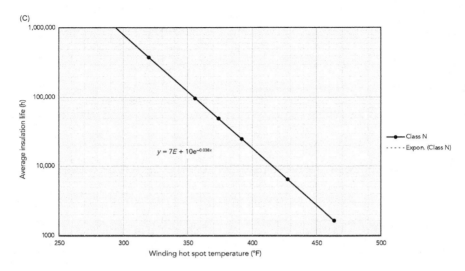

(C)

Figure 12.10 Continued.

Given the same conditions, a 400 series, 70 frame ESP motor is selected. Again, using the test data the operating temperature at the optimum voltage can be calculated at 362°F (183°C). Fig. 12.10C predicts the average life of class N insulation at this operating temperature to be around 74,000 hours or 8.5 years average insulation life.

Given the same conditions, a 400 series, 60 frame ESP motor is selected. Again, using the test data the operating temperature at the optimum voltage can be calculated at 409°F (209°C). Fig. 12.10C predicts the average life of class N insulation at this operating temperature to be 12,454 hours or 1.4 years average insulation life.

The 70 or 80 standard HP motor would be a choice between the best cost compared to insulation life.

12.2.7 Electric submersible pump motor insulation life— sensitivities

For the projected insulation lives calculated earlier, the operating conditions were assumed to be constant over the life of the well. The flow past the motor was assumed to be 1 ft./s or higher. However, well productivity changes with time and it usually declines. The ratio of the oil, gas, and water also changes with time. The HP loading may increase.

Assume that after 8 months, the water cut has increased to 80%, the HP loading has increased to 120 HP, the fluid velocity is still greater than 1 ft./s, and the free gas passing the motor is 30% of the total fluid. The 70 HP motor now has a hot spot temperature of 409°F with an average insulation life of 1.4 years. The 80 HP motor now has a hot spot temperature of 365°F with an average insulation life of 7.6 years. With this new information, the 80 HP motor would be a better choice.

Changing the free gas passing the motor to 70%, will increase the operating temperature of the 80 HP motor to 403°F with an average insulation life of 1.8 years. Under the same conditions, a 100 HP motor will have an operating temperature of 350°F with an average insulation life of 13.2 years.

This points up the need for longer term considerations when selecting the motor frame size.

12.3   Electric submersible pump seals

Bolted to the top of the motor is the base of the ESP seal as designated by Valiant Artificial Lift Solutions and Baker Hughes Centrilift. It is called the "protector" by Schlumberger and Halliburton and the "seal chamber section" by the American Petroleum Institute (API). (see Fig. 12.1A and B).

The ESP seal performs the following five functions.

1. It seals the motor oil from the well fluids.
2. It provides pressure equalization between the well bore and the oil in the motor and seal.
3. It provides chambers for the expansion and contraction of the motor oil, resulting from heating of the motor.
4. It provides a thrust bearing to carry the load of the pump shaft and impeller thrust, thereby isolating the motor thrust bearing from any pump thrust loading.
5. It provides a mechanical connection between the pump and motor.

12.3.1   The labyrinth seal

A labyrinth seal is cross-sectioned in Fig. 12.11A. There is a rotating mechanical shaft seal "mech seal" located in the "head." See Fig. 12.12. The head of the seal will be bolted to the bottom of the pump intake. Well fluid moving into the pump intake will be on the top of this shaft seal. The shaft seal is designed to prevent the well fluid from migrating down the seal's shaft to the secondary shaft seal below the bearing. This rotating shaft seal is designed to seal at pressure differentials up to 5 psi. Thus, the need for pressure equalization of the oil in the seal and the well fluid in the well bore. The seal's shaft is protected from the fluid in the top labyrinth chamber by the shaft "guide tube."

When the seal is serviced before being run into the well, the seal and motor will be filled with the high-dielectric strength oil used in the motor for electrical insulation, lubrication, cooling, and pressure equalization. The oil also has a specific gravity lower than the gravity of the oil and water being produced. If the well producing oil is of lower gravity than the motor oil gravity, then a bag-type seal should be used.

As the seal is lowered into the well, it will enter the well fluids and the well fluids will enter the pump intake and flood the head of the seal. Well fluids will displace the motor oil in the top labyrinth camber by migrating down the breather tube, as shown in Fig. 12.11A. If it is a vertical well bore then the well fluid can only fill the

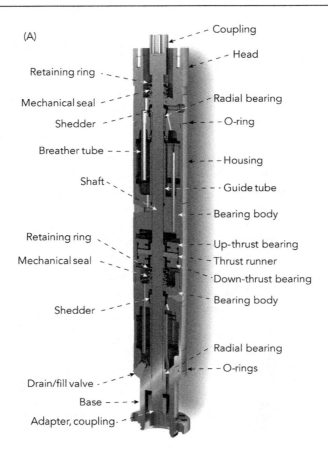

(A)

Coupling

Head

Retaining ring

Mechanical seal

Shedder

Radial bearing

O-ring

Breather tube

Housing

Shaft

Guide tube

Bearing body

Retaining ring

Mechanical seal

Up-thrust bearing

Thrust runner

Down-thrust bearing

Bearing body

Shedder

Radial bearing

O-rings

Drain/fill valve

Base

Adapter, coupling

Figure 12.11 (A) Component description of a cross sectioned labyrinth seal. (B) Component description of a cross-sectioned bag seal.
Source: Courtesy Valiant ALS.

labyrinth chamber to the bottom of the breather tube. The lighter motor oil will keep the level from increasing. The contact of the well fluids and the motor oil provides the pressure equalization necessary for the rotating shaft seals to function.

When the ESP is landed and turned on, the heat generated by the motor losses will heat the motor oil and cause it to expand. The excess volume of motor oil generated through expansion will be forced out the breather tube. The well fluid interface will remain at the bottom of the breather tube. When the ESP is shut down, the motor oil will begin to cool and will contract back down the standing tube shown on the right side of the labyrinth chamber in Fig. 12.11A. This contraction allows additional well fluid into the labyrinth chamber, raising the well fluid/motor oil interface in the chamber. The volume of the labyrinth chamber is designed such that this interface will be well below the top of the standing tube. When the ESP is

(B)

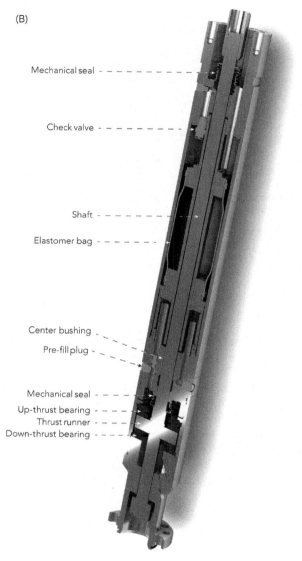

Mechanical seal - - - - - - - - -

Check valve - - - - - - - -

Shaft - - - - - - -

Elastomer bag - - - - - -

Center bushing -

Pre-fill plug -

Mechanical seal - - - -
Up-thrust bearing - - - -
Thrust runner - - - -
Down-thrust bearing - - - -

Figure 12.11 Continued.

powered up and the motor oil again expands, the well fluid will be forced back out the breather tube and the interface will return to the bottom of the breather tube.

If the ESP is started and stopped at regular intervals, the working level of the well fluid/motor oil interface will rise. The inside of the labyrinth chamber is "water wet" and will retain small amounts of water with each thermal cycle. This is one of several reasons that cycling the ESP should be avoided if possible.

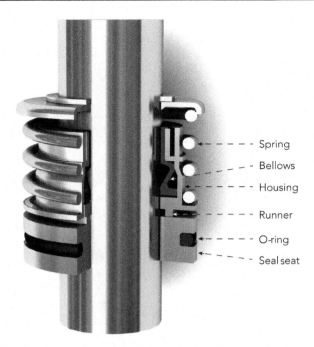

Figure 12.12 Component description of a quarter-sectioned rotating mechanical shaft seal. Source: Courtesy Valiant ALS.

With numerous cycles the well fluid level in the top labyrinth chamber could reach the intake of the standing tube and well fluids would be pulled down into the lower labyrinth chamber below the bearing. Continued cycling would eventually allow well fluid to enter the standing tube in the lower labyrinth chamber and into the motor collecting at the motor base.

The labyrinth seal's effective volume decreases as the chamber is inclined. Therefore they are not generally recommended at deviations greater than 30 degrees from vertical.

12.3.2 Positive barrier or bag seal

A bag seal is cross-sectioned in Fig. 12.11B. There is a rotating mechanical shaft seal, "mech seal," located in the "head" (see Fig. 12.12). The top labyrinth chamber is replaced with chamber enclosed with an elastomer bag. Motor oil is inside the bag and well fluid accumulates outside the bag through the breather tube. This provides the pressure equalization across the elastomer bag.

When the bag seal is lowered into the well, the bag is completely full of motor oil. As the oil heats up and expands, the pressure will build in the bag. The check valve shown in Fig. 12.11B is also a relief valve. The motor oil is on the topside of the valve and well fluid is on the bottom side of the valve. When the pressure tries to increase inside the bag, the check/relief valve will open and relieve the pressure. Then it will check to keep well fluid from entering the inside of the bag.

After the ESP is landed and energized, the oil will begin to heat and expand. Again, as the pressure tries to rise because of oil expansion, the relief valve will relieve the pressure until the pressure across the elastomer bag is equalized. Once the temperature stabilizes, the relief/check valve will check and the motor oil in the bag is again isolated from the well fluid.

When the unit is shut down, the motor oil will begin to cool and contract. As the oil contracts, the bag will collapse. The bag volume has been designed to account for the maximum contraction for the given motor and seal volumes. When again energized, the bag will expand as the motor oil heats up and expands.

Table 12.1 lists the ratings for the different O-ring and bag elastomer materials.

Should the top seal or top bag fail, there will be a secondary shaft seal and secondary bag or labyrinth chamber as a backup just as with the labyrinth seal.

Tandem seals are often run for additional protection. Instead of two or three shaft seals and two or three chambers available with a single seal, tandem seals will have four to six shaft seals and four to six chambers between the well fluid and the motor. Seals with multiple chambers in one housing are also available. Fig. 12.13 is an example of a seal with two bag chambers on the top connected in parallel instead of series. This allows for additional expansion and contraction that is required with larger motors with more oil capacity. Below the two parallel bags are the two labyrinth chambers in series and a high-load (HL) down-thrust bearing.

12.3.3 Seal thrust bearing

The thrust bearing location is depicted in Fig. 12.11A for the labyrinth seal type and in Fig. 12.11B for the positive barrier or bag seal type. The bearing consists of a thrust runner, a down-thrust bearing, and an up-thrust bearing, as shown in Fig. 12.14. The thrust runner is keyed to the seal's shaft, and its position on the shaft is fixed with snap rings. The thrust runner can move up and down between the down-thrust bearing and the up-thrust bearing by a fixed amount. For a 400 series seal the travel between the bearings is approximately from 1.55 to 1.65 in.

A down-thrust bearing, thrust runner, and up-thrust bearing are shown in Fig. 12.15. The down-thrust bearing in 400 series seal can carry 6000 lbs of thrust at 3500 RPM at oil temperatures up to 200°C (392°F). The up-thrust bearing is designed for much smaller loads since a special set of circumstances must occur to place the seal shaft into up thrust. Thrust will be discussed in detail in Section 12.5.

12.3.4 Seal horsepower requirement

The seal horsepower requirement increases with bearing loading. Fig. 12.16 plots the HP requirement versus total dynamic head or lift, in feet, for a 400 series (4 in. diameter) seal. Fig. 12.17 plots the HP requirement versus total dynamic head or lift, in feet, for a 500 series (5.13 in. diameter) seal.

Table 12.1 O-ring and bag elastomer ratings

Compound	Max operating temp (°F)	Max bht (°F)	Oil/H_2O	H_2S	CO_2	[a]Amines	Polar chemicals
HSN	275	175	Excellent	Excellent	Good	Fair	Good
Viton	400	300	Good	Not recommended	Not recommended	Not recommended	Fair
Aflas	450	350	Excellent	Excellent	Good	Excellent	Fair

Ratings: Excellent—long-term resistance and retention of high physical properties with low swell. Good—long-term resistance and retention of physical properties with moderate swell. Fair—long-term resistance and retention of physical properties dependent upon time and temperature, good short-term resistance and retention of physical properties.
[a]Amines—organic derivatives of ammonia (NH_3) containing carbon.
Source: Courtesy Valiant ALS.

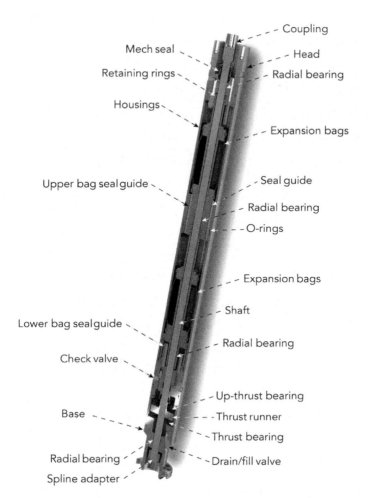

Figure 12.13 Seal from top to bottom, two bag chambers connected in parallel, in series with two labyrinth chambers connected in series, (BPBSLSL). Three shaft seals, with the bearing below the bottom labyrinth chamber.

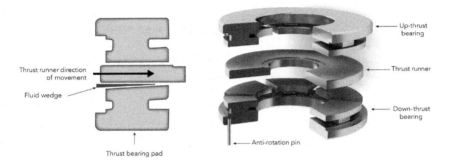

Figure 12.14 Typical seal thrust bearing assembly.
Source: Courtesy Valiant ALS.

Figure 12.15 Left to right, seal down-thrust bearing, seal thrust runner, and seal up-thrust bearing.
Source: Courtesy Valiant ALS.

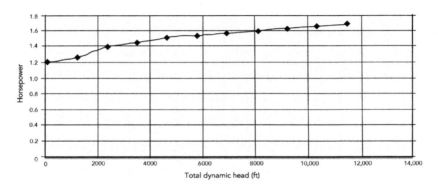

Figure 12.16 4 in. diameter (400 series) seal HP requirements versus total dynamic head.
Source: Courtesy Valiant ALS.

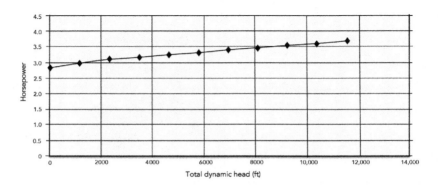

Figure 12.17 4.13 in. diameter (513 series) seal HP requirements versus total dynamic head.
Source: Courtesy Valiant ALS.

12.4 Electric submersible pump intakes

The ESP intake in Fig. 12.1A and B functions as a suction manifold feeding well fluid to the pump. Depending on the fluid gradient, it can be a standard intake with inlet holes (Fig. 12.18A and B) or a gas separator (Fig. 12.19A and B). The use of a gas separator may help mitigate potential problems inherent in wells where excessive free gas at the intake pressure may enter the pump and cause the pump to stop pumping, "gas lock,"[12] and create heating problems. The effectiveness varies with each design, well conditions, and differences between the gas bubble rise velocity and the liquid velocity at the intake of the pump.

12.4.1 Standard intake

The standard intake has several large ports allowing fluid to flow into the lower section of the pump and enter the bottom stage in the pump. The holes are approximately 1 in. in diameter. The intake in Fig. 12.18A and B is equipped with a screen to keep debris out of the pump. The screen is optional, and it is often not used. The bottom of the intake is bolted to the top of the seal and the top of the intake is bolted to the bottom of the bottom pump. The intake shaft is connected with a coupling to the seal shaft and the bottom pump shaft.

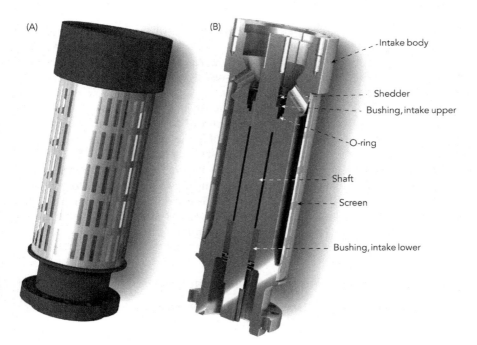

Figure 12.18 (A) ESP standard bolt on intake with screen for debris. (B) Cross section of a standard bolt on intake.
Source: Courtesy Valiant ALS.

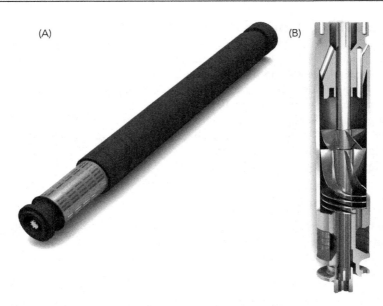

Figure 12.19 (A) ESP built in standard intake with screen. (B) ESP Vortex separator intake. Source: Courtesy Valiant ALS.

There is a Ni-resist bushing at the top and bottom of the intake to support the shaft. For abrasive applications the bearings should be abrasion resistant (AR) and will be tungsten carbide (TC) or silicon carbide. All ESP manufacturers recommend AR bearings in the intake in place of the Ni-resist bearings for all applications. The AR radially stabilized intake will have reduced wear with time. Wear causes the intake shaft to vibrate, which will damage the rotating seal located just below the intake. For long run lives, hardened bearings should be considered in all cases. The intake uses the same hardened bearing technology as the AR pumps. A graphical presentation of a radially stabilized intake and seal can be seen in Fig. 12.20.

The standard intake would be selected when the well has a very low free gas to liquid ratio (FGLR) at intake pressure and temperature or when motor or pump shrouds are used below the producing interval. The term "very low free gas to liquid ratio (FGLR)" has historically been defined as no more than 10%−15% free gas by volume, or a GVF of 0.1−0.15 as a rule of thumb.

12.4.2 Determining the gas volume fraction

Reservoir pressure−volume−temperature (PVT) correlations can be used to determine the GVF at pump intake pressure and temperature.[13,14]

Calculate the in situ water formation volume factor, bbl/stb = (B_w) (often assumed to be 1)
Calculate the in situ solution gas to oil ratio (GOR), scf/stbo = (R_s)
Calculate the in situ oil formation volume factor, bbl/stb = (B_o)

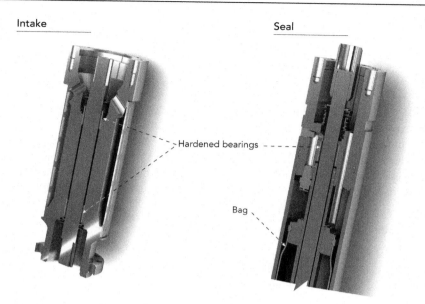

Intake   Seal

Hardened bearings

Bag

Figure 12.20 ESP bolt on intake and the seal head.
Source: Courtesy Valiant ALS.

Calculate the in situ gas formation volume factor, bbl/mscf = (B_g)

$$GVF = \frac{Mscf/d \times B_g}{(bbl_o/d \times B_o) + (bbl_w/d \times B_w) + (Mscf/d \times B_g)}$$ (12.4)

where Mscf/d is the 1000 standard cubic feet per day; bbl_o/d is barrels of oil per day at standard conditions; bbl_w/d is barrels of water per day at standard conditions.

After the GVF has been determined, the natural separation efficiency can be calculated.

12.4.3 Estimating natural separation efficiency

The next step is to calculate an estimate of the natural separation efficiency so that we can have an estimate of the free gas that will be entering the pump. Alhanati's[15] gas separation efficiency correlation is relatively easy to implement and can be run on a spread sheet.

$$E = \frac{V_t}{V_t + V_{sl}}$$ (12.5)

where E is the efficiency of natural separation, fraction; V_{sl} is the superficial velocity of the liquid phase (ft./s); V_t is the terminal bubble rise velocity (ft./s).

And

$$V_t = 0.79 \times \sqrt[4]{\frac{\sigma_l(\rho_l - \rho_g)}{\rho_l^2}} \qquad (12.6)$$

where σ_l is the interfacial tension (dyne/cm); ρ_l is the liquid density (lb/ft.^{3}); ρ_g is the gas density (lb/ft.^{3}).

Gas moving through a water and/or oil mixture will usually have a terminal velocity (V_t) between 0.5 and 0.6 ft./s. Superficial liquid velocity (V_{sl}) can be found by calculating the area, given the ESP intake O.D. and the casing I.D. and then determining the velocity using the in situ oil and water production rate at the intake as follows:

$$\text{Annular area (ft.}^2) = \frac{\frac{\pi}{4} \times ((\text{casing I.D. (in.)})^2 - (\text{intake O.D. (in.)})^2)}{144 \text{ in.}^2/\text{ft.}^2} \qquad (12.7)$$

$$Q_{o+w}(\text{ft}^3/s) = \frac{Q_{o+w}(\text{bbl/day}) \times 5.61458 \text{ ft.}^3/\text{bbl}}{24 \text{ h/day} \times 3600 \text{ s/}h} \qquad (12.8)$$

where Q_{o+w} (bbl/day) is the oil and water in situ flow rate at the pump intake. Then,

$$V_{sl} \text{ (ft./s)} = \frac{Q_{o+w} \text{ (ft.}^3/s)}{\text{annular area (ft.}^2)} \qquad (12.9)$$

For Q_{o+w} (bbl/day) = 800 bbl/day, intake O.D. = 4 in., casing I.D. = 4.892 in.

$$(12.10)$$

$$\text{Annular area (ft.}^2) = \frac{\frac{\pi}{4} \times (4.892 \text{ in.}^2 - 4.00 \text{ in.}^2)}{144 \text{ in.}^2/\text{ft.}^2} = 0.04326 \text{ ft.}^2 \qquad (12.11)$$

$$Q_{o+w} \text{ (ft.}^3/s) = \frac{800 \text{ bbl/day} \times 5.61458 \text{ ft.}^3/\text{bbl}}{24 \text{ h/day} \times 3600 \text{ s/h}} = 0.052 \text{ ft.}^3/s \qquad (12.12)$$

$$V_{sl} \text{ (ft./s)} = \frac{0.052 \text{ ft.}^3/s}{0.04362 \text{ ft.}^2} = 1.2 \text{ ft./s} \qquad (12.13)$$

$$E = \frac{0.5 \text{ ft./s}}{0.5 \text{ ft./s} + 1.2 \text{ ft./s}} = 0.294\% \text{ or } 29.4\% \qquad (12.14)$$

Notice that the natural gas separation efficiency is not a function of the GVF at intake conditions. However, natural separation is affected by the flow regime at the intake. Fig. 12.21A depicts modified flow regimes for gas and liquid flow in a vertical pipe. Alhanati's gas separation efficiency calculation works well in bubble flow

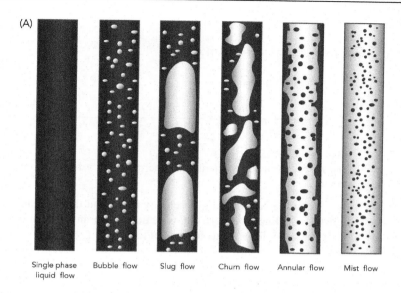

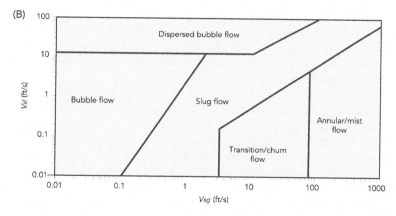

Figure 12.21 (A) Flow patterns for 2 phase vertical flow. (B) Superficial gas velocity (Vsg) versus superficial liquid velocity (Vsl) flow map for vertical flow.
Source: Ref. SPE 72998, by AS Kaya, C Sarica, and JP Brill, Mechanistic Modeling of Two Phase Flow in Deviated Wells and Courtesy Valiant ALS.

but a more complex model, if available, will be required for the more complex flow regimes.

Using the example calculation and Fig. 12.21B, an estimate of the percentage of free gas that can be handled before slug flow develops can be calculated.

1. The superficial liquid velocity (V_{sl}) calculated in the example is 1.2 ft./s.
2. Use Fig. 12.21B to determine the approximate superficial gas velocity (V_{sg}) before slug flow or 0.5 ft./s. Note: V_{sg} is not the same as V_t.
3. Use the annular area to calculate the gas flow in ft.^{3}/s.

$$Q_g(\text{ft.}^3/\text{s}) = V_{sg}(\text{ft./s}) \times \text{annular area (ft.}^2) \qquad (12.15)$$

$$Q_g(\text{ft.}^3/\text{s}) = 0.4 \text{ ft./s} \times 0.04326 \text{ ft.}^2 = 0.017 \text{ ft.}^3/\text{s} \qquad (12.16)$$

4. Calculate the GVF using ft.^{3}/s units.

$$\text{GVF} = \frac{Q_g(\text{ft.}^3/\text{s})}{Q_g(\text{ft.}^3/\text{s}) + Q_{o+w}(\text{ft.}^3/\text{s})} \qquad c = \frac{0.017 \text{ ft.}^3/\text{s}}{0.017 \text{ ft.}^3/\text{s} + 0.052 \text{ ft.}^3/\text{s}} = 0.246 \qquad (12.17)$$

5. Or 24.6% free gas by volume.
6. The GOR, R_s and B_g can be used to determine if the measured values will exceed the 24.6% free gas limit.

Fig. 12.22A shows that the superficial gas velocity (V_{sg}), and therefore the GVF, can be much higher in inclined or horizontal sections than in vertical sections because only the mist or spray flow needs to be avoided.

If the ESP is set in the horizontal or inclined section of the well, then a bottom feeder intake should be selected. For most of the modified flow regimes in inclined or horizontal sections, shown in Fig. 12.22B, incorporating an ESP intake that will adjust the intake opening to the pump, so that it takes the fluid from the bottom of the casing, will avoid much of the free gas production.

During system installation, the ESP may be rotating all the way down and the exact orientation of the intake once the system is landed cannot be predicted. A Baker Hughes Gravity Cup or bow spring intake will assist with avoiding gas, regardless of system orientation in the well.

Fig. 12.23A is a bow spring intake. With the bow spring intake, eight bow springs resting on the casing bottom activate mechanically and open intake ports to divert production from the casing bottom. Bow springs on the top keep these ports closed. Since ingested gas can reduce pump capacity and minimize production, bow spring intakes reduce the negative impacts of free gas. Fig. 12.23B is a gravity cup intake. With the gravity cup intake a series of internal cones fall with gravity, creating an opening near the intake bottom. The intake diverts production and avoids gas regardless of orientation or well deviation.

Fig. 12.24 is a rotating can or "sleeve" gas avoiding intake. This type of intake is designed with a rotating slotted sleeve over the intake designed to rotate the sleeve opening at the bottom of the casing by gravity. This allows the intake to be exposed to the liquid flow before the gas flow (see Fig. 12.25).

Now that the amount of free gas into the pump has been estimated, the next calculation should determine the effect the gas into the pump might have on the ability of the pump stage to produce the water curve head.

12.4.4 Estimating the probability of stage head degradation

The Dunbar[16] correlation uses the flow rate of the oil, water, and gas into the pump at the pump intake pressure to determine the minimum pump intake pressure (P_{ip_m}) before the pump stage performance will be degraded. This correlation was

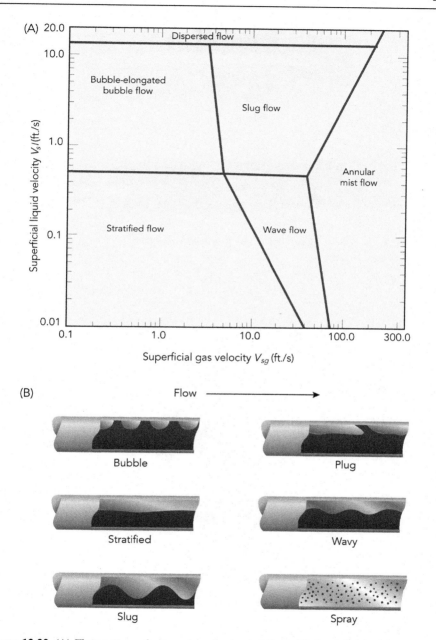

Figure 12.22 (A) Flow patterns for two phase horizontal flow. (B) Superficial gas velocity (V_{sg}) versus superficial liquid velocity (V_{sl}) flow map for horizontal flow.
Source: Courtesy Valiant ALS.

(A)

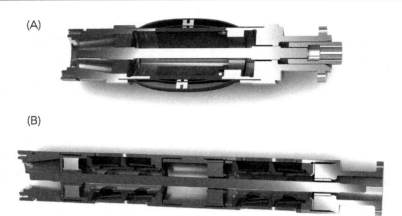

(B)

Figure 12.23 (A) A bow spring intake. (B) A gravity cup intake
Source: Courtesy Valiant ALS.

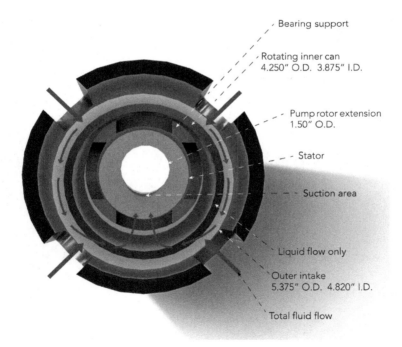

Figure 12.24 Gas avoider intake.
Source: Courtesy Valiant ALS.

developed using laboratory and field data. Dividing by the P_{ip_min}, a correlation constant, Φ (PHI), is calculated. For $\Phi \leq 1$ the stage head should not be effected by the free gas. As Φ approaches one, the probability of gas interference with the development of head increases. Any Φ greater than one should be avoided. The

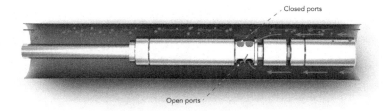

Figure 12.25 ESP bottom feeder intake for ESP's set in horizontal section of non-conventional wells.
Source: Courtesy Valiant ALS.

head produced will be less than the water head defined in the manufactures catalog. $\Phi \leq 1$ for the minimum intake pressure that can be obtained for a given FGLR into the pump, to apply the homogenous flow model with no head degradation.

The Dunbar equation is

$$P_{ip\_\min} = 935 \left(\frac{\text{bgpd}}{\text{bopd} + \text{bwpd}} \right)^{1/1.724} \tag{12.18}$$

or

$$1 = \frac{935}{P_{ip\_m}} \left(\frac{\text{bgpd}}{\text{bopd} + \text{bwpd}} \right)^{1/1.724} = \Phi \tag{12.19}$$

If the estimated pump intake pressure is unacceptably high, then additional methods to avoid the gas or remove additional gas should be considered.

12.4.5 Avoiding the gas—intake below the production interval—motor shrouded intake

Motor shrouds, depicted in Fig. 12.26, can be installed where casing sizes allow. The motor shroud is designed to provide cooling to the motor when fluid velocities are below minimum. A motor shroud is installed to increase the fluid velocity past the motor in wells with large diameter casing or to allow the lowering of the motor below perforations. In the latter case, the fluid must come out of the perforations, travel down to the bottom of the shroud, then travel up between the shroud and the motor and into the intake of the pump. This allows proper cooling of the motor.

One case where this type application may be used is in high-GLR wells or when maximum draw down is desired. Free gas bubbles will rise at about ½ ft./s. If the liquid velocity moving down toward the shroud intake is less than the velocity of the gas bubbles moving up the annulus, then separation occurs because the well fluid is flowing down to the shroud opening then up past the motor.

The shroud is attached above the pump intake and the motor flat runs through the shroud. To be effective the shroud must be sealed at these two locations. This is

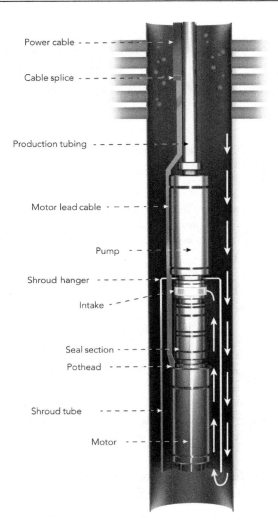

Power cable
Cable splice
Production tubing
Motor lead cable
Pump
Shroud hanger
Intake
Seal section
Pothead
Shroud tube
Motor

Figure 12.26 Shrouded motor.
Source: Courtesy Valiant ALS.

usually accomplished with the shroud hanger. The shroud can be manufactured
from thin wall casing, stainless steel, or fiberglass.

12.4.6   Avoiding the gas—intake below the production interval—recirculating system

The ESP recirculation system,[17] Fig. 12.27, is a solution for the motor cooling problem associated with placing an ESP below the perforations. The system design
includes a recirculation pump and special tubing to redirect fluid flow past the

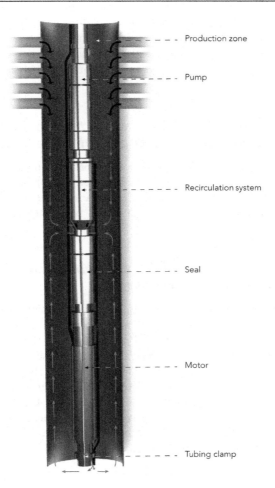

Figure 12.27 Fluid recirculation system for motor cooling.
Source: Courtesy Valiant ALS.

motor, ensuring the necessary motor cooling. The recirculation tubing may be designed to carry pumped fluid to the bottom of the motor and then the pumped fluid returns to the pump intake, or the tubes may be connected to the intake of the recirculation pump and the fluid is returned to the intake through the recirculation tubes. The system eliminates the need for a motor shroud, allowing operators to often use larger diameter, higher efficiency, and less costly motors.

12.4.7 Avoiding the gas—intake below the production interval— permanent magnet motor without cooling

The PMM is much more efficient than the IM, as discussed in Section 12.2. The rotor has no losses, therefore the heat generated is primarily due to I^2R losses in the stator. At the time of writing this book, the author was familiar with only one test installation using a PMM below the production interval without a shroud.[18] While the test case was a success, the authors of the paper also noted the need for additional modeling information for the estimation of PMM heat rise and more test cases.

12.4.8 Avoiding the gas—intake above the production interval— motor shrouded intake or pod with a tail pipe or dip tube

The shroud may be fitted with a crossover at its base to convert to tubing thread. The tubing can then be lowered into the deviated or horizontal section of the well, or into a small diameter liner. If the tubing tail pipe becomes too long, the pressure loss may require additional pressure which would negate the reason for installing the tail pipe.

12.4.9 Avoiding the gas—intake above/below the production interval—encapsulated system

One of the limitations of a conventional shroud hanger is that it does not provide a positive seal. An alternative is to use an encapsulated system, as shown in Fig. 12.28. With this system the entire ESP assembly is enclosed in a shroud. These units typically have a differential pressure rating of 5000 psi.

In an encapsulated system the shroud hanger is placed above the pump, so that there is room to use a power feed-through to feed the power to the ESP motor. The pump is suspended from a tubing sub below the base of the hanger while the shroud tube is attached to the hanger with a positive seal. A threaded crossover is attached to the base of the shroud tube to convert it into a tubing thread.

The tubing tail pipe may end with a packer seal assembly.

12.4.10 Avoiding the gas—intake above the production interval —pump shrouded intake—upside-down shroud

The pump shroud or upside-down shroud, shown in Fig. 12.29, also can provide free gas separation from well fluids. The use of a pump shroud requires that the motor is set above the producing interval, so that fluids moving past the motor can dissipate heat from the motor and provide adequate cooling.

The effectiveness of the shroud type design requires that the velocity rise of the gas bubbles be greater than that of the liquids falling back into the shroud. The length of the shroud is critical for a successful gas separation. Separation will occur just above the shroud intake. Therefore the intake should be above the pump head, so that the maximum annular area is available to reduce the liquid superficial velocity to a minimum.

It is thought that the pump shroud may also provide a liquid reservoir to continue to feed the pump should gas slugging occur.

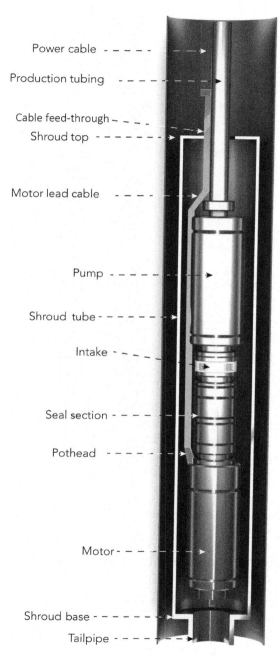

Power cable

Production tubing

Cable feed-through
Shroud top

Motor lead cable

Pump

Shroud tube

Intake

Seal section

Pothead

Motor

Shroud base

Tailpipe

Figure 12.28 ESP enclosed in a pod.
Source: Courtesy Valiant ALS.

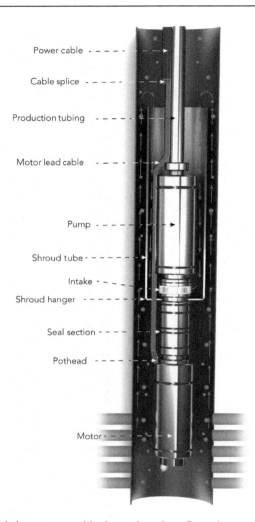

Power cable

Cable splice

Production tubing

Motor lead cable

Pump

Shroud tube

Intake

Shroud hanger

Seal section

Pothead

Motor

Figure 12.29 Shrouded pump or upside-down shroud configuration.
Source: Courtesy Valiant ALS.

12.4.11 Removing the gas—gas separators—rotary gas separator

The rotary gas separator, Fig. 12.30, will separate free gas with an efficiency of 75%−98%[19,20] under most conditions and is generally preferred in viscous applications.[21] The rotary gas separator may be considered where the free gas into the intake causes the Dunbar PHI to exceed 1. If abrasives are present, then the vortex design separator should be with abrasive resistant (AR) bearings.

Rotary gas separators allow the entry of fluids and gas at the base of the separator into a rotating centrifuge with inducers and straight vanes. As the liquids and gas rotate in the centrifuge, the denser liquids, oil and water, move to the outside of

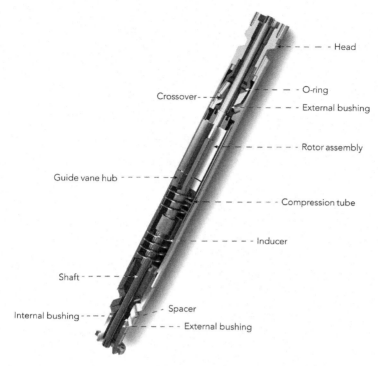

Figure 12.30 Rotary gas separator.
Source: Courtesy Valiant ALS.

the centrifuge and the lighter components, the free gas, move to the center. A crossover diffuser allows the gas to be discharged into the annulus and the liquids can enter the pump.

Since there is a large mass rotating, shaft stabilization is very critical. Seal failure and then motor failure due to excessive gas separator vibration can occur if the shaft is not stabilized with AR (hardened) bearings. Earlier models used only two bushings to stabilize the long shaft. Modern designs incorporate three bearings. Hardened bearings are an option that should be incorporated into the gas separator.

Even though the rotary separator is very efficient, there will be cases where the pump will gas lock. If the pump intake pressure gets so low that slug flow develops, then there will be moments of time when nothing but gas is present at the intake. During this interval, only gas will enter the pump.

12.4.12   Removing the gas—gas separators—vortex gas separator

The vortex gas separator, Fig. 12.31, also uses centrifugal force to separate the fluids according to the fluid gravity. The denser, high-specific gravity fluids, are

Figure 12.31 Vortex gas separators.
Source: Courtesy Valiant ALS.

forced against the separator wall, and lighter, low-specific gravity fluids, are left around the shaft. The vortex separator creates a vortex in the separation chamber with a spinning paddle set above the inducer or high-angle vane auger. One advantage is that separation occurs without the additional weight of the spinning chamber. The vortex separator will also have a higher maximum throughput rate than the rotary. Fig. 12.32 gives a general idea of liquid throughput versus percentage of free gas that can be separated. Notice that the separation efficiency is 0% at the maximum liquid rate. There is no room for free gas to enter the separator. If the total flow of oil, water, and gas is greater than the maximum liquid rate, then this will begin to cause a choking effect on flow through the ESP and the P_{ip} will increase, reducing the free gas, until a stable operation point is found.

The vortex separator will handle abrasives better than the rotary chamber separator, and the rotary chamber separator will degas viscous oil better than the vortex separator.

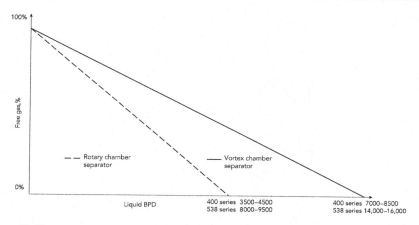

Figure 12.32 Approximate operating characteristics of rotary and vortex gas separators. Source: Courtesy Valiant ALS.

12.5 Electric submersible pumps

The ESP is located above the intake, as shown in Fig. 12.1A and B. The intake is bolted to the bottom of the pump, and the pump discharge head is threaded onto the bottom of tubing string.

ESP is a multistage centrifugal pump and is built in a center tandem configuration. The pump may be a single pump housing, as shown in Fig. 12.1A, or it may be made of two or more pump housings bolted together, as shown in Fig. 12.1B. Fluid enters the pump through this pump intake. The intake is a separate component that bolts onto the bottom of the bottom pump housing.

The submersible pump is a multistage centrifugal pump that produces head and fluid flow by converting the motor shaft horsepower to fluid hydraulic horsepower. A typical pump cross-section is shown in Fig. 12.33. The major components comprising a submersible pump along with their respective metallurgies are shown in Table 12.2.

Using corrosion resistant materials, cast Ni-resist impellers and Ni-resist diffusers with Monel K-500, Nitronic 50, and Inconel shafting, pump wear and corrosion can be minimized. However, unless specified otherwise, the housings, heads and bases of the pumps, seals, and motors will be carbon steel. In corrosive applications the equipment should be coated or specified with special ferritic steel housings and 416 stainless steel heads and bases.

12.5.1 The pump stage

A pump stage consists of an impeller, a diffuser and thrust washers. The impeller develops a fluid velocity pressure through either a centrifugal force (radial flow) or an axial force (mix flow) design. The diffuser converts this velocity to head of pressure to initiate the lifting of fluids from the well. Fig. 12.34A cross-sections a radial flow stage design, and Fig. 12.34B cross-sections a mixed flow stage design.

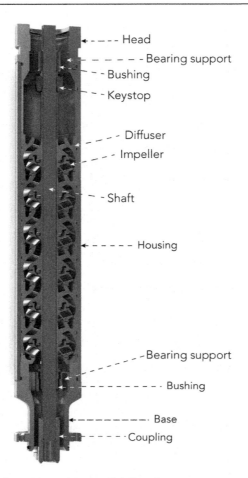

Figure 12.33 Detail of a cross-sectioned radial flow floater pump.
Source: Courtesy Valiant ALS.

The impeller is keyed to the shaft and rotates at the motor operating RPM. Centrifugal force causes the fluid to move from the eye of the impeller outward. These forces impart kinetic or velocity energy to the fluid. The vanes in the impeller create channels through which the fluid is directed. Given a fixed diameter and RPM, the volume between the top and bottom shroud determines the volume per unit time (or barrels per day (bpd)) that can be produced at peak efficiency.

The diffuser is stationary, and its function is to allow the fluids to flow efficiently from one impeller to another and to convert a portion of the velocity (kinetic energy) into pressure (potential energy).

The stages (an impeller—diffuser combination) are placed onto a keyed shaft and then loaded into a steel housing. When the threaded head and base are screwed into the housing, they compress against the outside edge of the diffusers. This compression holds the diffusers in stationary. If, for any reason, this compression was lost,

Table 12.2 Pump component metallurgy and associated operating temperature range.

Component	Metallurgy	Temperature range
Head	1020 steel	N/A
	Ferritic steel	
	416 SS	
Housing	Buna-N	180°F (82°C)
O-Ring	Viton	400°F (204°C)
	AFLAS	450°F (232°C)
Diffuser	Ni-resist	N/A
Impeller	Ryton	200°F (93°C)
	Ni-resist	N/A
Pump shaft	Monel k-500	N/A
	Nitronic 50	
	Inconel 718	
	Inconel 718 HS	
Pump housing	1020 steel	N/A
	Ferritic steel	
Shaft bearing	Bronze	N/A
	Ni-resist	
	Alloy bronze	
	Tungsten carbide	
Base	1020 steel	N/A
	Ferritic steel	
	416 SS	

Source: Courtesy Valiant ALS.

the diffusers would be free to rotate. This rotation would cause the pump to lose almost all its ability to produce any head (or lift).

The stages are of a fully enclosed curved vane design, whose maximum efficiency is a function of impeller design and type. There are two basic types of stages used in oil well submersible pumps.

12.5.2 Pump radial flow stages

The radial flow design, shown in Fig. 12.34A, develops pressure (or head) through the radial centrifugal force. This is generally used where a low flow, less than 1200 bpd, is required. This design is accomplished by allowing the fluid to be accelerated at a 90 degree angle from the eye of the impeller. Nominal rates are in the range from 100 to 1200 bpd with pump efficiencies in the order of 60%. The radial stage is a flat stage and is the most efficient design for these lower flow rates. The general subjective rule, or rule of thumb, is that this stage design can handle a GVF of 0.10−0.15 before head performance degradation starts to occur. The more objective method would be to apply the Dunbar correlation using the stage input pressure, temperature, liquid rate, and free gas rate to determine the minimum pressure before degradation begins to occur.

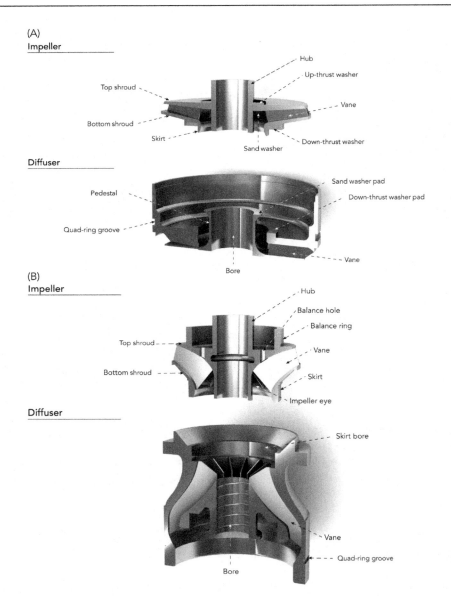

Figure 12.34 (A) Radial flow stage design. (B) Mixed flow stage design. (C) Helico-axial stage design — a type of gas handler stage.
Source: Courtesy Valiant ALS.

(C)
Helico-axial design

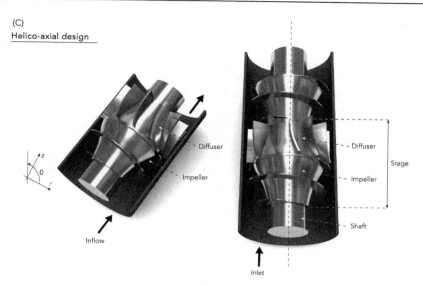

Figure 12.34 Continued

Dunbar correlation:

$$P_{ip\_min} = 935\left(\frac{GVF}{1-GVF}\right)^{1/1.724} \tag{12.20}$$

where GVF is the calculated gas void factor at the stage input; P_{ip_min} is the minimum pump intake pressure for the GVF.

Dividing both sides by P_{ip},

$$\frac{P_{ip\_min}}{P_{ip}} = \frac{935}{P_{ip}}\left(\frac{GVF}{1-GVF}\right)^{1/1.724} \tag{12.21}$$

Setting,

$$\Phi = \frac{P_{ip\_min}}{P_{ip}} \tag{12.22}$$

$$\Phi = \frac{935}{P_{ip}}\left(\frac{GVF}{1-GVF}\right)^{1/1.724} \tag{12.23}$$

Therefore when $P_{ip} < P_{ip\_min}$, then PHI < 1 indicating that stage head degradation should not occur and when $P_{ip} > P_{ip\_min}$, then PHI > 1 indicating that stage head degradation should occur.

12.5.3 Pump mixed flow stages

The mixed flow stage design, shown in Fig. 12.34B, develops pressure (or head) through radial centrifugal force and axial thrust force. It is generally used where a high flow, greater than 1200 bpd, installation is required. The angle at which the fluid is accelerated from the eye of the impeller is less than 90 degrees. Nominal rates are generally in the range from 1200 to 70,000 bpd, but with pump efficiencies greater than 60%. This is the most popular of the two designs due to fact they handle gas and solids more effectively.

This stage design can handle a GVF of 0.15−0.25 before head performance degradation starts to occur. The Dunbar correlation calculation can be used to estimate the minimum pressure before degradation begins to occur.

12.5.4 Pump gas handler stage

The advanced gas handler (AGH), sold by Schlumberger, is a modified mixed flow stage with a unique design to alter the pressure distribution of the impeller, creating a homogenized mixture with reduced gas bubble size. This conditioned fluid behaves as a single-phase fluid before entering the lifting stages of the pump.

The multiple vane pump with extended range (MVPER), sold by Baker Hughes, is also a modified mixed flow stage to prevent free gas from accumulating in the stage and causing a gas lock condition and a system shutdown. It is also designed to alter the pressure distribution of the impeller, creating a homogenized mixture with reduced gas bubble size.

Both AGH and MVPER stages are supplied with AR ceramic bearings for extended flow ranges. The stages can handle a higher GVF than the conventional mixed flow design, but will not handle the GVF that the helico-axial stage can handle, as described later.

12.5.5 Pump gas handler helico-axial stage

The "helico-axial" stage, shown in Fig. 12.34C, was developed for pumping mixtures of oil, natural gas, water, and sand with GVFs from 0% to 97% or even higher.[22,23] The impellers have a very high hub to impeller tip ratio, which increases from the inlet to the outlet. The blading resembles that of an inducer with very low blade height. Downstream of the impeller, a diffuser decelerates the flow and deflects it into an essentially axial direction. The diffuser hub ratio decreases from inlet to outlet to match the impeller inlet of the subsequent stage.

Camilleri et al.[24] reported successful operation, at a calculated GVF of 0.75, at the inlet stage of a 15 stage, 500 series, helico-axial pump (gas handler), after gas separation.

12.5.6 Pump performance curve, mixed and radial flow stages

The pump performance curve in Fig. 12.35A plots the water performance of a 4 in. diameter, mixed flow pump (or stage), rotating at 3500 RPM and pumping specific gravity 1 water. The left vertical axis is scaled in feet (or meters) for head (or lift). The bottom horizontal axis is scaled in bpd (or meters cubed per day). The curve−labeled head defines the lift (or head) the impeller can produce as the flow rate changes. For example, at 1750 bpd the single stage VC1750 shown in

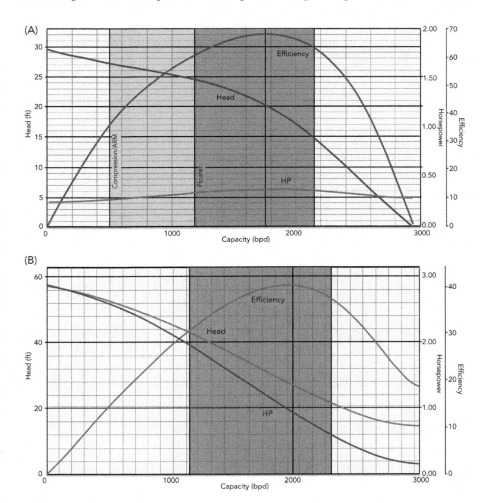

Figure 12.35 (A) VC1750 single stage performance curve. (B) VEGH4500 gas handler single stage performance curve. (C) VC1750 single stage variable frequency performance curve. (D) Graph of a 150 stage, VC700 pump VSD curve with the well productivity curve plotted on the graph.
Source: Courtesy Valiant ALS.

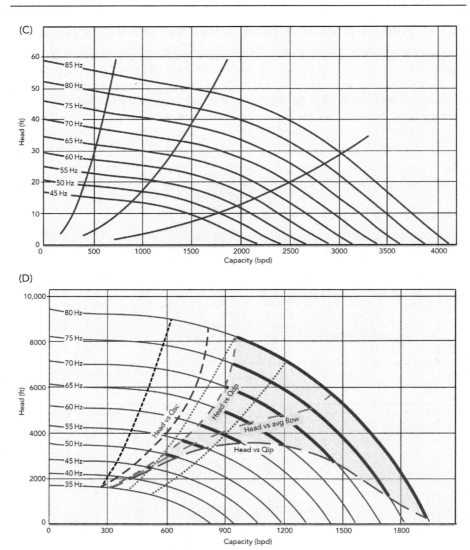

Figure 12.35 Continued

Fig. 12.35A will produce 20.14 ft. of head. Centrifugal pump performance is defined by the head produced, not pressure. The 20.14 ft. of lift in the aforementioned example represents 8.72 psi for specific gravity 1 fluid. However, the impeller will produce the same 20.14 ft. of head with a specific gravity 0.85 fluid which calculates a pressure of 7.41 psi. The centrifugal force acting on the fluid is the same regardless of the fluid's density.

Fluid density does affect the power required to lift the fluid. The curve in Fig. 12.35, labeled HP, plots the power requirements for this impeller at various flow rates. The first vertical axis on the right is scaled in horsepower motor load. This horsepower is based on pumping water with a specific gravity of 1. The horsepower will vary proportionally with the gravity of the pumped fluid.

As an example, at 1750 bpd the one stage pump[25] in Fig. 12.35A will require 0.381 HP if the fluid has a specific gravity of 1. For fluid having a specific gravity of 0.85, the pump will only require 0.324 HP (0.381×0.85).

The rightmost vertical axis of Fig. 12.35A is scaled in percentage of efficiency. Sometimes the curves will not agree with the calculation due to errors in reading and reproducing the curves. Because of this the API has established that mathematical coefficients should be used to determine an impeller's head, horsepower, and efficiency.

The curves are also RPM dependent, and the RPM for the curve will be listed. Changing the RPM of the impeller will affect the flow, head, and horsepower according to the pump and fan affinity laws.

12.5.7 Pump performance curve, helico-axial stage

The pump performance curve in Fig. 12.35B plots the water performance of a 5.38 in. diameter, helico-axial stage (gas handler pump), rotating at 3500 RPM and pumping water with a specific gravity of 1.

12.5.8 Pump performance curve changes with changes in impeller RPM

Pump performance changes with RPM and diameter. Diameter is a variable that cannot be changed when deliquefying a gas well. However, the RPM can be changed using a VSD instead of a constant frequency switchboard. The VSD applies variable frequency and voltage to the motor, which change the motor RPM. Most VSDs are set up for manual speed adjustment. Although the VSD can be configured with a feedback control loop to maintain a constant pressure or a constant flow. The user would adjust the desired pressure or flow instead of the frequency.

The pump performance can be calculated using the pump and fan affinity laws, stated later.

With the impeller diameter held constant:

1. Flow (Q) is proportional to shaft RPM (N)

$$\frac{Q_1}{Q_2} \cong \frac{N_1}{N_2} \tag{12.24}$$

2. Head (H) is proportional to the square of the shaft speed (N)

$$\frac{H_1}{H_2} \cong \left(\frac{N_1}{N_2}\right)^2 \tag{12.25}$$

3. Power (P) is proportional to the cube of the shaft speed (N)

$$\frac{P_1}{P_2} \cong \left(\frac{N_1}{N_2}\right)^3 \tag{12.26}$$

For ESP applications, usually frequency is used in place of RPM. It is assumed that RPM and frequency are proportional. This is strictly true for the PMM, but a small error is introduced with the IM since full-load slip is constant at all frequencies (Hz). The error is small and within the error range of the other variables that affect motor loading, therefore exact calculations are not necessary to determine the new flow, head, and power as the Hz are varied to an IM.

The pump curves in Fig. 12.35A were developed using an IM operating at 60 Hz or 3500 RPM. Now the aforementioned proportionalities can be set to equalities.

1. Equation to determine a new flow at a new Hz.

$$Q_{\text{Hz}} = Q_{60\,\text{Hz}}\frac{\text{Hz}}{60\,\text{Hz}} \tag{12.27}$$

2. Equation to determine a new head at a new Hz.

$$H_{\text{Hz}} = H_{60\,\text{Hz}}\left(\frac{\text{Hz}}{60\,\text{Hz}}\right)^2 \tag{12.28}$$

3. Equation to determine a new head at a new Hz.

$$P_{\text{Hz}} = P_{60\,\text{Hz}}\left(\frac{\text{Hz}}{60\,\text{Hz}}\right)^3 \tag{12.29}$$

where $Q_{60\text{Hz}}$ is a flow rate from a point on the 60 Hz curve; $H_{60\text{Hz}}$ is the head for the above flow point on the 60 Hz curve; $P_{60\text{Hz}}$ is the power for the above flow point on the 60 Hz curve; Hz is the new operating Hz for the above flow point; Q_{Hz} is the flow rate at the new Hz using the flow point on the 60 Hz curve; H_{Hz} is the head at the new Hz flow rate using the head at the flow point on the 60 Hz curve; P_{Hz} is the power at the new Hz flow rate using the power at the flow point on the 60 Hz curve.

Dividing the 60 Hz curve into 20 or 30 flow points, a computer can step through all the points for operating Hz from 45 to 85 Hz in 5 Hz increments to develop the variable speed performance curve in Fig. 12.35C.

These curves define the stage performance at different frequencies. The operating point on the curves will be defined by the well's productivity curve at the in situ flow rates through the stage. Fig. 12.35D is an example of a well's productivity curve plotted into the pump's VSD curves. For each frequency the intake, rate (Q_{ip}), the discharge rate (QDP), the average rate (avg. flow), and the standard condition liquid flow rate are plotted versus head. The darkened portion of the head capacity curves represents the stage by stage change in flow and head as the produced fluids move from the intake stage to the discharge stage. The plot assumes that the stages are producing the water curve head even though at Hz greater than 60 Hz, the intake fluids would have a very low average density and the stage would likely suffer head degradation making the actual production rates much lower. Above 70 Hz the pump would probably gas lock making the head capacity curves at the higher Hz theoretical.

12.5.9 Pump stage thrust

All centrifugal pumps develop axial thrust. The impeller thrust has four components and Fig. 12.36 shows a description of each component, net hydraulic thrust, and total thrust. Thrust is an important factor in establishing the recommended operating range of a centrifugal pump. Fig. 12.37 depicts a typical thrust curve versus

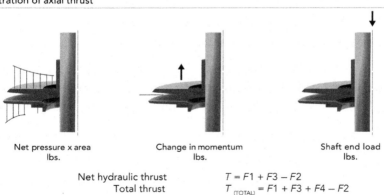

Total thrust of an impeller has four contributors:

1. The pressure acting upon the upper shroud minus the pressure ↓ **F1**
   acting upon the lower shroud. (acts downward)

2. The change in axial fluid momentum (acts upward) ↑ **F2**

3. The weight of the impeller in fluid (acts downward) ↓ **F3**

4. Pump discharge pressure acting on the end of the pump shaft (acts downward) ↓ **F4**

Illustration of axial thrust

| Net pressure x area lbs. | Change in momentum lbs. | Shaft end load lbs. |

Net hydraulic thrust $\quad T = F1 + F3 - F2$
Total thrust $\quad\quad\quad T_{(TOTAL)} = F1 + F3 + F4 - F2$

Axial thrust is one of the most important factors to be considered to establish "operating range" for multistage centrifugal pumps.

Figure 12.36 Thrust forces in the submersible pump.
Source: Courtesy Valiant ALS.

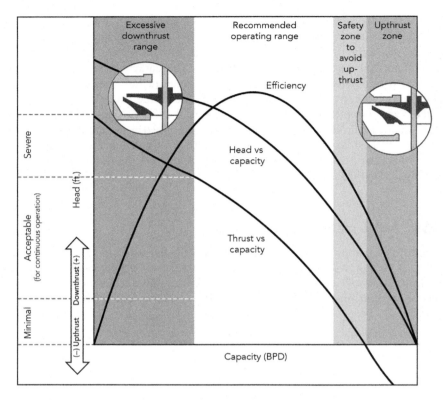

Figure 12.37 Pump stage thrust and head versus flow.
Source: Courtesy Valiant ALS.

capacity over the entire range of the ESP pump. It shows where up thrust and down thrust occurs in relation to the head capacity curve.

Every centrifugal stage is designed to produce at a certain flow rate range. There is a best efficiency point (BEP) for every stage. Every stage type has a recommended range. In Fig. 12.35 the recommended range is indicated by a shaded area. The recommended range for the VC1750 is from 1200 to 2150 bpd for floater (FLT) construction and from 500 to 2150 for compression (CMP) or abrasion resistant modular (ARM) construction. Across the operating range, the forces on either side of the impeller shroud are such that they minimize the amount of thrust that would be transmitted to the diffuser pads. When an impeller operates at a higher than recommended volume, it may be operating in up thrust, as shown in Figs. 12.36 and 12.37. In up thrust, the down-thrust force, because of the difference in intake and discharge pressure, is small, and the up-thrust force, due to fluid momentum change, is large resulting in a net upward thrust. Conversely, when the impeller is producing less than the design capacity, the down-thrust force, because of the difference in intake and discharge pressure, is large, and the up-thrust force,

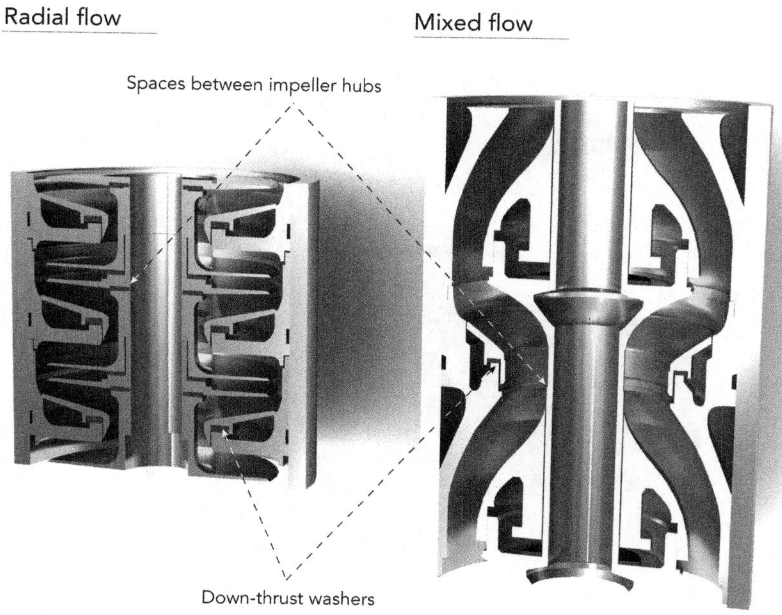

Radial flow

Mixed flow

Spaces between impeller hubs

Down-thrust washers

Figure 12.38 Cross sectioned radial and mixed flow stages in a floater pump.
Source: Courtesy Valiant ALS.

due to fluid momentum change, is small resulting in a net downward thrust. Down-thrust wear, or operating out of recommended range, is one of the main reasons for pump failure.

The longest pump life will occur if the stage is operated to the right of the BEP. That is, it is best to operate the impeller with lower pressures across the stage and at higher flow rates, while staying within the recommended operating range. However, this means that the pump will require more stages for the same amount of lift than the next larger flow pump and the pump may not be quite as efficient. Each application should be reviewed with this longevity versus efficiency problem in mind. Thrust generated by a multistage centrifugal pump is supported in one of three construction methods:

1. Floater construction (FLT)
2. Compression (CMP)
3. ARM construction

12.5.10   Floater pump construction

This is the original and simplest constructed model. Each impeller in a floater pump is free to move up and down on the shaft within the confines of the diffuser, as depicted in Fig. 12.38. The thrust washers, shown in Fig. 12.39, between the impeller and diffusers support the stage thrust up or down. The amount and

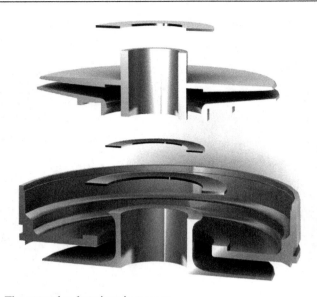

Figure 12.39 Thrust washer locations in a stage.
Source: Courtesy Valiant ALS.

direction of thrust varies with the head developed and the amount of fluid being pumped.

A floater pump should always be operated within the recommended operating range. The seal section thrust bearing supports the shaft thrust, F2, the discharge pressure of the pump acting on the cross-sectional area of the shaft, shown in Fig. 12.40. Even though the impeller is called a floater, the down-thrust washer is in contact with the diffuser pad from shutoff to 10%−15% past the maximum flow of the recommended operating range. The thrust washer also functions to seal and minimize recirculation of fluid within the stage. This construction's main advantage is that it does not transfer impeller thrust, F1, to the seal thrust bearing shown in Fig. 12.40. The impeller thrust, F1, is transferred through the thrust washers to the diffuser, which then transfers the thrust to the pump housing, then the housing thrust is transferred to the tubing.

12.5.11 Compression pump construction

The impellers in a compression pump are not free to move up and down on the shaft. The hubs of each impeller are in contact with the ones next to it. The down-thrust washer does not support any stage thrust. The seal section thrust bearing supports all the thrust, both the stage thrust and the shaft thrust, as shown in Fig. 12.41. The compression pump may operate outside of the recommended range without seriously damaging the stage parts. However, caution should be exercised, especially with high-volume pumps, not to operate more than 10% outside of the

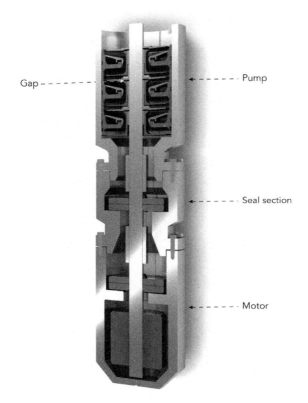

Figure 12.40 Floater pump, seal, and motor shaft settings.
Source: Courtesy Valiant ALS.

operating range on the right-hand side of the curve or operate at too low a rate on the left side of the curve which might affect motor cooling.

The compression pump is more difficult to assemble than a floater pump due to the closer tolerances required for the impellers and diffusers, shown in Fig. 12.42. Thrust bearings with higher load carrying capabilities made this type of construction popular even though the cost is high. This type of construction usually endures abrasives longer than a floater pump.

12.5.12   *Abrasion resistant modular construction*

Fig. 12.43A shows the TC-sotted bushing that is pressed into the diffuser bore. The TC beveled sleeve radial and down-thrust bearing rotates on and inside the slotted bushing. This is an AR radial and down-thrust bearing. The impeller down thrust is carried in this TC bearing as shown in Fig. 12.43B. Generally, there will be three or four additional stages above the stage with the TC bearing that will also transfer

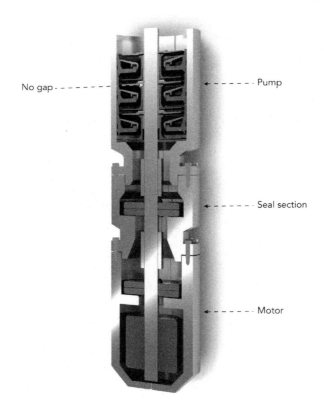

Figure 12.41 Compression pump, seal, and motor shaft settings.
Source: Courtesy Valiant ALS.

thrust to that bearing. The module TC bearing to stage ratio is generally 1:X where X may be from 1 to 8, or perhaps higher, depending on the application. The 1:1 ratio is the most robust and the most expensive and would only be considered where the application is severe and/or the cost to intervene and replace a failed pump is exceptionally high.

The weight of the pump shaft and the weight of the fluid on the pump shaft is carried by the seal bearing. Construction is very similar to the floater construction except AR bearings that are used for radial and down-thrust support instead of the Ni-resist for radial support and thrust washers for the impeller down thrust. Thrust washers may be included for improved head development.

12.5.13   Designing a pump for gas handling

When natural separation and mechanical separation do not reduce the GVF to an acceptable level, as defined by the Dunbar correlation, then the pump should utilize some form of a tapered pump design.

Radial flow   Mixed flow

No spaces between impeller hubs

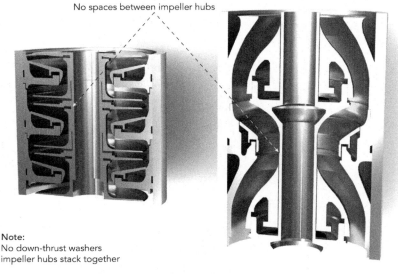

Note:
No down-thrust washers
impeller hubs stack together

*Some radial flow stages may use down-thrust washers
to improve head characteristics

Figure 12.42 Cross-sectioned radial and mixed flow stages in a compression pump.
Source: Courtesy Valiant ALS.

Tapered pump design

The tapered pump is used when the intake flow rate into the first stage of the pump, the pump intake, is much higher than the flow rate at the last stage of the pump, the pump discharge. As the pump intake pressure is lowered, the GVF at the pump intake will increase, lowering the natural and mechanical separation efficiency and increasing the GVF through the pump.

As the pressure increases from stage to stage, the free gas in the fluid is compressed and some of the free gas may go back into the oil as solution gas. As a result, the first stage of the pump may be in range but as the fluid moves from stage to stage, the flow rate keeps decreasing and the flow may move out of the recommended range for the selected stage.

The solution to the problem is to select the next smaller stage (by full housing) and place it above the larger stage (by housing). Housings are individual pumps with a specified number of stages that can be bolted together to form the total pump. Tapered pumps can be two or three different stage sizes. Four or more different stage types could be bolted together. However, this author has no knowledge of one ever being run or designed.

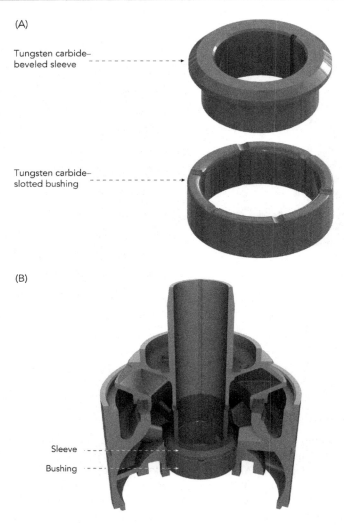

(A)

Tungsten carbide–
beveled sleeve - - - - - - - - - - - - →

Tungsten carbide–
slotted bushing - - - - - - - - - - - →

(B)

Sleeve - - - -
Bushing - - - -

Figure 12.43 (A) Tungsten carbide (TC)–slotted bushing and TC-beveled sleeve. (B) A
stage with the TC bushing and sleeve.
Source: Courtesy Valiant ALS.

Often a mixed flow stage will be used as the intake pump in a tandem configuration. This is sometimes called a booster pump to help compress the free gas that
enters the pump. This will assist the radial stage pump above the booster pump to
handle more gas.

Fig. 12.44 displays the curves of a staged tandem pump.

The lower, intake pump, is a mixed flow stage booster pump designed for an
intake rate of 1638 bpd, 543 bpd water, 517 bpd oil, and 587 bpd of gas at intake
conditions of 531 psi and 165°F. This is a GVF of 0.353. The Dunbar PHI is 1.24,

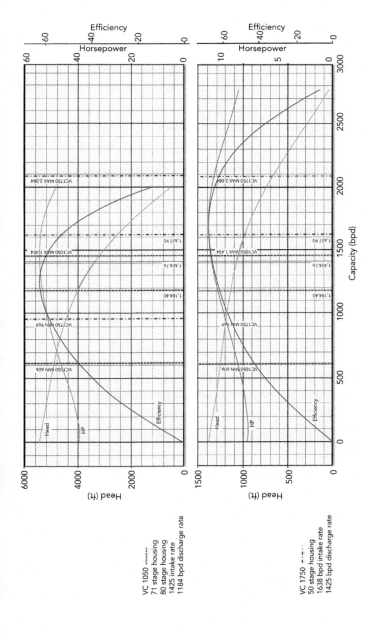

Figure 12.44 Flow diagram for a tapered pump.
Source: Courtesy Valiant ALS.

greater than the 1.0 PHI that is the estimated maximum to avoid head degradation in the VC1750 booster pump.

The original PHI correlation was developed by Turpin et al.[26] from laboratory test data.

$$\Phi = \frac{667}{P_{ip}} \times \frac{\text{bgpd}}{\text{bopd} + \text{bwpd}} = \frac{667}{P_{ip}} \times \frac{\text{GVF}}{1 - \text{GVF}} \tag{12.30}$$

For this example case the Turpin PHI is

$$\Phi = \frac{667}{531} \times \frac{0.353}{1 - 0.353} = 0.70 \tag{12.31}$$

Turpin states in the conclusions that for "PHI $\sim < 1$, correlations (are) applicable and good head production with gas present, (for) PHI $\sim > 1$, correlations (are) not applicable, but pump performance is generally so poor that operation in these regions is to be avoided."

The "correlations" refer to head correction versus GVF for the two stages that were being tested by Turpin.

Based on the above, we can postulate the following guidelines for pumps containing mixed flow and radial flow stages.

1. If the Dunbar PHI < 1, then there should be no head degradation.
2. If the Dunbar PHI > 1 and the Turpin PHI < 1, then the stage will suffer some head degradation.
3. If Turpin PHI > 1, then the stage may experience unstable head performance, surging, or gas lock.

Using the aforementioned guidelines with this example tandem pump design, it should be assumed that the VC1750 booster pump will likely suffer some head degradation. This will lower the total design head, which will lower the design discharge pressure, force the pump intake pressure up, increase the producing pressure and reduce the flow rate and the amount of free gas, leading to a stable operating point.

If the aforementioned scenario occurs, then the head degradation could be compensated by increasing the pump RPM, assuming a VSD is part of the installation. However, care must be exercised so that the RPM is not increased to the point that the GVF increases, the P_{ip} decreases, the Turpin PHI becomes > 1 and the pump gas locks.

The VSD curve set for the tandem VC1750 and VC1050 pumps is shown in Fig. 12.45. The bottom curve is the VC1750. Line A is at the inlet rate of the VC1750 of 1938 bpd, at a pump intake pressure of 531 psi and a calculated Turpin PHI of 0.70. Line B is the VC1750 outlet rate and the VC1050 inlet rate of 1425 bpd, a pressure of 815 psi, an accumulated head of 1028 ft., and a Turpin PHI of 0.28. Line C is at the discharge of the VC1050 with an outlet rate of 1184 bpd, at a pump discharge pressure of 2073 psi, an accumulated head of 4851 ft., and a

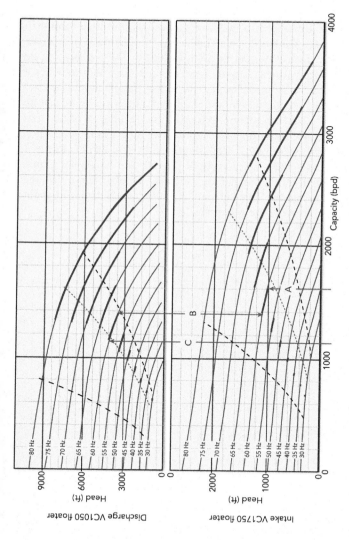

Figure 12.45 VSD flow diagram for a tapered pump.
Source: Courtesy Valiant ALS.

Turpin PHI of 0.04. The program that plotted the curve is designed to provide the above information as the operating frequency in the design is changed.

Tapered pump including the helico-axial stage (gas handler) design

When removing liquids from a gas well to increase production, the lower the producing pressure the greater the production and lower the producing pressure the larger the GVF. GVFs greater than 50% and as high as 80% or 90% would not be unusual at very low producing pressures. A two or three pump taper that includes a gas handling staged housing as the bottom pump or the intake pump, should be considered, along with a tandem gas separator. Minimum pump intake pressures can be calculated and plotted using the information provided by Dunbar, Turpin, and Camilleri, as shown in Fig. 12.46. If the pump is staged without a gas handler, then the Dunbar curve sets the limit for zero head degradation and the Turpin = 1 (Turpin1) curve sets the limit for the maximum head degradation before the pump gas locks. If the pump is staged with a gas handler, then the Dunbar curve sets the limit for zero head degradation and the Turpin = 3 (Turpin3) curve sets the limit for the maximum head degradation before the pump gas locks.

Turpin PHI = 3 (Turpin3) derivation.

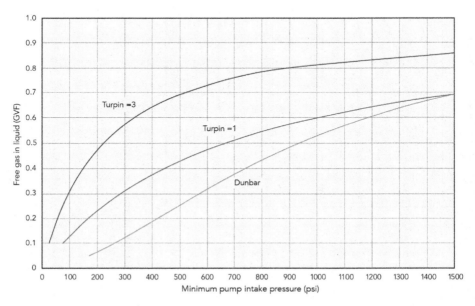

Figure 12.46 GVF versus the minimum pump intake pressure for Dunbar, Turpin1, and Turpin3 Φ.
Source: Courtesy Valiant ALS.

Turpin1 correlation for radial and mixed flow stages:

$$P_{ip\_min} = 667 \times \frac{GVF}{1 - GVF} \tag{12.32}$$

Camilleri et al. found that using a helico-axial stage for the intake pump, allowed the pump intake pressure to be three times lower before gas locking a mixed flow or radial flow stage.

$$P_{ip\_min\_T3} = \frac{667}{3} \times \frac{GVF}{1 - GVF} \tag{12.33}$$

Solving for the Turpin3 PHI in terms of P_{ip}:

$$\Phi_{T3} = \frac{223}{P_{ip}} \times \frac{GVF}{1 - GVF} \tag{12.34}$$

Example 1, no gas handler.
If the desired pump intake pressure is 300 psi, then from a GVF of 0–0.12, Dunbar predicts no head degradation. At 300 psi, from the Dunbar curve to the Turpin = 1 curve (GFV = 0.12 to GVF = 0.31) the pump stage head degradation will go from 0% to the maximum before the stage gas locks. Using the data available in the Turpin reference, we might assume the maximum head degradation before gas lock to be 20% for computational purposes.
Example 2, with a gas handler.
Again, if the desired pump intake pressure is 300 psi, then from a GVF of 0–0.12, Dunbar predicts no head degradation. At 300 psi, from the Dunbar curve to the Turpin = 3 curve (GFV = 0.12 to GVF = 0.58) the pump stage head degradation will go from 0% to the maximum before the stage gas locks. Again, we might assume the maximum head degradation before gas lock to be 20%.
Based on the Turpin et al. tests, Lea and Bearden tests,[27] and field data gathered by Camilleri, the following can be used as a method for estimating the head and flow degradation correction factor (HQCF) to apply in the pump design in high-GVF situations. The HQCF is defined as percent deviation from the catalog water curve. A HQCF of 20% (or 0.20) would be implemented by multiplying the catalog water head by 1-HQCF or 0.80 and the catalog water flow by 0.80 across the curve.
Method of calculation with pump configured with no gas handler.

1. Calculate the pressure and temperature at the intake of the stage.
2. Calculate the GVF at the intake of the stage.
3. Calculate the Dunbar PHI.
4. If the Dunbar PHI is less than or equals one, then HQCF = 0%.
5. If the Dunbar PHI is greater than 1, then calculate the Turpin1 PHI.
6. If the Turpin PHI is less than 1, then calculate the HQCF.

$$HQCF_f = \frac{GVF - GVF_{Dunbar=1}}{(GVF_{Turpin1=1} - GVF_{Dunbar=1})} \times 0.20 \qquad (12.35)$$

where GVF is the calculated GVF at the calculated stage intake pressure and temperature; $GVF_{Dunbar=1}$ is the calculated GVF at the stage intake pressure if the Dunbar calculation = 1; $GVF_{Turpin1=1}$ is the calculated GVF at the stage intake pressure if the Turpin1 calculation = 1.

Else, calculate the correction factor and then show the error "The Turpin1 PHI > 1, Probable Gas Lock."

Method of calculation with pump configured with a gas handler.

1. Calculate the pressure and temperature at the first stage after the last helico-axial stage.
2. Calculate the GVF at the intake of the stage.
3. Calculate the Dunbar PHI.
4. If the Dunbar PHI is less than or equals one, then HQCF = 0%.
5. If the Dunbar PHI is greater than 1, then calculate the Turpin3 PHI.
6. If the Turpin PHI is less than 1, then calculate the HQCF.

$$HQCF_f = \frac{GVF - GVF_{Dunbar=1}}{(GVF_{Turpin3=1} - GVF_{Dunbar=1})} \times 0.20 \qquad (12.36)$$

where GVF is the calculated GVF at the calculated stage intake pressure and temperature; $GVF_{Dunbar=1}$ is the calculated GVF at the stage intake pressure if the Dunbar calculation = 1; $GVF_{Turpin3=1}$ is the calculated GVF at the stage intake pressure if the Turpin3 calculation = 1.

Else, calculate the correction factor and then show the error "The Turpin3 PHI > 1, Probable Gas Lock."

Consider the design example shown in Fig. 12.47 using the helico-axial stage (gas handler). The ESP curves are designed to produce 376 oil, 172 water, and 645 mscf/d at standard conditions. This design used the catalog water curve head to calculate the required stages. The calculated results for flow, GVF, Dunbar PHI, and Turpin3 PHI at the input to each stage type and at the discharge of the pump are listed in Table 12.3.[28,29]

The first 20 stage types of the pump are helico-axial. The in situ rates at the first stage of the helico-axial pump after natural and mechanical separation are 408 bpd oil, 177 bpd water, and 784 bpd gas. The GVF at the intake to the first stage is 0.573. This is assumed to be an acceptable GVF for the helico-axial stage design. As stated by Camilleri et al.,[24] "As a result, an ESP equipped with a Poseidon helico-axial booster pump can handle GVF of 75% in steady state without gas locking and up to 100% in transient conditions such as gas slugs."

The flow passes through 20 helico-axial stages before entering the first VC1750 stage (BEP = 1750 bpd). The in situ rates into the first VC1750 stage are 408 bpd oil, 177 bpd water, and 731 bpd gas. The GVF at the intake to the first stage of VC1750 is 0.555. The calculated Dunbar PHI is 3.49, which indicates that the stage

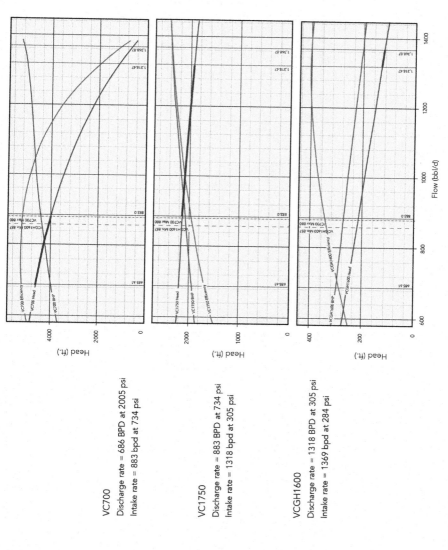

Figure 12.47 VSD flow graph for the gas handler 3-taper pump.
Source: Courtesy Valiant ALS.

VC700
Discharge rate = 686 BPD at 2005 psi
Intake rate = 883 bpd at 734 psi

VC1750
Discharge rate = 883 BPD at 734 psi
Intake rate = 1318 bpd at 305 psi

VCGH1600
Discharge rate = 1318 BPD at 305 psi
Intake rate = 1369 bpd at 284 psi

Table 12.3 Intake pressure, liquid flow, gas flow, GVF, and PHIs for the 3-taper pump.

Stage BEP (bpd)	Stage type	Stages	Pump construction	Intake pressure	Liquid (bpd)	Gas (bpd)	GVF	Dunbar PHI	Turpin3 PHI	GVF at Dunbar = 1	GVF at Turpin3 = 1	HQCF
1600	Helico-axial	20	Compression	284	585	784	0.573	3.90	1.05			
1750	Mixed flow	92	AR modular	305	585	733	0.556	3.49	0.92	0.13	0.58	0.19
700	Radial flow	159	Floater	734	583	300	0.340	0.87	0.16			
Discharge				2005	579	106	0.155	0.17	0.02			

Source: Courtesy Valiant ALS.

will suffer head degradation. The calculated Turpin3 PHI is 0.91, which indicates the stage should not gas lock.

The earlier method is now used to estimate the actual head after accounting for the effects of head degradation because of the low-density fluid in the stage. The GVF at the Dunbar PHI $= 1$ is calculated to be 0.13.

$$GVF_{Dunbar=1} = \frac{(P_{ip}/935)^{1.724}}{\left((P_{ip}/935)^{1.724} + 1\right)} \qquad (12.37)$$

$$GVF_{Dunbar=1} = \frac{(305/935)^{1.724}}{\left((305/935)^{1.724} + 1\right)} = 0.13 \qquad (12.38)$$

The GVF at the Turpin3 PHI $= 1$ is calculated to be 0.58.

$$GVF_{Turpin3=1} = \frac{P_{ip}/223}{(P_{ip}/223 + 1)} \qquad (12.39)$$

$$GVF_{Turpin3=1} = \frac{305/223}{(305/223 + 1)} = 0.58 \qquad (12.40)$$

HQCF for the first stage of the CV1750 can be now calculated.

$$HQCF_f = \frac{GVF - GVF_{Dunbar=1}}{\left(GVF_{Turpin3=1} - GVF_{Dunbar=1}\right)} \times 0.20 \qquad (12.41)$$

$$HQCF_f = \frac{0.556 - 0.13}{0.58 - 0.13} \times 0.20 = 0.19 \qquad (12.42)$$

Using the catalog water curve, the VC1750 stage will produce 21.15 ft. of lift at the 1318 bpd flow rate into the stage, when the Dunbar PHI is less than or equal to one, as shown in Fig. 12.48. When the 0.19 HQCF (or -19%) is applied, the head produced at 1318 bpd becomes 15.11 ft., as shown in Fig. 12.49. The additional head required to compensate the degradation caused by the low-density fluid is 40%.

The stage-by-stage computation using the water curve shows that the Dunbar PHI is lowered to less than 1 after 81 stages of the VC1750 stage. Analysis of the correction required per stage shows that the average correction would be 20% per stage. To compensate for the head degradation, the first 81 stages of VC1750 stages should be increased by 16 stages. The pump should be designed using 108 stages of VC1750 instead of the water curve design of 92.

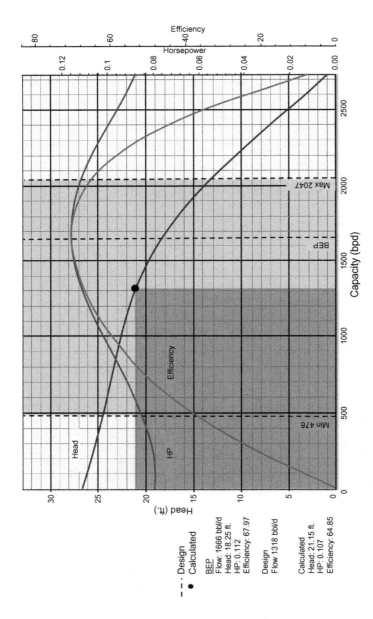

Figure 12.48 VC1750 catalog water curve at 1318 bpd.
Source: Courtesy Valiant ALS.

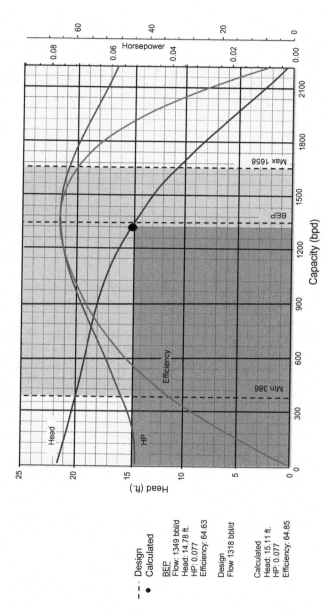

Figure 12.49 VC1750 corrected curve for gas degradation because of low fluid density at 1318 bpd.
Source: Courtesy Valiant ALS.

12.6 Summary

12.6.1 ESP motors

The ESP motor has historically been an induction motor; however, the ESP permanent magnet motor, new in the last 20 years or so, is gaining acceptance as a viable alternative because of their higher efficiency and the ability to run at lower temperatures than the equivalent size induction motor. The ESP motor is cooled by the produced fluids. When used for gas well deliquification, there will generally be a higher free gas-to-liquid ratio (BPDG/BPDL) around the motor. The specific heat (SpH) of the cooling fluid should be calculated. Water (SpH = 1) can remove more heat per pound than oil (SpH = 0.4) and oil can remove more heat per pound than gas (SpH = 0.15). Winding temperature calculations should be made to verify that the motor insulation will have an acceptable life based on the motor load, motor efficiency, ambient temperature, the velocity of the cooling fluid, and the average specific heat of the cooling fluid.

Potential changes in operating conditions, both short and long term, should be included when estimating the motor life.

12.6.2 ESP seals

When selecting the seal type (labyrinth or bag) the produced fluids and any treating chemicals should be considered. They should be compatible with the O-ring and the bag elastomers. Bag-type seals should be used when the well produces very light oils (API > 40) because the oil in the motor and protector can start to mix with the produced oil, reducing the electrical and lubrication properties of the motor and seal oil. Bag-type seal should also be used when the seal is set at an angle of 45 degrees or greater from the vertical to maintain well fluid and seal and motor oil segregation.

12.6.3 ESP intakes

The pump intake type and location will determine the GVF that will enter the pump. A standard intake would be required if the pump is set below the producing interval with a shrouded motor or a recirculation system. A standard intake would be used above the producing interval with an upside-down shroud (shrouded pump). Gas separation efficiency in the annulus is dependent on the gas terminal bubble rise velocity being greater than the superficial liquid velocity (V_{sl}) downward. V_{sl} should be less than 0.5 ft./s for good natural separation efficiency.

If the intake is set in a nonvertical section, then an intake that automatically adjusts its intake ports to allow only intake from the bottom may be preferred.

As the GVF at the pump intake increases, then a vortex separator or tandem separators should be considered to remove the gas before it enters the pump.

Calculations for determining GVF and the probability of stage head degradation due to the free gas (Dunbar correlation) are presented.

12.6.4 ESP pumps

Radial, mixed flow, AGH, MVPER, and helico-axial stage designs and their ability to produce head as a function of GVF versus the recommended flow range must be considered when designing for gas well deliquification. Floater, compression, and abrasion resistant modular pump construction types and their effect on thrust need to be selected so that the pump does not overload the seal thrust bearing.

Rules for quantifying the stage head degradation have been developed and used in two examples. An example using a mixed flow and radial flow stage in a tandem pump and an example using a helico-axial stage, a mixed flow stage, and a radial stage in a triple tandem pump design are presented.

References

1. Nyberg C. *The effects of high or low voltage on the performance of a motor.* <www.easa.com>; 2000.
2. Cowern E. Understanding induction motor nameplate information. *EC&M* May 2004.
3. *Voltage controller saves energy, prolongs life of motors.* <https://ntrs.nasa.gov/archive/nasa/casi.ntrs.nasa.gov/20080003940.pdf> [accessed March 2018].
4. Nola FJ. *Motor power control circuit for A.C. induction motors.* U.S. Patent 4,439,718; March 27, 1984.
5. Cashmore DH. Induction motor performance for a variable terminal voltage. In: *Presented at the 1992 ESP workshop, SPE, Gulf Coast Section, Houston, Texas.*
6. Lea JF, Liu J. *ESP motor voltage adjustment for maximum profits.* Society of Petroleum Engineers; 1997 Available from: https://doi.org/10.2118/37450-MS.
7. Knapp JM. *Systems and methods for adjusting operation of an ESP motor installed in a well.* US Patent 9,429,002.
8. Kang J, Yaskawa. *General purpose permanent magnet motor drive without speed and position sensor.* <https://www.automation.com/pdf_articles/yaskawa/WP.AFD.05.pdf>; 2009.
9. NEMA MG-1-12.45.
10. Toshiba. *"Temperature rise and Insulation", Application Guideline #105.* <http://toshont.com/wp-content/uploads/2017/06/Motor-Temperature-Rise.pdf>.
11. Siemens. *Application manual for motors, "Temperature Rise standards", Section 5, Part 6, 05/2008.* p. 566 and 567.
12. Bannwart A, Arrifano Sassim N, Estevam V, Biazussi J, Monte Verde W. *Gas and viscous effects on the ESPs performance.* Society of Petroleum Engineers; 2013 Available from: https://doi.org/10.2118/165072-MS.
13. Ahmed T. *Equations of state and PVT analysis.* Gulf Publishing Company; 2007.
14. Lake LW, et al. *Petroleum engineering handbook,* vol. I. Society of Petroleum Engineers; 2006. p. I-310−31.
15. Alhanati FJS. *Bottomhole gas separation efficiency in electrical submersible pump installations.* PhD dissertation, The University of Tulsa, Tulsa, OK; 1993.
16. Dunbar CE. Determination of proper type of gas separator. In: *Presented at the micro-computer applications in artificial lift workshop,* SPE Los Angeles Basin Section, Los Angeles, California, USA; 1989.

17. Mack JJ, Wilson BL. *Recirculating pump for electrical submersible pump system.* US Patent 5,845,709; December 8, 1998.
18. Guindi R, Storts B, Beard J. *Case study, permanent magnet motor operation below perforations in stagnant fluid.* Society of Petroleum Engineers; 2017 Available from: https://doi.org/10.2118/185273-MS.
19. Lea JF, Bearden JL. *Gas separator performance for submersible pump operation.* Society of Petroleum Engineers; 1982 Available from: https://doi.org/10.2118/9219-PA.
20. Sambangi SR. *Gas separation efficiency in electrical submersible pump installations with rotary gas separators.* MS thesis, The University of Tulsa, Tulsa, OK; 1994.
21. Lackner G. *The effect of viscosity on downhole gas separation in a rotary gas separator.* PhD dissertation, The University of Tulsa, Tulsa, OK; 1997.
22. Marsis EGR. *CFD simulation and experimental testing of multiphase flow inside the MVP electrical submersible pump.* A dissertation, submitted to the Office of Graduate Studies of Texas A&M University in partial fulfillment of the requirements for the degree of doctor of philosophy in mechanical engineering; December 2012.
23. Bozorgmehrian M. *Sizing and selection criteria for subsea multiphase pumps.* A master's thesis, presented to the faculty of the Department of Engineering Technology, University of Houston; May 2013.
24. Camilleri LAP, Brunet L, Segui E. *Poseidon gas handling technology: a case study of three ESP wells in the Congo.* Society of Petroleum Engineers; 2011 Available from: https://doi.org/10.2118/141668-MS.
25. Valiant Artificial Lift Solutions. *Electric submersible pump systems catalog*; January 10, 2018.
26. Turpin JL, Lea JF, Bearden JL. Gas-liquid flow through centrifugal pumps — correlation of data. In: *Proceedings of third international pump symposium*; 1987.
27. Lea JF, Bearden JL. *Effect of gaseous fluids on submersible pump performance.* Society of Petroleum Engineers; 1982 Available from: https://doi.org/10.2118/9218-PA.
28. Oliva GBFF, Galvão HLC, dos Santos DP, Silva RE, Maitelli AL, Costa RO, et al. *Gas effect in electrical-submersible-pump-system stage-by-stage analysis.* Society of Petroleum Engineers; 2017 Available from: https://doi.org/10.2118/173969-PA.
29. Pessoa R, Prado M. *Experimental investigation of two-phase flow performance of electrical submersible pump stages.* Society of Petroleum Engineers; 2001 Available from: https://doi.org/10.2118/71552-MS.

Coal bed methane (CBM) and shale

13

David Simpson, P.E. has been working in oil and gas since 1980 and is currently the Proprietor and Principal Engineer of MuleShoe Engineering (www.muleshoe-eng.com). Based on the San Juan Basin of Northern NM, MuleShoe Engineering addresses issues in unconventional gas, low-pressure operations, gas compression, gas measurement, oil field construction, gas well deliquification, and project management. A P.E. with a BSIM and MSME, Simpson has numerous articles in professional journals and has spoken at many conferences around the world. He is the author of Practical Onshore Gas Field Engineering, ISBN: 978-0128130223, published by Gulf Publishing, 2017.

13.1 Introduction

Coal bed methane (CBM) and the organic portion of shale plays are fundamentally different from the pore-volume storage in conventional and tight-gas plays. The three overriding differences are as follows:

1. The gas is adsorbed to the surface of the organic material and this is not subject to the gas laws that we are all familiar with.
2. The pathways from a storage site to a flow path are very limited.
3. The reservoirs are "capped" by the hydraulic pressure of liquid in the microfracture system rather than a "cap rock" (in fact, both coal and shale are very common cap rocks for conventional reservoirs).

Fig. 13.1 shows the impact of these fundamental differences. The pore-volume storage reservoirs do not have an internal driving force for flow other than the pressure of the fluids within the pore volume. This pressure along with fluid friction within the flow paths and the inherent compressibility of hydrocarbon gases results in the bottom line in Fig. 13.1. Anyone looking at this curve would think "when pressure gets low, gas in place is disproportionately lower" and would be reluctant to spend significant capital on lowering pressures late in the field life. At 50% of initial reservoir pressure, only 33% of the original gas in place (OGIP) is still in the reservoir and at 10% of original pressure, the remaining gas is less than 6% of OGIP.

In CBM and shale, there is an additional energy source called "desorption energy" that can push the gas toward the wellbore and you find that by the time you pull the reservoir pressure down to 10% of initial conditions you still have significantly more than 10% of the OGIP in both CBM and shale reservoirs, and flow rates can be very high at these "low" pressures.

Gas Well Deliquification. DOI: https://doi.org/10.1016/B978-0-12-815897-5.00013-5

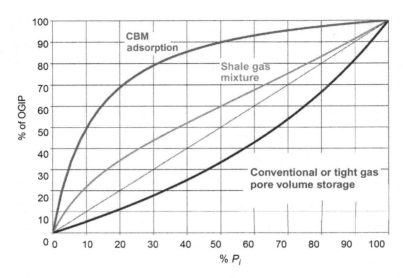

Figure 13.1 Gas in place versus reservoir pressure.[1]

CBM has approximately zero pore-volume storage because the coal matrix tends to be reasonably homogenous with minimal inclusions of sand or carbonates. This nature of CBM makes oil production and heavier hydrocarbon gas production rare.

Shale tends to be far from homogenous. In several shale plays, core analysis shows that upward of 60% of OGIP (and 100% of original oil in place) is stored in the pore volume of interbedded sandstone and carbonate-based rock. The remaining 40% of OGIP is adsorbed to the organic portion of the shale matrix. The most important ramification of this heterogeneous nature is that the potential for oil production is very high in shale, and the adsorbed gas provides an energy source to push the oil to the wellbore. Fig. 13.1 shows the shale line to be much closer to the 45-degree line than either CBM or conventional/tight gas. In early days, virtually all of the produced liquid and gases come from the pore-volume storage. At about 70% of initial reservoir pressure (Fig. 13.3), the lines crossover and desorption dominates production (with concurrent reduction in the oil cut).

Many companies have looked at versions of Fig. 13.1[1] and reached the erroneous conclusion that since CBM wells must spend so much of their productive life at very low pressures, money can be saved by providing very low-pressure equipment and piping for CBM fields. This conclusion has led to significant underperformance in several CBM fields that will be discussed later.

The producers in the shale plays, currently under development, can see the very high initial pressure and provide surface equipment consistent with those pressures. This decision is certain to improve the ultimate recovery from these wells.

The key to produce any reservoir is to understand what conditions are under your control, and what target parameters should be monitored to maximize ultimate recovery. Ask the question "I can control flowing bottom-hole pressure, what value should I shoot for at any given time?" It matters.

13.1.1 History

The first commercial hydrocarbon well was a shale gas well drilled 27 ft. into the Devonian shale in Fredonia, NY in 1821 (30 years before the first oil well in Titusville, PA). The well provided gas to the town's gas lights for many years.

Methane adsorbed to the surface of coal is a very old issue with some new commercial ramifications. This explosive gas has made underground coal mines dangerous both from the risk of explosion and the possibility of an oxygen-poor atmosphere that will not support life. The miner's main concern with CBM has been how to get rid of it.[1] Techniques to deal with CBM in mines have ranged from the "classic canary in a cage" that can (fatally) detect an oxygen-poor atmosphere prior to human distress, to huge ventilation fans to force the replacement of a methane-rich environment with outside air, to drilling CBM wells in front of the coal face to try to degas the coal prior to exposing the mine to the gas. All these techniques have met with some amount of success. None of the techniques to prevent CBM from fouling the air in an underground mine have been totally successful.

As the expertise in developing and CBM production spreads beyond the Black Warrior and San Juan Basins, it is becoming clear that CBM is a significant economic resource on its own, and capturing CBM for sale is often profitable even on coal seams that cannot be economically mined.

With CBM and shale-gas' unique method of gas storage the preponderance of the gas is available only to very low reservoir pressures. The coal face pressure is set by a combination of flowing wellhead pressure and the hydrostatic head exerted by standing liquid within the wellbore. Effective compression strategies can lower the wellhead pressure to appropriate values. Effective deliquification techniques can reduce or remove the back pressure caused by accumulated liquid. In other words, you have a great deal of control over flowing bottom-hole pressure.

Lowering the hydrostatic head can create suction-pressure challenges for most of the deliquification techniques presented in this book, and the successful operator must be very aware of both the minimum inlet pressure needs of their deliquification technique and the back pressure requirements of the well. Getting deliquification and compression "right" can result in recovery factors in excess of 90% of the OGIP, while getting them wrong can limit recovery to less than half of the OGIP. For a well with something like 20 billion standard cubic feet (bscf) OGIP, 50% can be worth upward of $120 million USD at $3 USD/MMBtu.

13.1.2 Economic impact

In 2003 CBM production was 8% of total gas production in the United States, 10% of gas reserves, and 15% of estimated undiscovered gas resources. As of 2003, the San Juan Basin of Northern New Mexico and Southern Colorado represented over

80% of worldwide cumulative CBM production.[2] Since 2003, CBM projects have been undertaken in several parts of the world, and the Bowen Basin of Queensland, Australia has surpassed both the Black Warrior Basin (Alabama) and Powder River Basin (Wyoming) in peak rate and cumulative recovery.

With the early dominance from the San Juan Basin, it is obvious that a significant portion of the data available for analysis of the lifecycle of a CBM well must come from this basin. Many things were learned in the San Juan Basin, which have proven to be unique to San Juan and not applicable to the other coal plays around the world. Other things were learned through difficult and expensive lessons that can be applied to developing basins.

In 2003 there were no shale reservoirs in the Energy Information Administration (EIA) top 10 basins based on either production or proved reserves of either oil or gas. In the 2015 "Top 100 U.S. Oil and Gas Fields" the largest oil field (more than twice the production of number 2) was the Eagle Ford Shale in south Texas. Three of the top ten oil fields were shale. For gas fields, shale operations make up the top three, and are four of the top ten. The four shale gas plays are 71% of the top 10, over half of the top 100, and nearly half of total US gas production.

13.2 Organic reservoirs

13.2.1 Reservoir characteristics

Rock formations that are typically important to oil and gas production fall into three categories.[1]

- *Source rock* is a formation containing organic matter whose decomposition has resulted in the formation of complex hydrocarbon products.
- *Reservoir rock* is a formation with the ability to store commercial quantities of hydrocarbons.
- *Cap rock* is a formation that is largely impermeable to the flow of liquids and gases and is located such that fluids that approach it from lower or adjacent formations cannot migrate further.

In conventional reservoirs, oil or gas is created in the source rock, it migrates to the reservoir rock, and its migration is stopped by cap rock. CBM and shale reservoirs do not follow this pattern. They both meet the criteria of source rock since the very matrix is rich in organic matter and in fact both are frequently the source rock for conventional reservoirs. They meet the definition of reservoir rock even though the pore volume of a coal bed is an order of magnitude less than conventional reservoirs. Both are often the cap rock for conventional reservoirs.

CBM and the organic portion of shale gas is adsorbed to the surface of the organic matrix. The adsorption sites can store commercial quantities of gas as part of the solid. This must not be confused with conventional pore-volume storage. Gas within a pore volume acts as a gas and the traditional pressure/temperature/volume relationships hold. Adsorbed molecules are not gas. They do not conform to the shape of the container. They do not conform to the modified ideal gas laws (i.e., $PV \neq ZnRT$), and they take up substantially less volume than the same mass of gas would require within a pore volume at a given pressure.

One effect of CBM and shale not being stored in pore volume is that most of the conventional equations that describe hydrocarbon flow within a reservoir are either outright invalidated or require extensive modification. A simple example is the *Bureau of Mines Method of Gauging Gas Well Capacity*[3] shown in Eq. (13.1).

$$q = c_p \cdot \left(\overline{P}^2 - P_{BH}^2 \right)^n \tag{13.1}$$

Eq. (13.1) is also known as the absolute open flow equation or sometimes the "back pressure equation" and can be used to predict how a well's production rate (q) will change with a change in flowing bottom-hole pressure (PBH) assuming that reservoir pressure ($\overline{P}$) is relatively constant in the short term. Both the nonlinearity term (n) and the flow constant (c_p) are reservoir properties and are constant for all pressure/temperature combinations.[4] It has been shown[2] that either c_p or n must be changing dramatically over time to allow changing reservoir pressure to match exhibited unconventional gas performance. Some of the changes recorded have been on the order of 30% increase per month. There is no value that can be assigned to any of the parameters in Eq. (13.1) that would result in an early-life incline and a late-life decline.

Although the ideal gas law does not apply to organic storage while the gas is adsorbed to the matrix, there are mathematical relationships that do apply. The most important relationship to describe the way that gas is adsorbed is the *Langmuir Isotherm* (Eq. (13.2)), which assumes that the reservoir is at a constant temperature and defines the quantity of gas adsorbed as

$$OGIP = 0.031214 \cdot A \cdot h \cdot V_m \cdot y \cdot \rho_{lang} \cdot \frac{b \cdot P_i}{1 + b \cdot P_i} \tag{13.2}$$

where OGIP = original gas in place (scf); A = drainage area in ft.^{2}; h = thickness of the coal in ft.; V_m = gas content of coal (scf/ton); y = mineral-matter free mass fraction of total coal (fraction); ρ = density (g/cc); b = Langmuir shape factor (psi^{-1}); and P_i = initial reservoir pressure (psia).

Most of these parameters are either acquired during drilling, logging, and coring, or are estimated from analogs. The equation can also yield remaining gas by substituting current reservoir pressure for initial reservoir pressure and OGIP minus cumulative production for OGIP since the other parameters are reservoir characteristics that do not change substantially over the life of a well. If you plot gas in place as a percentage of the OGIP versus reservoir pressure as a percentage of initial reservoir pressure, you will get a curve as in Fig. 13.2.

For this well the "declining pressure" period ends somewhere around 65% of initial reservoir pressure remaining. At this point, you have produced 7% of OGIP.

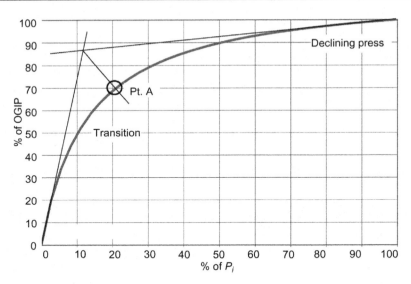

Figure 13.2 Annotated GIP versus P_i for CBM.[5]

This period is characterized by relative insensitivity to flowing bottom-hole pressure. Often in the declining pressure region, wells will not respond to wellsite compression or to deliquification techniques.

The transition region is a very large portion of a well's life and it is difficult to characterize. For the well in Fig. 13.2, the transition period starts with 93% of OGIP remaining and ends with about 30% of OGIP remaining. During this time the reservoir pressure has dropped from 65% of initial pressure to 7% of initial pressure—for this well the transition period corresponds to a pressure drop from 1100 to 130 psia. The transition period is difficult to characterize, because at the start the well will perform much like it did during "declining pressure," and at the end it is acting like a late-life "declining reserves" well.

Determining necessary equipment can be facilitated by inscribing a straight line on the data for both the first and last regions—at the point they cross, draw another line from the intersection back to a point normal to the curve (point A). With point A defined, you have set the place where the well will tend to stop acting like a "declining pressure" well and start acting like a "declining reserves" well.

The "declining reserves" region is characterized by declining reserves with fairly stable reservoir pressure. It is clear that successfully recovering this gas requires very careful management of flowing bottom-hole pressure. Seemingly, infinitesimal changes in either wellhead pressure or fluid level can cause significant changes in production.

The well in Fig. 13.2 is a San Juan Basin "Fairway" well that had an OGIP of 28.2 bscf and an initial reservoir pressure of 1820 psia. At the point where it left the declining pressure region, it had produced for only 7 months and had made 180 MMscf. It stopped acting like a declining pressure well at 5 years-5 months (cumulative production 8.9 bscf). This point coincided with a peak production rate of 10.5 MMscf/day, and shortly after this peak both wellhead compression and

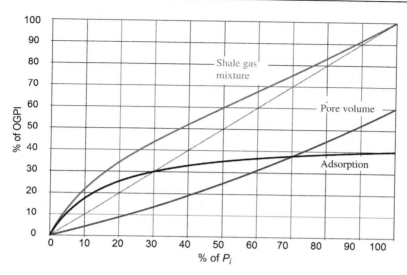

Figure 13.3 GIP versus P_i for shale gas.[5]

deliquification were required to sustain production rates. It entered the declining reserves region at 11 years-3 months (cumulative production 16.9 bscf) and the average gas rate was still 2 MMscf/day. With the last data available at the time of publication, the well had been producing for last 26 years and was at 1% of initial reservoir pressure (18 psia), it had produced 22 bscf (76% of OGIP), and the production rate was 360 Mscf/day.

The well in Fig. 13.2 is only unique in the San Juan Fairway because of the amount of high-quality data collected over its entire producing life. Other than that it is a fairly average San Juan Fairway well but is not necessarily representative of CBM projects in other basins. When other basins have moved farther down their isotherms, it will be possible to verify the late-life procedures that will be necessary in those fields.

You can do a similar analysis with shale gas, but you have to do it in two parts and combine the results (Figs. 13.3 and 13.5). The "Point A" analysis shown in Fig. 13.2 can be done on either the adsorption line or the combined line in Fig. 13.3 without a significant change in the decision points, but if you do it on the combined line, you do not have to convert partial pressures of the two storage types. The well in this figure is based on data from the Barnett Shale in central Texas and is very similar to wells in other basins. There is considerable variance in the expectations of the magnitude of OGIP, so I will refrain from putting actual pressure or GIP numbers to this figure until some future edition of this book.

13.2.2 Flow within an organic reservoir

One defining characteristic of conventional reservoirs is "permeability." Permeability is a measure of a reservoir's ability to flow. Permeability is measured

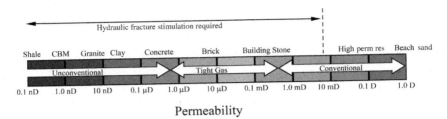

Permeability

Figure 13.4 Permeability continuum.[5]

in "Darcy" (or milli-Darcy, micro-Darcy, etc.). Conventional gas wells generally see permeability from 500 milli-Darcy (mD) to 1.0 Darcy (D). CBM can be expected to have a permeability in the region of 1.0 nano-Darcy (nD, 10^{-9} Darcy), CBM after all is a common cap rock for conventional reservoirs. Shale gas would be expected to be about 1/10th the permeability of CBM. Fig. 13.4 shows the expected values for permeability.

It is common practice to try to develop an "effective perm" or "cleat perm" value for CBM and shale. This process takes the matrix permeability (effectively zero) and adds the "permeability" in the channels (effectively infinite), then divides by 2 and gets a value around 30 mD. This made-up value is then used in numerical simulations to predict well performance over time—with universally horrible long-term results.

Flow within an organic reservoir depends on a system of channels that only require the gas to flow a very short distance (not more than a few feet) through the matrix to reach a near-infinite-permeability channel. Models that honor this characteristic of organic reservoirs tend to do a decent job of predicting performance, models that ignore this characteristic tend to do a consistently poor job of predicting performance.

CBM is mechanically much weaker than shale. Commercial flows within a CBM reservoir rely on the system of "channels" that have been created within the coal seam either through geological movements or designed stimulations. Wells where the stimulations have been aligned to cross the maximum number of face cleats have consistently outperformed wells that ignore the lay of the coal stresses.

The channels are analogous to pipe flow, and the pressure traverse from the wellbore out to considerable distances can be essentially constant. Since desorption is a pressure-swing phenomenon, the extent of these channels has an overriding effect on flow rates. If the system of channels and channel branches is extensive and robust, then the flow rate will be high. On the other hand, without these channels the well's ability to project low wellbore-pressures deep into the reservoir will be limited and the production rate and/or total recovery will be low. Conventional stimulation techniques have had very mixed success in CBM fields (see later). The problem is that the best coal seams are completed either open hole or with unce-mented liners and it is impossible to control where the stimulation will go once it is outside the pipe.

Shale reservoirs have significantly more mechanical strength than CBM reservoirs. The primary effect of this increased strength is that extended-lateral shale wells are strong enough to maintain hole-integrity while casing/liners are set. With pipe cemented down a lateral, hydraulic-fracture stimulation can be precisely located for maximum effect.

13.2.3 Adsorption site contamination

Adsorption sites can hold any molecule that fits. The typical adsorption-site "diameter" in organic fields is perfectly suited for CO_2 molecules, but methane and nitrogen also fit reasonably well. Heavier hydrocarbons do not fit well on the adsorption sites and it is rare to see significant hydrocarbon concentration heavier than methane adsorbed to the organic matrix (although interbedded sand-lenses are reasonably common and they can hold any gas or liquid that their pore structure can accommodate).

CO_2 is a very common contaminant. Estimating both peak CO_2 and a CO_2 production profile is a difficult, but useful, exercise. In the Fairway of the San Juan Basin CBM, there were a very limited number of pressurized cores taken during initial field development, and all of them showed a bulk CO_2 in the neighborhood of 18% (this number is a function of the amount of oxygen that was present when the organic material was sealed away from the atmosphere). Initial production from these wells mostly had 6%−8% CO_2 and the question was "where is the missing 10%−12%?" There is no definitive answer to that question, but close observation of the CO_2 levels over most of the life of several wells has yielded some clues. The CO_2 increased gradually over time toward 25%−29%, and then started to decrease. This would suggest that the actions of drilling, completing, and producing the gas have uncovered a very large number of unsaturated adsorption sites near the wellbore by breaking the coal. These new sites would preferentially adsorb CO_2 because it is a better fit, and you would be effectively filtering CO_2 from the combined 82% methane, 18% CO_2 stream. As pressures come down, the disproportionate CO_2 content in the near-wellbore will cause CO_2 production to increase with time. As operators anticipated very low-pressure and vacuum operations, there was a real concern that the CO_2 would continue to increase until a large quantity of gas would have to be abandoned in place because it would become uneconomic to treat it.

A careful material balance in the San Juan Fairway has indicated that the "missing" CO_2 in the early days has a total mass that can be determined. Also, if the original portion of CO_2 in the reservoir is known then the total mass of CO_2 in the reservoir is also known as a fraction of the OGIP. The analysis showed that there is a point where the missing CO_2 has been recovered, and the production stream should drop back toward the in situ CO_2 level of 18%. This peak was predicted at 25%−29% and has been seen in a number of wells.

Expectations are that shale will have a similar CO_2 production profile, but it will likely take many decades to reveal itself and we may not be certain of the profile until the 22nd century.

13.2.4 Coal mechanical strength

An important point that is common to all CBM plays is that the coal is soft. The friability (i.e., the ability of the material to resist shear forces) of coal has been reported as low as 15 psi in the San Juan Basin Fairway; other coals are comparatively stronger, but still very weak. This characteristic is responsible for the observed fact that significant quantities of coal solids are produced along with the CBM. It also accounts for the observed tendency of a coal bed to "heal" around an inclusion. For example, you can force a shovel into a coal face, but a short time later it will be very difficult or even impossible to remove the shovel. This self-healing characteristic has been observed in coal mines since the 1800s.

In the San Juan Basin Fairway the cavitation technique has been used very successfully to allow high-velocity gas-flow to sculpt large downhole cavities and a robust channel system by causing the coal to fail and then transporting the failed coal to surface. Success of cavitations outside the San Juan Basin Fairway has been limited, probably because the friability of the coals in other basins allows the coal to resist failure much longer than it can resist with the Fairway coal.

Every sort of hydraulic fracturing has been tried on CBM wells, and most of them have had both successes and failures. It is likely that many of the failures have relied on small liquid volumes and large proppant volumes, and the coal was able to heal itself around the proppants. On multiseam CBM wells, there is strong evidence that stimulation has been limited to a single zone with no stimulation in the other zones.

13.3   Organic reservoir production

During the time that a well's reservoir pressure is above Point A in Fig. 13.2, it is fairly forgiving. Flowing bottom-hole pressure needs to be below the pressure corresponding to Point A in Fig. 13.2 (400 psia for the well in Fig. 13.2), but efforts to drop it significantly below the Point A pressure have had minimal impact on production rate. For the well in Fig. 13.2, early-life flowing-tubing pressure was around 160 psia so the well could tolerate about 500 ft. of water above the formation and still produce approximately its maximum rate. During the declining pressure period, deliquification and/or compression may be required to get the flowing bottom-hole pressure below Point A, but often it is installed unnecessarily because the operator says "that is the way we do it."

After Point A in Fig. 13.2, managing flowing bottom-hole pressure is critical to maximizing ultimate recovery. Since most wells reach Point A with more than 70% of OGIP remaining, it is worthwhile to build initial wellsite facilities with the *anticipation* (but not the requirement) of low-pressure operations. Separation equipment should have "blowcases" to allow low-pressure liquid to be boosted easily to pressures required to enter tanks or pipelines. Gas pipe should have appropriate manifolds to allow inexpensive installation of wellsite compression. The wellsite should be set up to facilitate whatever sort of deliquification strategy you have adopted. For example, running a single line from the wellhead to the separator will require that any pumped liquids be commingled with produced gas at the wellhead for the run to the separator—that sort of multiphase flow is very energy inefficient.

Single-digit flowing bottom-hole pressures are achievable in CBM wells, but these low pressures require you to understand and minimize every tiny pressure drop up the wellbore and across the location.

Prior to any well-specific decisions, you should have a couple of detailed plans—you need to know how you are going to get water off the formation when the reservoir pressure is under 100 psig and you need to know what your gathering system pressures are going to be (and how they are going to be maintained). These strategies should be clearly documented and available to everyone making drilling, completion, deliquification, or facilities decisions.

13.3.1 Deliquification plan

If your operation is in a field like Horseshoe Canyon in Alberta, Canada where you can reportedly be confident that a well might never need help getting liquid to surface, then it is reasonable to design wells with small casing. If you think there is a reasonable chance the well will need deliquification equipment sometime in its life, then you will need more real estate downhole. Doing this analysis prior to spudding the first well goes a long way toward a wellbore design that will work for the life of the well. For example, one of the major operators of the San Juan Basin Fairway completed a large number of wells with cased-and-frac'd completions using small casing, and the wells significantly underperformed relative to offsets. The company revisited their plans and decided to sidetrack many of the wells and redrill many others with larger casing and cavity completions. Production increased significantly. Late in the life of the wells the sidetracked holes presented a difficult deliquification problem.

The deliquification plan should have three parts:

- initial deliquification;
- mid-life deliquification;
- late-life deliquification.

Initial deliquification

Recall that adsorbed gas is held onto the organic surface by hydrostatic pressure in the cleats and that any desorbed gas must have an exit path.

It has become very common for CBM producers to design surface facilities for very low operating pressure and install deliquification equipment on Day 1. This is rarely very effective since it leads us to choking the wellhead pressure while pumping at a rate inconsistent with actual reservoir inflow.

If surface facility design-pressure and temperature are consistent with full reservoir conditions, then it is quite reasonable to rapidly dump built-up pressure into the separator and gas/water-gathering system and let the pressure drop to below gathering pressure and then shut the well back in to build pressure back up. Operators who have done this have found that every shut-in cycle is shorter than the previous shut-in, and every flowing period is longer than the previous flowing period until the well is able to flow all the time. This approach has two very real benefits: (1) the rapid onset of flow tends to break coal and move coal pieces—improving the effectiveness of the flow channels and (2) it tends to remove liquid at a rate consistent with the needs of the reservoir.

In short the initial deliquification plan for CBM should consist of pressure build-up and rapid depressurization events without deliquification equipment. After the well begins flowing full-time, it is very effective to use a tubing-flow-control scheme that is made up of:

• a flow meter on the tubing (V-Cone meters have proven to be very effective for this);
• a flow-control valve on the casing;
• a process logic controller (PLC).

In the tubing-flow-control scenario, some value is used for "critical rate" (usually the Turner equation is used for initial deliquification), and the PLC watches the flow meter on the tubing. If the tubing flow is above the critical rate and increasing, then the PLC bumps the flow-control valve toward open. If the tubing flow is above the critical rate and decreasing, then the PLC bumps the flow-control valve toward shut. If the tubing flow approaches critical, then the PLC shuts the flow-control valve. This scenario can certainly take the well to the end the declining pressure region, and with some modification should be effective beyond Point A in Fig. 13.2.

If the maximum allowable working pressure (MAWP) of the surface equipment is inadequate for full reservoir pressure, then this technique is unlikely to be effective (and the well is unlikely to recover a significant portion of the OGIP). In the case of unavoidable low-pressure equipment and gathering pipe (generally due to company policies inconsistent with the needs of the reservoir), some sort of mechanical deliquification equipment will be required. For start-up, it is unlikely that the well will provide a reliable fuel source to power the equipment, and securing a reliable energy source is the most important part of the deliquification plan. This energy source can be grid-power, a sales-tap from the gathering system, or on-site propane tanks. The choice of equipment is a series of bad trade-offs.

Any pump technology has about the same chance of working as any other (i.e., not very likely). With any technology, it is important to avoid suction strainers/separators that can become clogged with coal very rapidly (the jet pump throat in Fig. 13.5 was subjected to a clogged suction line for less than 3 h).

Companies have gotten results that they seemed satisfied with using Electrical Submersible Pump (ESP), Progressive Cavity Pump (PCP), conventional rod pumps, and linear rod pumps. Many people have tried jet pumps, but the results have rarely been positive.

Because of the amount of pore-volume storage in a shale well, the intermitting process that works well for CBM is rarely as effective. Several companies (recognizing the value of oil vs gas) have had good success setting a packer in the vertical section of the wellbore and using gas lift to assist a continuous-run (i.e., internal bypass) plunger for initial deliquification. Other types of deliquification equipment have had problems with slugging in the lateral that has resulted in early pump failure.

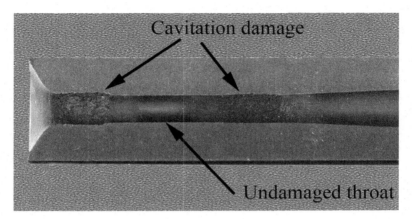

Figure 13.5 Damaged jet pump throat.[5]

Mid-life deliquification

During the transition period in Fig. 13.2 the tubing-flow-control technique described earlier can be successful for both CBM and shale. It is common as you enter the transition period to change the critical flow rate from Turner to Coleman, and that is probably not a bad approach. As the reservoir approaches Point A in Fig. 13.2, actual critical rate should be determined as follows (see Fig. 13.6):

- Install data logging equipment (or change the sample frequency on the automation equipment to 1 min if the automation equipment can do that) to allow tubing flow and casing pressure to be captured each minute.
- Shut a manual valve on the casing to override the PLC control of casing flow.
- Flow the tubing.
- As long as liquid level in the casing is near the bottom of the tubing, gas pressure in the casing will assist flow up the tubing (and consequently drop steadily). When the fluid level rises into the tubing/casing annulus, the annulus volume will become isolated from the tubing. Gas entrained in the liquid will evolve into the casing, raising the casing pressure.
- On the data log find the flow rate where the decline in casing pressure begins to stop, that is your actual critical rate that should be programmed into the PLC.

Before Point A in Fig. 13.2, it can take several days to determine critical rate, after Point A, it should take a few hours.

Tubing-flow control could easily be the only deliquification technique needed through most of the transition period for wells that have surface equipment consistent with reservoir pressure/temperature.

If surface equipment is inconsistent with reservoir conditions, then the mid-life deliquification plan should describe equipment to be used on wells that are liquid loading. Equipment used in the mid-life period should always be variable speed with flexibility to change the speed (either manually or through some clever algorithm). Mechanical deliquification in mid-life is expensive, difficult to match with reservoir needs, and very prone to failures. Shutting-in mid-life wells for extended

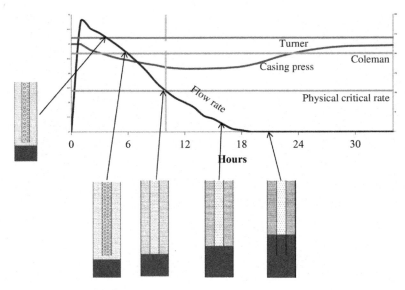

Figure 13.6 Determining actual critical rate.[5]

periods waiting on a rig will cause established channels to fill in and should be avoided. The plan should set goals for rig-access, replacement parts, and postappraisal requirements.

Late-life deliquification

Recovering reserves from a low-pressure reservoir requires very low flowing bottom-hole pressure. A consequence of very low pressures is that a lot of water will evaporate and move as water vapor. In many wells, evaporation will be adequate by itself to remove all the water inflow. Late in the life of all organic-storage wells, evaporation will represent a significant water volume. If a pump is set up for a given water rate and half that rate moves by evaporation, then the pump will begin to experience difficulties such as gas locking or cavitation. Again the issues can be successfully managed as long as they are fully understood and anticipated.

Typically the late-life plan will specify significantly smaller fluid-handling capacity and gentler equipment. For example, it is reasonable to use a standard oil field beam unit to drive a rod pump in the initial deliquification period, but late-life deliquification seems to do better with pneumatic or hydraulic surface equipment that can provide slower strokes and can take longer to reverse rod direction. It is not clear why the gentler pump action works better, but it is reasonably clear that it does.

13.3.2   Wellsite and gathering plan

What operators get wrong more often than anything else happens to be the thing that they absolutely must not get wrong. If you build facilities that are rated for full reservoir pressure and temperature, then every decision you make over the life of the field can start with "what does the reservoir need me to do if I want to maximize my return on investment." In other words, you can consider the needs of the reservoir and long-term profitability. It is preferable to say that you are looking toward the reservoir.

If you build facilities for late-life operations (and low pressure and/or low temperature), then you have to protect the artifacts from reservoir pressure from Day 1. This requires chokes, emergency shut down valves, and extensive procedures. I like to say that this approach is looking toward the artifacts and ignoring the needs of the reservoir. If you install a piece of equipment that is not rated for full reservoir pressure, you have limited the ultimate recovery from the field and this can be significant.

If initial reservoir pressure is like the well in Fig. 13.2 (i.e., 1820 psia), then adherence to always looking toward the reservoir would require ASME B16.5 Class 900 equipment (MAWP of 1960 psig at 100°F), which significantly limits the available equipment. Since the facilities will only see full reservoir pressure after a shut-in of several days, it is reasonable to slightly shave this requirement and install Class 600 equipment (MAWP of 1440 psig at 100°F) and emergency shut down valves off the wellhead tubing and casing line to shut if the pressure approaches the MAWP of the equipment. These valves can likely be removed within 3 years as reservoir pressure declines rapidly.

Wellsite equipment should be Class 600, separators should be vertical and have internal blowcases (both of which are designed to accommodate late-life, low-pressure operation), piping out of the separator should include a compressor 3-ball manifold with a check valve on the bypass, and provisions should be made for powering late-life compression and deliquification.

A producer who manages and controls the gathering system that the wells produce into will always do better than the same producer flowing to a third party.[5] There are two reasons for this: first the economic analysis for system modifications will use gas sales prices instead of gathering fees and second the incentives for steady pressures in the system are obvious and tangible to the well operator and they are not quite so immediate to a third party. The first reason can be very significant. A project that will add 3 MMscf/day of gas selling for $6/Mscf has a lot more attraction than a project that will add $1050/day at $0.35/Mscf—the payouts are much shorter and their ranking on net present value will be much higher.

The second point is really an alignment of the well operator's goals with the gathering system. In an ideal world the field techs who operate the wells would also be responsible for pigging and have some involvement in compressor operations. This alignment helps the tech identify when line pressure at a particular well is creeping up due to liquid in the gas line and allows the operator to organize pigging the line that is starting to have a problem. Pigging lines is hard and dirty work that no one likes. Third-party gatherers will generally run pigs on a rigid schedule or never run pigs.

These points are valid only when the gathering system is operated by the same people as the wells. In situations where a large company has a separate division that operates gathering systems, the benefits are completely lost and performance is typically worse than a true third party.

If the well operator also operates the gas-gathering system, then prior to field development a staged gathering-system plan should be developed. If the wells produce into a third-party gas-gathering system, then the producer needs to develop a compression strategy.

Initial system layout

As a field is developed, the only thing that is certain is that the production forecasts will be wrong. Occasionally the entire field production profile can be estimated reasonably closely, but any particular well is subject to having significantly more or less production rate than forecast. The wells cannot produce until they have a route to market. You cannot know what a well will produce until it has passed its initial deliquification period.

The only reasonable approach to initial system layout in an unconventional field is to assume that every well will have the "average" production rate for both water and gas and design the piping to accommodate those rates. It is certain that most wells will not flow at this rate, but that cannot be helped. What is also certain is that late in the field's life, significant quantities of liquid will flow up the wellbore as water vapor and much of this vapor will subsequently condense in the gathering system. Every line needs a technique to remove condensation. Simple "pigging valves" are effective on lines that are 6 in. and smaller. Larger lines require more elaborate pigging facilities. In any line, removing the water will improve the efficiency of the flow and will reduce the horsepower that must be deployed to overcome parasitic pressure drops. In steel lines, removing the standing water will prevent the formation of corrosion cells and can significantly increase the life of the system.

Water strategy

Virtually, all wells will produce some amount of water during their entire lifecycle. Some produced water is quite suitable for surface discharge into rivers and streams or for irrigation. This is an environmental and regulatory consideration, but if the water is suitable for surface discharge, then it is better not to aggregate wellsite water but to discharge it as close to the wellsite as practical. Many small introductions of foreign water to a stream will have a much smaller impact on the stream's biology than pumping an aggregated volume at one point.

Most produced water is not suitable for surface discharge and must be disposed of. Disposal options are outside the scope of this document, and they must be developed in consultation with environmental, legal, and engineering experts. For any disposal option, the water must be transported to the disposal facility. The trade-off that must be considered within a gathering plan is "do I spend capital dollars to aggregate wellsite water or do I spend expense dollars to haul it?" The answer to this question is never simple. One approach that seems to minimize the difficulties is to install "transfer points" and haul the water from the wellhead to these points while piping it to disposal facilities. This technique allows efficient use of water hauling while reducing capital.

Government regulators are beginning to "strongly encourage" the use of water-gathering systems instead of water hauling. Water trucks have a significant negative impact on roads, create very real risks to the public, and are very fuel-inefficient. Although water-gathering lines typically leave a more or less permanent mark on the landscape, the mark has a lower total impact than ongoing water-truck traffic.

13.4 Pressure targets with time

Early-life wells are reasonably easy to produce. They may not need any deliquification help, and reservoir pressures are high enough to flow into moderately high line pressure. There is a point in the life of every well that it changes from easy to very difficult.

A pressure analysis over the life of the well can show that in the early days the well can reach mainline pressure with a fairly small gathering system and a central delivery point designed for 10 compression ratios. As the well approaches Point A on the Langmuir Isotherm (Fig. 13.2), you may need another stage of compression to get 40 compression ratios. Later in the well's life, you could easily need five stages of compression to get from required wellhead pressure to mainline pressure.

Since pipe loses efficiency as pressures decline, it is generally suboptimum to try to achieve very low wellhead pressures from distant central compressor stations. One approach that has been very effective in several operations is to build an initial gathering system for all the wells being average with the piping funneling toward a single compressor station. Produce the field for a year or so and develop a set of debottlenecking projects to try to equalize the wellhead pressures. As those debottlenecking projects are designed, pick sites for straddle or booster compression stations. As wellhead pressures begin approaching Point A as shown in Fig. 13.2, start designing the straddle sites. After the straddle sites are in service for a year or so, begin implementing your late-life strategy. This will be a combination of wellhead compression and (possibly) some sort of mechanical deliquification.

13.4.1 Wellbore

For any operation that anticipates ever operating at low pressure, the operator should look at each component of the downhole equipment and ask, "Why is this here, and is this the best equipment/size to do that job?" That includes everything from X-nipples to the tubing itself. For example, if your primary deliquification method is going to be evaporation, then any tubing at all will increase your velocities up the wellbore and unnecessarily add to the pressure drop due to friction.

Horizontal and highly deviated wells are becoming more common every year. Success with unlined horizontal laterals in CBM has been limited due to the frequency of lateral collapse caused by the weak mechanical strength of the coal. It can be demonstrated that the hoop strength of a coal bore decreases with increasing bore diameter, so unlined horizontal laterals should be as small a bore-hole as production velocities will allow.

Removing liquid from low-pressure horizontal wells is a serious problem that has not been adequately solved. Some fields have had good success with orienting the lateral up dip and normal to the face cleats. This allows pumps or foamers to be set in the horizontal portion of the wellbore without imposing a large hydrostatic force on the formation.

Some very encouraging work is being done in treating the lateral as a high-perm reservoir-section and transporting gasses and liquids to a pump in the vertical section through a packer set in the horizontal section. There are commercial products using this techniques with success, and there is a very good chance that this technique will be viewed as a "solution" within the next few years.

13.4.2 Flow lines

It can be said with a great deal of confidence that a single-phase flow line will be more efficient than a multiphase line. Consequently a separate flow line for each potential flow stream is nearly always a very good idea. A well that is planned for mechanical pumping should have a flow line from the tubing and a separate line from the casing. They may both end at a wellsite separator or may diverge, but they should not be joined except in a piece of equipment which will separate them permanently.

13.4.3 Separation

CBM operations generally call for a two-phase separator to remove liquid from the gas, but typically do not need to further separate the liquid into oil and water. For a small added cost, it is a good idea to provide two inlet nozzles to the separator to allow the tubing flow line to enter separately from the casing flow line. This minimizes the mixing of the two streams prior to the separator and will generally result in better liquid removal.

In anticipation of very low separator pressures late in the well's life, it is also a good idea to anticipate forcing the water out of the separator. If your late-life compression strategy includes wellhead compression, then a separator with an integral blowcase is a good choice. The blowcase will accumulate liquids, and periodically the compressor-discharge pressure is used to blow out the liquid. In the absence of wellhead compression a pump can be used in conjunction with a dump valve to pump out a chamber. The pump-chamber can physically be a blowcase, so the initial wellsite separator can easily be the same vessel for either strategy.

Shale operations nearly always need to consider both water and hydrocarbon liquids, so it is very common to see three-phase separators on shale wells (Fig. 13.7). Since paraffin is quite common in shale wells, these units often fall into the heater/treater category (i.e., the liquid section of the separator is directly fired instead of the entire vessel being indirectly fired).

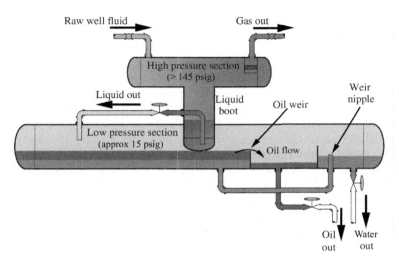

Figure 13.7 High-low pressure (HLP) separator.[5]

Some operators use a vertical 2-phase separator in front of the vessel to remove most of the liquid from the gas and dump the extracted liquid into the low-pressure section on a vessel as shown in Fig. 13.7. This approach allows the high-pressure section to only see a fraction of the wellbore liquids to reduce the frequency that the high-pressure dump valve operates (upon adding one well this vertical preprocessor vessel reduced the cycle frequency on the high-pressure dump from 6/min to 3/day and significantly reduced failure frequency).

13.4.4 Compression

Transportation lines typically have nominal operating pressure on the order of 1000 psig. Late-life wells require flowing wellhead pressures under 10 psig. Transporting these two required pressures from one to the other requires compression. A combination of machines to provide compression ratios of above 70 and sometimes significantly higher (i.e., from 15 inHg to 1000 psig is 144 compression ratios) indicates a range of technologies.

For example a two-stage reciprocating compressor with a suction pressure of 100 psig and a discharge pressure of 1000 psig is a very efficient piece of equipment. Efficiency drops considerably if you lower the suction pressure to 40 psig (increasing compression ratios from 9.3 to 20.6 at sea level), and requires a three-stage compressor. Multistage reciprocating compressors work best within a very narrow suction-pressure range, so the variability caused by water sloshing in the gathering system or liquid-level changes in wellbores will create problems for a reciprocating compressor.

An effective strategy is "staged compression." With this strategy, you start your field production with central compressor station suction at a fixed value. For example, you can set the station up for 100 psig suction (pressure upstream of the site inlet pressure-control depends on the needs of the reservoir, but the pressure inside the site is fixed). This station can be expected to run with this suction pressure for the entire life of the field. At some point the wells will need lower pressures and this can often be provided with single-stage reciprocating compressors located at strategic "straddle compressor sites" that were described in the gathering strategy. The straddle sites can be designed for an inlet pressure of around 40 psig and a discharge pressure consistent with 100 psig inlet to the central stations. Later in the life, very low wellhead pressures can be provided with flooded-screw compressors that handle varying suction pressure very well and work effectively with line pressures consistent with the straddle site design suction pressure.

Since flooded-screw compressors can develop 20 compression ratios and reciprocating compressors should be designed with about 4 compression ratios per stage, this staged-compression scenario results in obtaining 1280 compression ratios (Eq. (13.3)) or raising the pressure of the gas from 1 psia to above 1000 psia (there are always losses in flow lines, coolers, and compressor valves) in most gathering scenarios.

$$(20)_{\text{screw compressor}} \cdot (4)_{\text{single stage}} \cdot (4)^2_{\text{2-stage}} = 1280 \text{ ratios} \tag{13.3}$$

13.4.5 Deliquification

Deliquification of low-pressure wells uses the techniques described in the rest of this book. The key to success is understanding the minimum Net Positive Suction Head Required (NPDH-r) for the pump you want to use. For example, a hydraulic jet pump requires approximately 300 psig pump intake pressure to prevent cavitation. Consequently, whereas an early-life jet pump can be effective, after Point A in Fig. 13.2 jet pumps do not have a place in organic reservoir operations.

One deliquification technique that is unique to a very low-pressure operation is "evaporation." Refer to Figure 20.3 in the *GPSA Field Data Book* to see how much liquid water will evaporate at low pressures and moderately high temperatures. This will often be higher than the liquid inflow rate and it is possible to rely on evaporation to satisfy all the well's deliquification requirements. Phase-change scale issues that are discussed in other chapters are an important consideration, but these issues vary from field to field.

References

1. Simpson DA, Lea JF, Cox JC. Coal bed methane production, SPE 80900. In: *Presented at SPE production and operations symposium*; March, 2003.
2. Simpson DA, Kutas M. Producing coalbed methane at high rates and low pressures, SPE 84509. In: *Presented at SPE annual technical conference and exposition*; October, 2003.
3. Bradley HB. *Petroleum engineering handbook.* 3rd ed. Society of Petroleum Engineers; 1992.
4. Stephens MM. *Natural gas engineering.* 2nd ed. Mineral Industries Extension Services, School of Mineral Industries, The Pennsylvania State College; 1948.
5. Simpson DA. *Practical onshore gas field engineering.* Gulf Professional Publishing; 2017.

Production automation

14

Cleon Dunham is president of Oilfield Automation Consulting which was founded in 2000. He is also president and CEO of the Artificial Lift R&D Council (ALRDC) which was founded in 2001. Cleon has 30 years of experience in oilfield automation and optimization. He worked for 36+ years in Shell Oil Company (United States) and Shell International Exploration Production, with primary focus on production operations, engineering and technology, and gas-lift operations. He has designed and implemented automation systems for gas-lift systems, sucker rod pumping systems, electrical submersible pumping systems, water flood injection systems, and CO_2 and steam injection systems. He has been a coordinator of oilfield automation and artificial lift for Shell's worldwide producing operations.

Dunham provides consulting services to evaluate existing oilfield automation systems and to plan for new automation systems. Gas-lift training and awareness classes have been provided by him in Argentina, Bolivia, Singapore, Kuala Lumpur, and Vietnam. Currently, Cleon helps in defining and designing new oilfield automation products, and in testing and qualifying new products. He is also an author and coauthor of numerous automation and artificial lift articles.

Dunham holds a bachelor of engineering degree from Cornell University. He is awarded four patents and numerous awards from the Society of Petroleum Engineering, American Society of Mechanical Engineering, and others for teaching, making various presentations, leading conferences, and writing journal articles. He has many related recent projects and is active in numerous organizations.

14.1 Introduction

Many gas fields, oil fields, and fields with unconventional production have large numbers of wells, which are located in remote locations. Many of the gas wells experience liquid loading and require deliquification to obtain desired gas production rates and ultimate recoveries. Many of the oil fields and fields with unconventional production also require careful operation and surveillance. The combination of these conditions can make manual surveillance of gas wells, oil wells, and wells with unconventional production difficult, control challenging, and optimization almost impossible. For these and other reasons, many companies are turning to production automation systems to improve the management of their gas well, oil well, and unconventional production operations.

Gas Well Deliquification. DOI: https://doi.org/10.1016/B978-0-12-815897-5.00014-7

As used here, the term "unconventional production" primarily refers to production from highly deviated and/or horizontal wells. Often, these wells are completed in shale formations and the horizontal sections are several thousand feet long. They may be drilled in a toe up, toe down, or undulating manner. Often, this configuration can lead to production of slugs of liquid and gas which can make artificial lift and production management challenging.

Surveillance. Production automation systems are used to monitor gas well, oil well, and unconventional production. This includes measuring gas and liquid production rates, gathering related information such as pressures and temperatures and monitoring the performance of artificial lift equipment. These measurements and monitoring are used to determine gas and liquid production volumes and provide the surveillance and problem detection needed for problems to be addressed.

Control. Automation systems are used for control. This is particularly pertinent when artificial lift systems are used for gas well, and management of production from oil and unconventional wells. Control systems are important, some would say essential, to use plunger lift systems, pumping systems, gas-lift systems, chemical injection systems, etc. It can be virtually impossible to manually perform the necessary control in the way needed and with the timeliness required.

Optimization. Automation systems are used for optimization. The goal of gas, oil, and unconventional well optimization is to maximize both current production rates and ultimate gas and oil recoveries while minimizing capital, operating, and maintenance costs. The question is: how can the minimum amount of energy and manual effort be expended to produce the maximum amount of gas and oil, on a sustained basis?

Production automation systems are being used by many companies, but in many cases the companies acknowledge that they are not gaining the benefits they expect. Some systems are not sufficiently reliable, some are underutilized, and some are too difficult to understand. In some cases, personnel are not properly trained to use or support them.

Another important factor is that companies are becoming more sophisticated in their selection of artificial lift systems to deliquefy their gas wells and produce their oil and unconventional wells. Some wells are better served by plungers, chemical systems, wellhead compression, pumping systems, gas-lift, etc. Each of these systems requires different surveillance, control, and optimization methods. Production automation systems are being called upon for surveillance, control, and optimization of a range of artificial lift systems. This must be in a way that is understandable and usable. There cannot be separate systems for each type of artificial lift; they must be integrated into one approach.

This chapter covers automation equipment, general applications that are available in most production automation systems, special applications that are designed specifically for each type of artificial lift, some issues that must be considered when planning a production automation system, and finally some case histories.

Note that production automation systems are sometimes referred to as SCADA (Supervisory Control and Data Acquisition) systems. In reality, true production automation systems contain much more than SCADA capabilities. They also contain information analysis, logic to diagnose problems, and production optimization capabilities.

This chapter is long, but it only "scratches the surface" of the overall topic of production automation for gas, oil, and unconventional production operations. If someone is interested in actually pursuing an automation project or enhancing an existing automation system, they should contact an automation service company or a consultant.

14.2 Brief history

Production automation systems, in one form or another, have been used since the 1950s. There was much development of systems for surveillance, control, and optimization of selected forms of artificial lift in the 1970—90s. However, significant advancements in automation of gas well, oil well, and unconventional operations have occurred only in the past few years. A brief history of this development is presented in this section.

14.2.1   Wellsite intelligence

Since the mid-1980s, dramatic developments have occurred in the area of wellsite intelligence. The development of microprocessor technology has made it practical to place devices at the wellsite with capabilities similar to those of personal computers (PCs). This has opened up opportunities for real-time monitoring, control, and optimization of artificial lift systems. Numerous application-specific field devices have been developed which allow for autonomous control and optimization of artificial lift systems. For instance, in the area of plunger lift, wellsite intelligence has evolved from simple time-cycle—based control to sophisticated condition-based control logic, to self-tuning algorithms which minimize the need for direct intervention from personnel. This has improved the viability of this lift method in remote operations. Similarly, pumping systems can now be equipped with motor controllers or variable frequency drives (VFDs) which can start, stop, speed up, or slow down the pump based on the real-time condition of the well. These devices can monitor the state of the well using both surface and downhole measurements, such as pressure, temperature, load, current draw, and other parameters. With such data the controllers are able to detect and diagnose abnormal operating conditions and take corrective action, thus protecting equipment from damage and possible failure.

14.2.2   Desktop intelligence

In the early days of automation, host systems were able to provide fairly simple functionality such as trending of data and basic control capabilities. Most systems were custom-built for the end user at significant expense. Over time, systems were developed with ever-increasing levels of sophistication. Such systems enabled operators to detect, diagnose, and address the problems by changing operational parameters or even redesigning artificial lift equipment to better suit field conditions. In addition, software was developed which allowed operators to manage by exception. This meant that instead of reviewing every well on a daily basis, operators could focus their efforts on those wells where there was a known problem or opportunity for improvement, greatly improving operational efficiency. With the development of PC technology, these concepts were incorporated in off-the-shelf commercial software which could be run on Windows-based PCs and be deployed with minimal if any need for customization. As a result, many operators who could not previously justify the expense of implementing an automation project were able to gain access to this technology and put it to use in their fields. Today, host systems continue to evolve. Where systems have traditionally focused on helping producers improve operational efficiency, new systems are being developed to help engineers maximize the value of the asset. Such systems are intended to link the reservoir, to the wells, to the gathering system, to the facility, to the sales point. By utilizing real-time enabled engineering tools, engineers can use such systems to uncover hidden performance trends and better manage the asset.

14.2.3   Communications

Early automation systems typically used hard-wired or telephone-based connections between remote terminal units (RTUs) located at the wellhead and host systems at a central location. In many cases, this came at considerable expense and posed serious logistical challenges. As systems developed, operators migrated to other communication technologies such as microwave, spread-spectrum, and licensed radios. This solved many of the logistical challenges faced in the field, but also required considerable capital investment. Over time, other communications options have opened up. These include cellular (CDPD/CDMA), satellite, and fiber optic. Each of these has proven to be a key enabler. Cellular has been particularly useful in automating remote fields with little infrastructure, due to its ease of installation, low cost, and broad coverage. Unfortunately, not all locations in the world have cellular coverage. For such areas, satellite communications have proven to be a useful tool, but also bear a considerable price. In assets where operators wish to transmit extremely large quantities of data and desire maximum bandwidth, fiber optic is proving to be another useful, yet expensive tool. While there are pros and cons to each of the communications options, with so many options available, it is now possible to find a fit-for-purpose communications solution for almost any application in the world.

14.2.4   System architecture

Early automation systems typically consisted of a number of simple RTUs deployed at wellsites and connected via hardwire to a central host system on a mini computer. These were custom-built installations that required considerable up-front development and support. Over time, technologies have evolved that have replaced the various components of these systems and offer a variety of options to operators. In many cases, operators deploy off-the-shelf RTUs which communicate via licensed or spread-spectrum radio to a host system on a Windows server. Users then interface with this system using client software or via company intranet using an Internet browser. In many corporate environments, operators deploy large-scale SCADA systems that utilize RTUs and programmable logic controllers (PLCs) that communicate directly with a distributed control system (DCS) and archive data in historians for future retrieval. Such systems may even communicate with other enterprise data sources to gain access to well testing, accounting, or other data. In many gas deliquification or oil-field projects—where wellsites are remote, minimal infrastructure is in place and it is important to minimize cost—operators are increasingly choosing to go another route. In such applications, web-hosting services are proving to be a popular option. Web-hosting allows operators to simply connect a cellular radio or satellite transmitter to the wellsite RTU and transmit data to a third-party service. Operators are then able to view their data in an Internet browser with a preconfigured interface. The entire automation infrastructure is managed by the third-party web-hosting service and operators pay a low monthly fee. While this is not a practical solution in every part of the world or in every corporate environment, it is proving to be a popular choice for operators of gas, oil-field production, and unconventional projects, particularly in North America.

14.3   Automation equipment

Automation equipment consists of the hardware and software used to implement production automation systems. There are many components and suppliers of these components. The purpose of this section is not to evaluate or judge the various brands or suppliers, but to provide information of the equipment that is available and some insights into what works well.

14.3.1   Instrumentation

The core components of any production automation system are the instruments used to measure gas and/or oil production variables of pressure, temperature, flow rate, etc. In addition, special instruments are needed to measure variables required for some artificial lift systems. For example, the load on the polished rod and the position of the beam are required measurements for sucker rod pumping systems. These special instruments are discussed in the appropriate sections.

Several technologies are available including analog current instruments, direct current voltage instruments, digital instruments, instruments designed to work with Foundation Fieldbus, and others. In general, all transmitters that are obtained from reputable companies are rugged and reliable, and all are reasonably priced.

Figure 14.1 Pressure transmitter.

Fig. 14.1 shows a typical pressure transmitter that measures gauge pressure and transmits a 4−20 mA (milli-amp), 1−5 Vdc (volt direct current), or digital output (DO) signal. It is used to measure tubing (production pressure), casing, line, separator pressure, etc.

The signal is transmitted to an RTU or a PLC (see Section 14.3.4) where it is converted into engineering units of psi, kPa, °F, °C, MCF/day, M^3/day, etc.

Fig. 14.2 shows a typical differential pressure transmitter. It is used to measure the pressure drop across an orifice meter or similar device. For gas well production, this is often used to measure the gas flow rate.

Fig. 14.3 shows a typical temperature transmitter. It is used to measure the temperature of the produced gas or oil. For gas well production the temperature of the gas must be known to accurately calculate the flow rate.

Fig. 14.4 shows a typical multivariable transmitter that measures pressure, differential pressure, and temperature with one device. This device may reduce overall cost since only one device must be installed to perform the functions of three separate measurements.

Transmitters can use analog signals, voltage signals, digital signals, Foundation Fieldbus, or other methods (see Table 14.1). Analog current and digital are the most common.

Analog and voltage instruments require a pair of wires between each instrument and the RTU or PLC. Digital and Fieldbus instruments can be "daisy chained" with several instruments on one cable. Some systems can use wireless communications between the instrument and the RTU or PLC. Since there are many choices, a trained instrument engineer should evaluate the options and recommend the right system for each application.

Figure 14.2 Differential pressure transmitter.

Figure 14.3 Temperature transmitter.

14.3.2   Electronic flow measurement

Often the custody transfer point in gas deliquification applications is at or near the wellhead. For this reason, there is an additional requirement in these applications to provide custody transfer quality gas measurement. To address this need the industry has adopted a measurement standard from the American Petroleum Institute, called API MPMS 21.1. This standard sets forth requirements for the measurement of gas flow rates as well as storage and transmission of such data. A range of RTUs exist in industry which are designed, tested, and certified to comply with this standard. These devices are commonly referred to as electronic flow measurement (EFM) devices.

Figure 14.4 Multivariable transmitter for P, DP, and T.

Table 14.1 Types of signal outputs from transmitters

Type of signal output	Description
Analog voltage	The output voltage is a simple (usually linear) function of the measurement
Analog current	Often called a transmitter. A current (4–20 mA or any other analog current output) is imposed on the output circuit proportional to the measurement. Feedback is used to provide the appropriate current regardless of line noise, impedance, etc. This output is useful when sending signals to long distances
RS-232/RS-485	The output of the transmitter sends out a serial communications signal
Parallel	A standard digital output (DO) protocol (parallel) such as a printer port, Centronics port, and IEEE 488
HART Protocol	HART (Highway Addressable Remote Transmitter) is a method of transmitting data via frequency shift keying on top of the 4–20 mA process signal to allow remote configuration and diagnostic checking. HART is a registered trademark of the HART Communication Foundation
PROFIBUS	PROFIBUS is an open Fieldbus standard for use in manufacturing and building automation, as well as process control

(Continued)

Table 14.1 (Continued)

Type of signal output	Description
DeviceNet	Utilizing CAN protocol, DeviceNet is a network designed to connect industrial devices such as limit switches, photoelectric cells, valve manifolds, motor starters, drives, and operator displays to PLCs and PCs
Foundation Fieldbus	Fieldbus or Foundation Fieldbus is a generic term used to describe a common communications protocol for control systems and/or field instruments
Ethernet	A very common method of networking computers in a local area network (LAN). Ethernet will handle about 10,000,000 bits-per-second and can be used with almost any computer
Analog frequency or modulated frequency	The output signal is encoded via amplitude modulation (AM), frequency modulation (FM), or some other modulation scheme such as sine wave or pulse train, but the signal is still analog in nature
Special digital (TTL)	Any DO other than standard serial or parallel signals. Simple TTL logic signals are an example
Switch/alarm	The "output" is a change in state of a switch or alarm
Other	Other unlisted, specialized, or proprietary outputs

System description

An EFM system consists of three major elements: *primary*, *secondary*, and *tertiary* devices. Primary devices refer to the meter itself. These could be orifice, turbine, venturi, or any other form of gas flow meter. Secondary devices include electromechanical transducers which convert the physical inputs of the meter (i.e., pressure, temperature, differential pressure) into an electrical signal. Tertiary devices refer to the flow computer which takes the electrical inputs from the secondary devices and uses them to calculate a flow rate.

Algorithms

In addition to differential pressure metering algorithms such as those defined in A. G.A. Report Numbers 3, 5, 7, and 8, EFM devices must also perform algorithms which account for the effects of sampling and calculation frequency during periods of fluctuating flow.

Sampling frequency

In general an EFM device must sample data from end devices once every second. However, there are exceptions. If the RTU collects data at a frequency that is greater than once per second, these inputs may be averaged using techniques specified in API MPMS 21.1. Also, if sampling frequency is slower than once per second, these values may be used if it can be demonstrated that the difference in uncertainty between the slower sampling rate and 1-second sampling rate is no more than 0.05%.

Data availability

EFM devices are required to collect and retain a minimum amount of data to ensure that gas flow rate calculations are performed accurately, and to provide an audit trail of system operation and quantity determinations. These devices are generally expected to retain hourly averages of all key values as well as the associated configuration parameters and totalized values for each gauge-off period. Typically a gauge-off period is 1 day.

Audit and reporting requirements

EFM devices provide an audit trail in the form of daily and hourly quantity transaction records, algorithm identification, configuration logs, event logs, corrected quantity transaction records, and test records for the metering equipment. This audit trail provides support for the current and prior quantities reported on the measurement and quantity statements as well as the ability to make reasonable adjustments when gas measurement equipment has stopped working, is deemed to be out of calibration, or in cases where parameters were incorrectly entered into the RTU.

Equipment installation

All EFM equipment is required to be installed in a manner that is consistent with the practices described in API MPMS 21.1. Affected equipment includes the transducers (or transmitters), gauge lines, RTUs, communications, peripherals, and cabling.

Equipment calibration/verification

An EFM system is required to be calibrated such that the system as a whole will provide no more than $\pm 1\%$ uncertainty over the expected range of temperatures and pressures for the installation. EFM components requiring calibration/verification include static pressure transmitters, differential pressure transmitters, temperature transmitters, pulse generators and counters, online analyzers, and densitometers. These calibrations must be performed once per quarter (i.e., at least once each 3 months).

Security

All EFM systems are expected to provide specific safeguards pertaining to access, integrity of logged data, algorithm protection, protection of original data, memory protection, and error checking.

14.3.3 Controls

Automatically controlled valves and accessories

Automatic control valves are used in a variety of gas deliquification and oil production applications including plunger lift, gas-lift, hydraulic lift, and well testing. These devices are generally classified as either fluid or electrically operated. In generally, fluid-controlled valves are either diaphragm operators or fluid cylinders. In automation applications, these devices are equipped with transducers or related equipment which allows them to accept the various inputs and protocols as described in Table 14.1.

Fluid-controlled valves

In generally, fluid cylinder operators are used in valves requiring a 90-degree bend, while diaphragm operators are used in valves that have angle, butterfly, globe, or Saunders-style valve bodies.

A variety of fluids may be used to actuate fluid-controlled valves. In generally, natural gas is used in oil-field applications. However, other fluids may be used in cases where a suitable natural gas source is not available. These include compressed air, nitrogen, or hydraulic fluid.

Fig. 14.5 shows a typical on-off-style globe valve. These are commonly used as dump valves on separators or flowline valves in plunger lift applications.

Fig. 14.6 shows a typical throttling-style globe valve. These are used in applications requiring variable control of throughput such as gas-lift injection.

Fig. 14.7 shows a typical electro-pneumatic transducer. These are used in conjunction with automatic control valves and use an analog input (AI) (generally 4−20 mA) and convert this to a proportional pneumatic pressure output in order to adjust the position of the valve.

Figure 14.5 On−off-style globe valve.

Figure 14.6 Throttling-style globe valve.

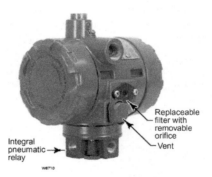

Figure 14.7 Electro-pneumatic transducer.

Fig. 14.8 shows a globe valve equipped with an electro-pneumatic transducer for actuation.

Electrically controlled valves

Two general forms of electric operators are commonly used in oil-field applications. These are generally classified as electric-solenoid and electric motor operators.

Figure 14.8 Globe valve and transducer.

Electric-solenoid operators are used to adjust the longitudinal motion of a valve stem and are generally limited to valves of 2 in. diameter and smaller. Electric motor operators are used in a variety of valve types, but generally require additional accessories to be installed such as torque limiters or limit switches to prevent damage to the unit. In addition, electric motor actuators require the use of a rack-and-pinion assembly to convert the motor's rotary movement to longitudinal displacement.

Production safety controls

A variety of devices are used in oil-field applications to ensure that equipment operates under fail-safe conditions. These devices are commonly referred to as production safety controls. Typical production safety controls include high-pressure/low-pressure safety shut-in valves, excess flow valves, pressure relief valves, pressure and temperature switches, and pump-off controls.

Figure 14.9 Artificial lift motor controller.

Motor controllers

Motor controllers are devices which regulate the operation of an electric motor. In artificial lift applications, motor controllers generally refer to those devices used in conjunction with switchboards or VFDs to control the operation of the prime mover. Motor controllers often include a manual or automatic means for starting and stopping the motor, selecting forward or reverse rotation, speeding up or slowing down, and controlling other operational parameters. In addition, motor controllers can provide protection for the artificial lift system by regulating or limiting the torque, and protecting against overloads and faults. Many motor controllers contain additional capabilities such as data collection and data logging as well as application-specific control logic.

Fig. 14.9 shows a typical motor controller for electric submersible pumping (ESP) applications. This device receives and displays data from downhole gauges and surface electrical parameters. This data can then be used to adjust the operation of the pump according to changing conditions.

Switchboards

The switchboard is basically a motor control device. The switchboards range in complexity from a simple motor starter/disconnect switch to an extremely sophisticated monitoring/control device.

There are two major construction types: electromechanical and solid state. Electromechanical construction switchboards provide basic over current and under current protection to the artificial lift system. Monitoring these features allows for protection of the artificial lift system from damage caused by conditions such as pump-off, gas lock, tubing leaks, and shut-off operations. The solid-state switchboards incorporate a solid-state motor controller that allows more elaborate and accurate protection from a much greater list of potential problems. In addition, most solid-state controllers incorporate data logging functions.

A valuable switchboard option, particularly in ESP operations, is the recording ammeter. Its function is to record, on a circular strip chart, the input amperage to the prime mover. The ammeter chart record shows whether the unit is performing as designed or whether abnormal operating conditions exist. Abnormal conditions

Figure 14.10 Motor switchboard.

can occur when a well's inflow performance does not match correctly with pump capability or when electric power is of poor quality. Abnormal conditions indicated on the ammeter chart record are primary line voltage fluctuations, low current, high current, and erratic current. Ammeter readings can also be sent to an RTU and on to a host automation system for analysis.

Fig. 14.10 is an example of a motor switchboard for an ESP.

Variable frequency drives

A VFD changes the capacity of the artificial lift system by varying the motor speed. By changing the power frequency supplied to the motor and thus the motor RPM, the capacity of the pump is changed in a linear relationship. Thus well production can be optimized by balancing flow performance with pump performance. This applies to both long-range reservoir changes and short-term transients such as those associated with high-GOR (gas—oil ratio) wells. This may eliminate the need to change the capacity of a pump to match changing well conditions or it may mean improved run life by preventing cycling of the system. This capability is also useful in determining the productivity of new wells by allowing evaluation and measurement of pressure and production values over a range of drawdown rates. The change in frequency can be made manually or automatically. The VFD can automatically adjust the operating frequency to maintain a target pressure, flow rate, current or other set points when operating in a "closed loop" mode.

Fig. 14.11 is an example of a VFD drive for progressing cavity pump applications. This device receives both electrical and other production data and can use this information to change pump speed to optimize performance or prevent damage to the artificial lift system.

Figure 14.11 Variable frequency drive.

14.3.4 Remote terminal units and programmable logic controllers

Remote terminal units

An RTU is an electronic device utilizing a microprocessor, which links objects in the physical world with an automation system. This is accomplished by transmitting telemetry data to the system and/or changing the physical state of connected objects based on control messages received from the automation system. RTUs share many common characteristics with PLCs, but in general, tend to be designed to handle a smaller number of points and will often contain application-specific control logic.

One way of looking at an RTU is as a small computer sitting at the wellsite that is ruggedized to handle field conditions and has input and output capabilities for communicating with the field equipment. Many RTUs have been customized with application-specific control logic to allow them to perform specific functions in the field. Examples of these include rod pump controllers (RPCs), plunger lift controllers, data loggers, and a variety of other application-specific devices.

An RTU comprises several major components. These include: (1) a communications interface, (2) a microprocessor, (3) nonvolatile memory, (4) environmental sensors, (5) override sensors, and (6) a bus, which is used to communicate with devices or interface boards. This bus is commonly called a field bus or device bus.

A variety of standards (or protocols) are used to communicate with RTUs. These include both generic and proprietary protocols. Perhaps the most widely used generic protocol is MODBUS. Others include ODBC, OPC, and ISO Controller Area Network (ISO 11898). Examples of proprietary protocols include Weatherford's Baker 8800 protocol and Allen-Bradley's data Highway.

Figure 14.12 Remote terminal unit.

RTUs can have a number of different types of interface boards. These interface boards can be either digital or analog and can come up with inputs only, outputs only, or a combination of the two. These main types of interface boards are often abbreviated as AI, AO (analog output), DI (digital input), or DO. Interface boards are connected to physical objects using wires.

RTUs often have application-specific logic programmed into firmware and/or software. This control logic, sometimes referred to as "wellsite intelligence," allows for autonomous control based on changing conditions, without the need for instructions from a human operator or host system. Such control logic is useful for executing functionality that is data-intensive, time-sensitive, or is required for fail-safe operation of equipment. Such functions would not generally be practical to carry out remotely from a host system. Examples of RTU-based control logic include proportional integral derivitive controller (PID) loops which measure and maintain a given flow rate by adjusting a valve's position, gas measurement, pump-off control, data logging, and a variety of others.

Fig. 14.12 is an example of a generic RTU. Typical of many RTUs, this device contains a microprocessor, multiple communication interfaces, support for eight AIs, two AOs, eight DIs, and eight DOs. This device is typical for a stand-alone, single well control application. Also pictured are the associated instrumentation and cabling, solar power array, battery backup, and radio.

Programmable logic controllers

A PLC is a digital computer that is specifically designed for the automation of industrial processes. PLCs have the same basic components as RTUs, yet differ in both form and function. While they share many characteristics with RTUs, PLCs tend to be more scalable, interface with more end devices, and are less likely to contain customized control logic. Typical oil-field applications for PLCs include tasks such as automatic well testing, scanning multiple end devices, and other process control tasks. In a typical oil-field automation system, PLCs may be used to interface with a number of RTUs to collect data or adjust set points in the controllers (operating the RTUs in a master—slave relationship) and, in turn, communicate that data to a DCS.

PLCs originated in the automotive industry in the late 1970s. They were a replacement for the relay logic used to control machinery. Their advantage over the relay logic was that they were programmable and that the program could be changed relatively easily; relay logic is hard-wired, takes up significant space, and is not easily changed. The ladder logic language, which is still popular in PLCs, is the same as the ladder logic drawings used for relay logic wiring and hence is well understood by electricians.

PLCs were originally very large, expensive, and suitable only for large manufacturing plants. They were capable of only binary logic (no analog) and were aimed at machine control. Early PLCs had minimal communications functionality beyond providing a port to plug in the programming terminal. The operator interface was mostly provided by switches and lights hard-wired to PLC I/O. Over time, PLCs were developed with communications ports to allow them to communicate to one another and to provide computer-based operator interfaces, giving rise to the MODBUS protocol. Today, both PLCs and RTUs have evolved to a point where, in many cases, the lines have blurred between the different devices, making the distinctions less meaningful.

Fig. 14.13 is an example of a typical PLC-based control panel. Typical of most PLCs, components are rack-mounted and are scalable, in that they allow for the addition of blocks of I/O or processors.

Figure 14.13 Control panel with PLC.

14.3.5 Host systems

The "host" computers in production automation systems provide many important functions. These are covered in Sections 14.4 and 14.5.

General automation systems

There are companies that make "host" systems for the general automation market. These systems, as they come from the factory, provide most of the general applications as discussed in Section 14.4 but do not contain the unique applications as described in Section 14.5. It may be possible to add some of the unique applications at a cost, but usually the supplier of the "host" system is not able or interested in doing this; it will be necessary to use an independent software supplier. The general applications will typically need to be configured to meet the specific requirements of each location, but they do not need to be developed and tested from scratch.

Equipment-specific systems

Other companies produce "host" systems specifically designed for the oil and gas production industry. These systems will typically contain most of the general applications and some (most will not contain all) of the unique applications. In some cases the unique applications are designed to work with specific RTU/PLC logic and capabilities and/or artificial lift systems that are provided by the same company; and they may or may not support RTU/PLC logic or artificial lift systems produced by other companies.

Home-grown systems

A third category of "host" systems are produced by the operating companies themselves. Very few operating companies develop their own systems. But when they do, they tend to focus on the specific types of production equipment, artificial lift systems, and production automation equipment that are important to them.

Generic oil and gas systems

There are a few companies with the primary business of developing "host" production automation systems. To make their systems attractive to a wide range of customers, they try to support as many different types of production equipment, production automation components, and artificial lift systems as possible. It is difficult for one company to be expert and provide good capabilities for all forms of production and artificial lift systems, but some do a reasonable good job of this.

It may be difficult for an operating company to know what type of system to choose. Often, it may seem attractive to use a "host" system provided for the general automation market. However, this may not be wise in that it may not be possible to obtain the unique applications that may be of significant value in gas and oil well production. Or, it may seem attractive to buy a "host" system that supports specific artificial lift equipment. This should be carefully scrutinized if there is a likelihood of using other types of artificial lift, or even artificial lift systems from other suppliers. Since most operating companies are not going to develop their own system, the best approach may be to work with a company that focuses on building generic "host" systems for oil and gas production.

14.3.6 Communications

Communications are required at several levels in production automation systems:

- Between the instruments and controllers and the RTU or PLC
- Between the RTU/PLC and the host automation system
- Between the host automation system and the general user community
- Between the host automation system and other computer systems
- Between the other computer systems and the user community

Instrument to remote terminal unit

Communication between the instruments and controllers and the RTU or PLC is normally over a twisted pair cable. This cable may be placed in conduit. Often, many pairs of wires are installed in a large cable that connects from the RTU to several instruments. Normally, as in the case of analog current or voltage signals, a single pair of wires goes directly from the instrument to the RTU or PLC. In some cases, as in the case of digital or Fieldbus transmitters, a single wire may be connected from the RTU or PLC to many instruments or controllers. In a few cases, wireless communications are used.

This communication is one of the weakest links in the production automation system. Wires may be cut, damaged, or shorted. Fortunately, it is easy to tell if there is a communication outage. For example, with analog current transmitters, a value of 4 mA represents a zero (0) value of the signal being measured. If the analog current signal goes to 0 mA, this signifies a communication outage. Similar indications exist for voltage, digital, and Fieldbus signals.

Remote terminal unit to host

There are many alternatives for communicating between the RTUs or PLCs and the "host" automation system. These include hardwire, radio, microwave, spread-spectrum, satellite, etc.

When low-speed communications (up to 9600 bps) are used, it is common for the host system to poll the RTUs for information. A typical polling frequency might be once every 15 or 20 minutes. Normally, the host would "ask" the RTU for its status. If all is OK, the response is short. If there are problems, the RTU may respond with a larger set of information that describes the situation.

With higher speed communications, it is possible to poll much more frequently. Also the possibility exists for the RTUs to "report by exception." In this case the RTUs do not need to wait until they are polled. If they detect a problem, they can initiate the communication and send the information to the host (Tables 14.2–14.5).

Method of communication	Brief description	Pros	Cons
Hardwire	Physically wired connection from RTU/PLC to host computer	Very fast Always open to communication	Wire may deteriorate or be damaged over time Can be expensive
Telephone	Telephone line hard-wired between RTU/PLC and host computer	Can handle communication over very long distances	Monthly communication cost Occasional lack of service Slow speed
Radio	FCC regulated UHF or VHF licensed frequencies	30 + miles line of sight communication. 19,200 kbps	Must have line of sight Need to rent or own tall towers to mount master and repeater stations
Microwave	High bandwidth wireless communication	Up to 100 Mbps throughput Ethernet addressability Very high bandwidth	High initial cost to install Must have AC power at all sites Expensive
Fiber optic	Communication between RTU/PLC and host computer over fiber-optic cable		May be difficult to maintain
Spread-spectrum	Nonlicensed wireless communication. Can be frequency hopping, direct sequence, or 802.11	Up to 1 Mbps throughput No fees or licenses Uses multiple repeaters Fits well with solar power	30-mile range Line of sight Shared frequency
Satellite	Wireless communication from remote site to satellite to ground station	Very high bandwidth Communicates to or from anywhere on earth	Very expensive Latency, time lag
Hybrid	Combining multiple communications devices to build one system	Allows multiple RTUs communicating by radio to talk to 1 satellite or phone line	Operators need to be proficient with multiple instruments
Cell phones	Wireless data over cell phone network	No infrastructure needed Easy installation Accessible from anywhere	Recurring monthly phone bills Not available in some areas No ability to repair or troubleshoot

These are the physical methods that may be used to communicate between RTUs/PLCs in the field and a "host" production automation system.

Table 14.3 Communication standards

Communication standards	Brief description	Pros	Cons
RS-232	Method of enabling serial communications	• Universal RTU communication protocol • Up to 20 kbps throughput	• 50 ft maximum between RTU/PLC and radio • Only one device per cable run • Requires trenching or conduit
RS-485	Method of serial communications that allows multiple drops	• Allows multiple devices on one cable run • Up to 10 Mbps throughput	• 4000 ft maximum • Limited to 32 devices • Requires trenching or conduit
Ethernet	Network standard using coaxial, twisted pair cable, or spread-spectrum radio	• Up to 1 Gbps throughput • Multiple conversations at one time	• Not all RTUs/PLCs support Ethernet • 300 ft limit on wired connections
Combination	Some radios can accept 232 and act as a terminal server or protocol translator to provide the computer with Ethernet (TCPIP) data		

These are the physical standards that are employed to connect the communication systems into the RTUs/PLCs and the "host" computers.

Host to users

As indicated in Section 14.3.5, most "host" computer systems use desktop PCs. The primary method of communication for users is direct on the PC using Microsoft Windows or similar tools. Some users are connected to the host computer via a network connection. Some use PC-to-PC links. Some use intranet or Internet access.

Host to computer systems

Many production automation systems expand their capabilities by connecting to other computer systems for storage of large volumes of information in various types

Table 14.4 Communication protocols

Communication languages	Brief description	Pros	Cons
Modbus	PLC language provides 256 addressable locations. It is very close to becoming the universal language between RTUs/PLCs and host computers	Addressability Common language for multiple vendors and equipment	Limited addresses, slower baud rates, typically 9600 or 19,200
Modbus RTU	This is a MODBUS protocol designed for use with RTUs/PLCs		
TCPIP	Ethernet language	Provides error checking, guaranteed data delivery, and polling of multiple units at once	More overhead Higher power consumption No idle mode Requires a lot of bandwidth
	Allows for multiple conversations at once by giving every packet its own IP address		

These are the "languages" that are employed so the RTUs/PLCs and the "host" computers can communicate with and understand each other.

of database systems, access to analysis, design, and simulation software, access to the worldwide web, etc. Often, these "extra" systems are not provided by the supplier of the production automation system, so some form of agreement must be negotiated.

Computer systems to users

There are many ways for users to communicate with the computer systems that are part of the extended production automation system. In some cases the information

Table 14.5 Methods of data security

Communication standards	Brief description	Pros	Cons
BCH	A BCH (Bose, Ray-Chaudhuiri, Hocquenghem) code is used for error detection. The side (RTU/PLC or host computer) that is transmitting data calculates a BCH code based on the data being sent and appends it to the transmission. The receiving side recalculates the code. If the two codes match, the data have been correctly received	Provides a very high degree of data transmission security In theory, BCH codes can be decoded to correct communication errors	Most RTUs/PLCs only use error detection and re-transmission They do not use error correction
128 bit AES (encryption)	American Encryption Standard, commonly accepted by multiple industries	Provides security on outbound messages Prevents attack from incoming messages	Requires bandwidth and computing power Can be "hacked" May need to be used in conjunction with other security
RADUIS (central authentication)	Allows system to have conversations with only devices known to be authenticated by system administrator. All others are blocked	Prevents "rouge" users from entering system Has a "time out" feature that removes devices from system if they stop transmitting	Any device that temporarily goes offline must be manually reentered into the authentication list before it can resume conversation
MAC address filtering	Allows each port to be secure. Only allows conversations with a list of known MAC addresses	Prevents attack at the port level Unknown MAC addresses cannot access a device or port on the device	Complexity in management New equipment will not be accepted without hands-on intervention
VLAN tagging	Virtual LAN allows multiple LANs inside one network. Keeps management data separate and segregated from SCADA data Each packet has a tag that identifies its LAN and routes information to the proper network	Allows for multiple secure conversations within the same network at the same time Users can only access their VLAN	Inconvenient if you need access to more than one of the VLANs Data management is more complex

from the computer system is transmitted back to the automation "host" system so the user can access it there. In other cases the information may be available via an intranet or Internet connection.

14.3.7 Database

Overview

Automation systems generate vast quantities of data. For this data to be of value to end users, it needs to be handled so that it can be easily stored, retrieved, and displayed at some time. For this reason, databases play an integral role in automation systems and exist in some form or another in virtually every component of the system. A database can be defined as a structured set of records that is stored in a computer so that a program can consult it to answer queries.

Database models and schema

For a given database, there is a structural description of the type of facts held in that database. This is known as a schema. A schema describes the objects in a database and the relationship among them. There are different ways of organizing schemas, called database models. The most common database model is the *relational model*. Relational databases arrange all information in tables of rows and columns, where relationships are represented by values common to more than one table. Other database models include the *hierarchical model* and *network model*, which represent relationships more explicitly.

Storage

Databases are typically stored in memory or on hard disk in one of many formats. Data are often stored by category (i.e., data by month, data by well), creating preconfigured views known as materialized views. In some cases, data may be *normalized* to reduce storage requirements and improve extensibility. In other cases, data may be *denormalized* to reduce join complexity and reduce execution time for queries.

Indexing

All databases take advantage of indexing to increase their speed and efficiency. Indexing is a means of sorting information. The most common form of index is a sorted list of contents in a particular table column with pointers to the row associated with the value. Indexes allow a set of rows matching certain criteria to be located quickly.

Real-time databases

Unlike conventional databases, a real-time database is specifically designed to meet the demands of a system where information is constantly changing. Where traditional databases are adequate for handling *persistent* data which are generally unaffected by time, a real-time database must be able to "keep up" with constantly changing conditions. Real-time databases are traditional databases that use an extension to give the additional power to yield reliable responses. They use timing constraints that represent a certain range of values for which the data are valid. This range is called temporal validity. A conventional database cannot work under these circumstances because the inconsistencies between the real-world objects and the data that represent them are too severe for simple modifications. An effective system needs to be able to handle time-sensitive queries, return only temporally valid data, and support priority scheduling. To enter the data in the records, often a sensor or an input device monitors the state of the physical system and updates the database with new information to reflect the physical system more accurately.

Fig. 14.14 illustrates the difference between how real-time data are processed with a conventional database versus a real-time database. Conventional database protocols, which generally schedule transactions on a first-come, first-serve basis, will let A lock the data and complete, allowing A to meet its deadline. B, on the other hand, will miss its deadline because A's lock on the data prevents B from starting early enough. In contrast, a real-time database with time-cognizant protocols would preempt transaction A and transfer data control to B because B's deadline is earlier. Transaction A would regain control after B completes, and both transactions would meet their deadlines.

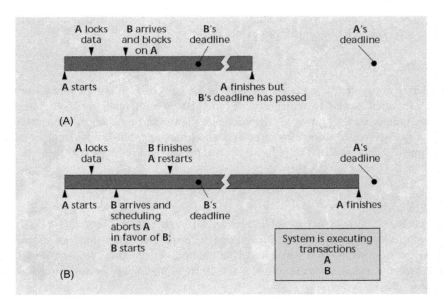

Figure 14.14 Processing of two transactions using (A) conventional database protocols and (B) time-cognizant protocols.
Source: After Stankovic JA, Son SH, Hansson J. *Misconceptions about realtime databases, cybersquare.* University of Virginia; 1999.

FIFO

FIFO is an acronym for first-in, first-out. This describes behavior similar to a queue in which people leave the queue in the order they arrived. In database terminology, this refers to a system in which there is a maximum amount of data which can be stored, and once that limit is reached, the oldest data are over-written as new data comes into the system. This structure is common in most components of real-time systems. RTUs, PLCs, and host systems all contain databases built around this principle.

The implication in these systems cannot archive data for an indefinite period of time. Depending on the size of the database and the quantity of data collected, such databases may be able to store as much as a year's worth of data or as little as a few hours. For example, consider a system where the database can hold up to 10 million records. If that system were to collect data from a field with 100 wells, each instrumented with 10 analog sensors at 1-minute intervals; the database would only be able to store 1 week's worth of data (100 wells $\times$ 10 AIs $\times$ 24 h/day $\times$ 60 min/h = 1,440,000 records). This is an important consideration when designing an automation system to support today's highly instrumented wells. Because of the large quantities of data collected by wells equipped with electronic downhole gauges and other instrumentation, an FIFO database may not be adequate to handle a system's long-term storage needs. For such an application, it may be necessary to augment the automation system with another type of database designed specifically for this purpose.

Historians

A data historian is a special class of real-time database that is designed to efficiently store and retrieve large sets of real-time data. These are commonly used as a repository for long-term data in automation systems. In addition to the normal characteristics of real-time databases, historians provide internal compression schemes to handle the extensive data storage requirements of an automation system. Further, historians provide tools that allow users to retrieve the data extremely quickly and even perform mathematical operations on the data. Historians are also built with extensibility in mind and can be easily integrated with other real-time data sources, host systems, or enterprise data sources. For this reason, historians often serve as the workhorse of an automation system. Several historians are in use throughout the industry today. The most commonly used historians are PI from OSISoft and Honeywell's Uniformance PHD.

14.3.8   Other

Some production automation systems extend beyond the traditional automation equipment and software of RTUs, PLCs, "host" computer systems, and databases. There are companies that are not in the automation business, but provide software systems for modeling reservoirs, well inflow, well outflow, nodal analysis, artificial lift systems, well test systems, etc. These software applications may be used for design, analysis, troubleshooting, and optimization of reservoir recovery, well inflow, and artificial lift system behavior. They may be more effective if they are provided with "live" data from an automation system. And the value of the automation system is enhanced if the results of system models and analysis can be fed back to it. For example, some alarms are based on a comparison of measured results versus results predicted by system models. Therefore some companies are integrating production automation with these other systems.

This integration is possible with the availability of standard software interface systems that allow automation systems and these other systems to communicate with one another, share information, share results, etc., without having to be written by the same company. The interface acts as a translator. The automation system can communicate using its language and data. The other systems can communicate using their approaches. The interface translates between the two systems.

One of the more common interfaces is the computer port or interface (COM) object interface system in Microsoft Windows. Many companies can communicate with each other's systems using this standard. COM lets data be exchanged using COM-supported software such as Microsoft Excel, Word, and PowerPoint.

Another common interface standard is the Petrochemical Open Standards Consortium (POSC) interchange format. Several companies including BP, Chevron, ExxonMobil., Shell, Halliburton, Invensys, OSISoft, Petroleum Experts, Schlumberger, Sense Intellifield, TietoEnator, and Weatherford are working on the PRODML project to develop a POSC Work Group Agreement. PRODML is a shared solution for upstream oil and gas companies to optimize their production.

The "good news" is that any application (database, simulation software, design program, surveillance program, optimization program, etc.) that is COM or POSC compatible can be interfaced to a production automation system. This is good news for the operating company; they can have access to the capabilities. It is also good news for the automation companies and the other service providers; they can offer their products without needing to develop interfaces with every different other system.

14.4 General applications

Automation systems are used in many industries for many purposes. Suppliers of these systems provide a wide range of general capabilities that are appropriate for use across a wide range of applications. This is "good news" for gas, oil, and unconventional well operators. For the most part, these general applications come ready for use "off the shelf." They must be installed and configured for each specific field and set of wells, but they do not need to be developed and tested, which has already been done.

If a field has only flowing gas or oil wells, these general applications, once properly configured for monitoring and reporting gas and/or oil well production, may

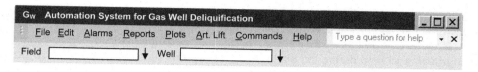

Figure 14.15 Example user interface for automation system for gas well deliquification.

provide 90% of the production automation requirements. Of course, this is rarely the case. Many fields require one or more forms of artificial lift. Some of the required special applications for artificial lift and unique gas and/or oil well needs are covered in Section 14.5. That is, these general applications are necessary, but they are not sufficient if special gas production requirements exist, or if artificial lift systems are used.

The purpose of this section is to discuss some of these general applications and how they can be used for management of gas production wells and systems.

14.4.1   User interface

Production automation systems start with a user interface. Fig. 14.15 shows an example of what a user interface might look like for an automation system for gas well deliquification. This is based on the graphical user interface standard developed by Microsoft Windows.

The purpose of the user interface is to allow the operator to easily navigate and select the specific reports, plots, or other needed functions or capabilities for specific wells or groups of wells. All production automation systems typically use an approach that is similar to the standard developed by Microsoft Windows. Each system will look slightly different, have different specific "pull down" options, have different methods for selecting specific field(s) and well(s), etc. All systems should have a "help" capability.

14.4.2   Scanning

A typical gas field may have tens or hundreds of wells. In many cases an RTU or PLC is installed at each well. A primary function of the "host" production automation system is to scan each RTU/PLC on a periodic basis to collect pertinent "real-time" information for alarming, reporting, plotting, analysis, etc. Real time means that the information is "live" and indicates the condition of the well "right now." The primary goal is to automatically collect pertinent information and have it available for processing and display, so the operator can focus on data analysis, not on data acquisition.

Typically the "host" system communicates with (scans) each RTU/PLC on a preset frequency. For example, if radio communications are used, it may scan each device once each 15–20 minutes. With other communication systems, it may scan more frequently. Typically a limited amount of information is "uploaded" from the RTU/PLC to the "host" on each scan. This might consist of a few data "words" to indicate the status of the well (e.g., are there any outstanding alarms?), the current gas flow rate and pressure, current oil production, etc.

If there is a problem (alarm), the "host" may then automatically upload more information to help define the alarm condition, or this function may be left to the operator.

In addition to periodic routine scans, the "host" system may also perform special scans at certain times of day, or in conjunction with certain events or conditions. For example, the system may upload a full set of data just before "morning report" time. Or, it may upload specified data in conjunction with a well test.

14.4.3   Alarming

A typical gas and/or oil field may have hundreds or thousands of instruments to measure pressure, temperature, flow rate, artificial lift equipment status, etc. In principle, alarms can be defined for every instrument. An alarm is an indication that there is or may be a problem. In theory the operators need to be made aware of all alarms, so they can initiate the necessary action to address the alarm condition.

While this is the theory, it may be counterproductive to actually implement all possible alarms. For example, most production automation systems allow configuration of:

- High and low alarms—an alarm exists if the process variable (e.g., pressure) is above the high alarm limit or below the low alarm limit.
- High, high and low, and low alarms—an alarm exists if the variable is above or below the high, high or low, and low limits.
- Rate of change alarms—an alarm exists if the value of the variable changes too fast or too slow.

If eight or ten alarm conditions are defined for each variable, the system may generate hundreds or thousands of alarms per day. No operator can properly deal with this many alarms. So, in reality, many of them are ignored. This is not good, especially if there was one or more "real" alarms in the mix.

Rather than using the "standard" types of alarms, as indicated earlier, it is preferable to "design" the alarm system to make it pertinent for gas and oil well operations. There may be specific alarms that are pertinent for gas or oil wells, and there are specific alarms that are pertinent for each type of artificial lift. The first type is discussed in this section. The second type is discussed in Section 14.5.

Before designing actual alarms, it is useful to consider three classes of alarms:

- Class I—simple alarms such as high and low alarms.
- Class II—combination alarms where combinations of variables are used to indicate specific alarm conditions that are pertinent for gas or oil wells.
- Class III—performance alarms where the values of measured variables are compared with values that are estimated or derived from models of well or system performance.

Alarm Name	Current State	Severity	State Change Time	Times In Alarm	Hours In Alarm	Yesterday Times in Alarm	Yesterday Hours in Alarm	Month Times In Alarm	Month Hours In Alarm	Consecutive Days In Alarm	Previous Month Times In	Previous Month Hours In
Scanning Disabled	Normal	295.00		0	0.00	0	0.00	0	0.00	0	0	0.00
Communication Failure	Normal	290.00		0	0.00	0	0.00	0	0.00	0	0	0.00
Test Unit Down	Normal	285.00		0	0.00	0	0.00	0	0.00	0	0	0.00
WT Engine Connection Failure	Normal	280.00		0	0.00	0	0.00	0	0.00	0	0	0.00
Switch Failure	Alarm	275.00	04/25/2004 08:05	2	64.57	0	0.00	2	64.67	0	0	0.00
Invalid Well	Normal	270.00		0	0.00	0	0.00	0	0.00	0	0	0.00
Programming Failure	Normal	265.00	04/21/2004 15:47	0	0.00	0	0.00	0	0.00	0	0	0.00

Figure 14.16 Current alarm display.

Class I alarms

Some Class I alarms are pertinent for gas and/or oil well operations. For example, a zero signal from an instrument may indicate that the instrument or the wiring to the instrument has failed. Fig. 14.16 shows a typical alarm display taken from a well test facility. The following information is shown on this panel:

- Alarm name.
- Current state.
- Severity—this number is assigned so alarms can be sorted by their severity.
- State change time—when the point last changed state from "normal" to "alarm."
- Times in alarm—number of times this point has been in alarm today.
- Hours in alarm—total time this point has been in alarm today.
- Similar information is shown for yesterday, so far this month, and the previous month.

Class II alarms

Class II alarms are specifically designed to indicate problems with gas or oil wells. They are based on a combination of field measurements. These alarms must be designed and configured for gas and oil well operations but once they are, they can be used in many gas and oil well production automation systems. Typically, these alarms are much more informative than simply determining that the pressure is too high or too low. Two examples are:

1. Gas or oil production line blocked or frozen. This may occur if the wellhead pressure is above normal and the production rate is below normal. If there is a wellhead or separator temperature measurement, this may also be used as part of the combination.
2. Gas or oil production line leaking or broken. This may occur if the wellhead pressure is normal, the flow rate is normal or higher than normal, and the line or separator pressure is below normal.

Class III alarms

Class III alarms are generated when the value of a measured variable differs from a value that is calculated or estimated by a model of the well or system. These are commonly used for artificial lift and are discussed in Section 14.5. An example for general gas well application is:

- Well flowing below critical velocity. This may occur if the calculated gas flow velocity, based on measured flow rate and pressure, is less than the calculated critical flow velocity.

14.4.4   Reporting

Production automation systems can produce several types of reports. The most common are:

- Current reports—reports of current information on individual wells or groups of wells.
- Daily reports—reports that summarize the wells' production and performance for a day.
- Historical reports—reports that summarize wells' production and performance for the past week, month, or longer.
- Special reports—reports that contain special information such as well tests.
- Reports for unique applications—reports that are unique for special applications, as discussed in Section 14.5.

Current reports

Current reports can show information as of the last scan of the RTUs, or a special scan can be forced so the report contains true current information. The reports can contain measured values such as production rate and pressure. They can contain calculated information such as gas—oil ratio, liquid—gas ratio, and critical flow rate. Information on the reports can be sorted by well name, by highest to lowest production rate, etc. Columns can be totaled to, for example, show the total gas or oil production rate for a group of wells. Columns can be averaged to, for example, show the average production rate. In addition to rates and pressures, reports can show alarm and status information, downtime information, actions performed by the operators, etc.

Daily reports

Most systems produce a set of reports at the end of each "production day." The production day may end at midnight or at some time early in the morning. These reports typically show the current production rate, the total production for the day just ended, the production for the previous day, and the cumulative production so far in the month. In addition to production, daily reports may show downtime for the most recent day and downtime so far in the month. As with current reports, these reports can be sorted, totaled, averaged, etc.

Historical reports

Most systems can produce historical reports. Usually, this is a report of the production rates, pressures, critical velocities, downtime, etc. for a well. The report may contain average values for the past few months and daily values for each day in the current month. There may also be historical reports of well tests.

Special reports

Well test reports are in the special category because they do not contain daily information. A well test is reported when it occurs, and usually this is periodic, for example, once per week, once per month, etc. Well test reports may show the current well test and a previous test for comparison.

Unique application reports

Reports are produced for most forms of artificial lift. These are discussed in pertinent places in Section 14.5.

Reports can be displayed on the automation screen, they can be printed, or they can be accessed by any system that is in communication with the automation system. Reports can be requested manually, scheduled automatically at some time, or produced in association with some event. For example, many locations schedule daily reports for automatic printing early in the morning, before people arrive, so they are available for the "morning meeting."

Reports can list all of the wells or conditions in an area. Or, they can list exception conditions. For example, a report may list only those wells where the production is too low, or those wells where the gas flow velocity or oil production rate is below the critical velocity or target value. Exception reports are popular since they allow the operators to focus on "problem" wells and not have to sort through information on wells that are producing normally.

14.4.5 Trending and plotting

In general, any measured or calculated variable that can be reported can be plotted with a trend plot. A trend plot is a plot of one or more variables versus time. Most production automation systems provide a general trending capability; variables can be plotted by configuring the trend plots. In addition, some other types of plots are possible, for example, the so-called xy plots where one variable is plotted versus another. And, many of the unique applications use various trends and other types of plots. These are discussed in Section 14.5.

Various adjustments are possible on trend plots. The timescale can be adjusted to show duration from minutes to months or years. The "y" axis can be adjusted to show one, two, or many variables on the same plot. The axis can be adjusted from 0.0 to a maximum value, or it can be "telescoped" so the range of data fills the plot.

Most trend plots have a "zoom" feature so the operator can "zoom in" on a smaller time window and/or vertical set of data. Most trend plots support a color and/or a line style coding system, so different variables can be color coded and can use different line styles if the plots are printed on a black-and-white printer.

Some trend plots are static, that is, they display data that have already been collected and/or calculated. Other plots are dynamic, that is, the data on the plot are automatically updated when a new set of data is scanned from the RTU. Some trend plots show only the data points; some show the data points connected by lines;

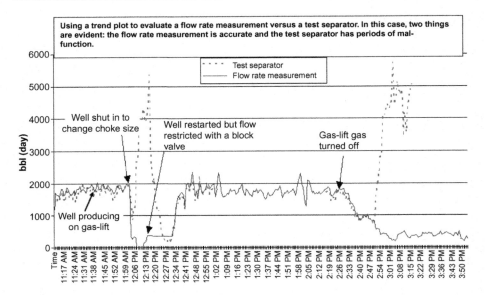

Figure 14.17 Trend plot of two variables.

some show only the lines. Some plots show only the measured or calculated data; some also show trends in the data; some show only historical data; some show both historical and projected future values.

Trend plots can be very useful in spotting changes in variables over time and in evaluating data. Fig. 14.17 is an example trend plot used to evaluate a flow rate measurement system versus test separator readings. The accuracy of the measurement was confirmed, and some problems were detected with the test separator measurements.

14.4.6 Displays

"A picture is worth a thousand words." This saying is attributed Fred R. Barnard in the December 8, 1921 advertising in the trade journal *Printers' Ink*. Production automation systems make use of this by featuring schematic displays of systems, facilities, and wells. The displays typically contain information about the operation of the item(s) being displayed, such as pressures, temperatures, flow rates, and alarm and status information.

Displays may be of several types: unique, generic, static, dynamic, and interactive.

Unique

Unique displays portray a given specific system or set of equipment. An example may be a specific production facility. The display may show a schematic of the facility with pertinent pressures, temperatures, flow rates, alarm and status information, etc.

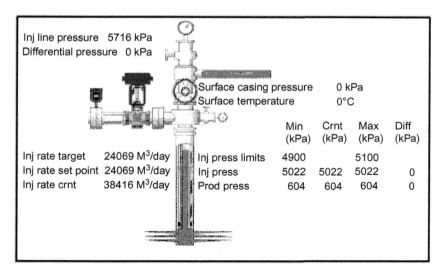

Figure 14.18 Example of a schematic display of a gas-lift well.

Generic

Examples of generic displays are for a gas well, an oil well, or an artificial lift system. Fig. 14.18 is an example of a generic schematic display of a gas-lift well. The display shows a typical gas-lift well with pertinent pressures, temperatures, flow rates, etc.

Static

A static display shows information that was collected on the last scan of the RTU(s) that provide information that is shown on the display.

Dynamic

The information on a dynamic display is updated each time new information is obtained from the pertinent RTU(s). Also, some dynamic plots have "live" graphics that show the equipment moving, liquid flowing, etc.

Interactive

An interactive display is one where the operator can enter parameters or commands on the display. The appropriate action is taken such as downloading the parameters to the appropriate RTU(s), issuing the desired commands, and performing the desired calculations.

Most production automation systems provide a "tool kit" for constructing the displays and populating them with the "tag numbers" for the information to be shown on the display. For unique displays, specific tag numbers are used. For generic displays, generic tag numbers are used so that the displayed information depends on the specific well selected for display.

14.4.7 Data historians

Some production automation systems collect huge amounts of data. If a system serves 1000 wells; collects pressure, temperature, and flow rate once per minute; and calculates critical velocity or pump performance once per minute, this is 5,760,000 pieces of data per day. Most systems collect, calculate, and store much more than four data items per well per scan.

Most automation systems are designed to focus on "real-time" information. They are primarily concerned with current rather than historical operations. If they store historical information, it is usually summary data, for example, hourly or daily average values. However, it is often valuable to store and be able to access detailed, minute-by-minute historical information, sometimes for months or years.

Data historians use special data compression techniques to store huge volumes of data and make this readily available. Real-time data are transmitted from the automation system to the historian. The historians typically provide special methods or techniques for access to the data. For example, one company provides a tool called a "processbook." Operators can create interactive graphical displays that extract information from the historian on an as-needed basis. The information can be saved and shared with other people who have access to the system via an intranet or Internet connection.

Specific examples of uses of historians are beyond the scope of this book. However, if there is interest, most automation companies can provide access to a historian system.

14.5 Unique applications for gas well deliquification and oil well production

As indicated in Section 14.4, production automation systems usually come with several general applications that can be applied for gas and oil well monitoring, control, and optimization. However, these general applications never address specific requirements for artificial lift or other unique gas and oil well capabilities. Therefore special or unique applications are required.

Again, there is "good news" for gas and oil well operators. Production automation systems for several types of artificial lift have already been developed for use in oil production. Systems for sucker rod pumping, progressing cavity pumping, ESP, and gas-lift exist have been developed, tested, and used extensively. In many cases, these can be applied for gas wells and unconventional production with only minor modifications.

However, these forms of artificial lift only address a fraction of gas well production. Other major production methods for gas wells include plunger lift, chemical injection, wellhead compression, etc. For these and others, new capabilities intended for use on gas wells are required. Some of these have been developed and some are in various stages of research and development.

The purpose of this section is to discuss these unique applications, how they can be used for monitoring, control, and optimization of gas and oil well operations, and some of their benefits and challenges. Some of these systems are very comprehensive. For example, to fully describe an automation system for plunger lift, sucker rod pumping, ESP, gas-lift, etc., a full-length book would be required for each. So the purpose here is to "hit the high spots." If more information is desired on automation for a specific form of artificial lift, it must be obtained from the authors or an appropriate service company or operating company that uses the system.

14.5.1   *Plunger lift*

Plunger lift is a low-rate artificial lift method, common in gas well deliquification applications but also in some oil applications. The method requires no outside energy source; it uses the well's natural energy to lift fluids (and the plunger) to the surface. The systems can be installed without a rig, provide easy maintenance, can be deployed at extremely low cost, are tolerant of deviation, and can produce a well to near depletion. Limitations of plunger lift include the system requires specific gas/liquid ratios to function; components are sensitive to solids; and the system can be labor-intensive, requiring surveillance to work properly. Historically the need to regulate flow as well as surveillance requirements and the labor-intensive nature of plunger lift have made it one of the most heavily automated lift methods in the industry.

Measurements

To effectively monitor, control, and optimize a plunger lift installation, the automation system must collect a number of surface parameters. These include the tubing-head pressure, casing-head pressure, flowline pressure, differential pressure across an orifice union, gas flow rate (calculated from differential pressure and flowline pressure), and an indication of plunger arrival.

Fig. 14.19 is a system illustration depicting the instruments and controls in a plunger lift system. These include: (1) a tubing pressure transducer, (2) a casing pressure transducer, (3) a differential pressure transducer, (4) an orifice union assembly, (5) a plunger lift controller, (6) an automatic control valve, (7) a flowline pressure transducer, and (8) closed contact switches.

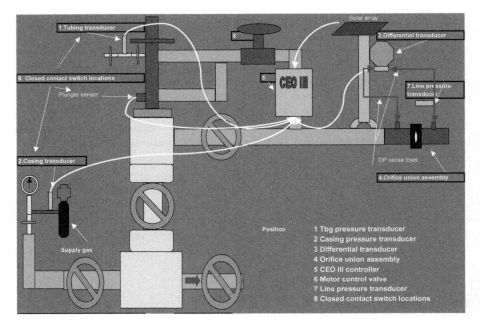

Figure 14.19 Plunger lift system illustration.

Control

Most plunger lift systems operate on time-based control algorithms utilizing control of one or two surface valves to control the movement of the plunger. When the system is closed, bottomhole pressure builds up. When the system is open, this pressure forces the plunger to move up to the surface, carrying fluid on top of the plunger. The plunger lift controller monitors and records the time between plunger cycles and makes adjustments to the time-based control algorithm.

Unique hardware

Plunger lift uses specialized RTUs called plunger lift controllers. These devices monitor and control the well based on internal control logic, and, in some cases, communicate with a host system to transmit data and accept commands. In addition, because plunger lift controllers are often used in gas deliquification applications, they can either communicate with an API 21.1-compliant EFM device or have EFM capabilities integral to the controller itself. In controllers equipped with telemetry, this flow measurement data can then be transmitted to a host system (Fig. 14.20).

Historically, plunger lift has been viewed as a labor-intensive form of artificial lift due to the significant requirement for surveillance and adjustments to the system to optimize production. For this reason, many operators tended to avoid this

Figure 14.20 Typical plunger lift controller.

artificial lift method. The development and evolution of plunger lift controller technologies has reduced the amount of labor required for normal operation and enabled this lift method to become more widely used throughout the industry.

Unique software

Plunger lift controllers contain special software that provides a computerized means for opening and closing the control valve based on programmed responses or sets of parameters. A variety of options are available for these devices, including: (1) actuation based on pressure and flow, (2) time-cycle control, (3) self-adjusting models, and (4) telemetry support.

Plunger lift controllers have evolved from simple time-cycle controllers, to time-cycle controllers with plunger arrival recognition, to auto adjusting time-cycle controllers. As a result, controllers have become self-managing plug and play devices; and plunger lift systems no longer require the constant supervision they once did.

Various forms of control logic are available in plunger lift controllers. The following are some commonly used operating parameters for plunger lift well control using a flow- and pressure-operated control system.

On pressure limit control
Controller initiates *on* cycle when the following conditions are met.

1.	Tubing pressure ≥ on pressure limit	(looks for tubing pressure to exceed a set point)
2.	Casing pressure ≥ on pressure limit	(looks for casing pressure to exceed a set point)
3.	Tubing-line ≥ on pressure limit	(looks for tubing pressure to build to a set point above line pressure)
4.	Casing-line ≥ on pressure limit	(looks for casing pressure to build to a set point above line pressure)
5.	Foss and Gaul calculations	(looks for casing pressure to reach a calculated value)
6.	Load factor	(looks for the casing-tubing/casing-line pressure ratio to be a factor of less than 40%)

Off pressure limit control
Controller initiates *off* cycle when the following conditions are met.

1.	Plunger has arrived	(used on oil wells)
2.	Casing pressure ≤ off pressure limit	(looks for casing pressure to fall below a set pressure)
3.	HW ≤ off pressure limit	(looks for flow rate to fall below a set differential in inches of water)
4.	Flow rate	(sometimes a calculated Turner rate)
5.	Casing-tubing ≥ off pressure limit	(looks for differential between casing and tubing pressure to increase)
6.	Casing to tubing sway	
7.	Casing to line ≥ off pressure limit	(looks for differential between casing and line pressure to increase)

Specialized alarms

Plunger lift systems provide a variety of alarms to notify users of adverse operating conditions or potential opportunities to enhance performance. Some of the more interesting alarms are:

- High plunger velocity—This could indicate a number of issues. More commonly, a high-velocity indication could be the result of either: (1) a failure of the plunger to reach bottom (i.e., due to a wellbore obstruction, such as hydrates) or (2) a plunger that is coming up dry (i.e., inadequate inflow or inadequate shut-in time).
- Number of runs per day—An excessive number of cycles per day could be an indication that the plunger is wearing out, resulting in reduced efficiency.
- High tubing-head pressure—In certain cases an extended shut-in period could result in an excessively high tubing-head pressure. If an on-cycle were initiated under these conditions, the resulting high-pressure slug could potentially damage the separator or cause other upsets to the surface facilities.

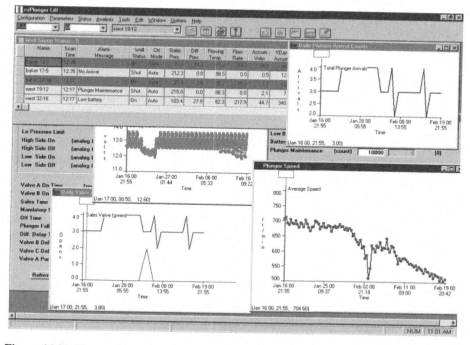

Figure 14.21 Plunger lift surveillance panel.

Surveillance

Plunger lift surveillance is performed by continuously checking for special alarm conditions and reviewing trend plots. Fig. 14.21 shows a typical surveillance panel from a plunger lift automation system. It contains status information, trend plots, and operating parameters for a plunger lift installation.

Analysis

Once data are captured in the plunger lift automation system, they are used to perform analysis to optimize the performance of the system. One of the key offline analysis methods is a determination of whether to switch from continuous plungers to conventional plungers. In addition, an evaluation of gas velocities and pressures can assist in determining whether the well is operating below the critical velocity for liquid loading. In addition to these offline analyses, some plunger lift controllers perform continuous real-time analysis of pressures and velocities to make continuous adjustments to system parameters and optimize production.

Design

Plunger lift is unique in that it lacks the same rigorous design requirements that are common to other lift methods. Application engineering techniques generally consist of: (1) evaluation of flowing well conditions to determine if the well is a candidate for plunger lift, (2) evaluation of the most suitable type of plunger to use, and (3) selection of equipment that is most appropriate for the conditions.

Optimization

Through surveillance of key operating parameters, it is possible to determine if opportunities exist to improve system performance. For example, by monitoring casing pressure, one might identify an opportunity to reduce casing pressure and, in turn, increase formation drawdown. Another key parameter is gas flow rate. In many cases, there is an upper limit to the rates that can be accurately measured by the gas meters at the sales point. If the gas flow rate exceeds this amount, the operator may not be compensated for all the gas that is transferred to the pipeline. So, instantaneous flow rates should not exceed this threshold. Finally, through evaluation of key parameters, users can adjust the shut-in time so sufficient pressure builds to ensure plunger arrival while ensuring that the well is not shut in for an excessive duration.

Safety

There can be a safety issue if a plunger arrives at the surface too fast. This could damage the plunger or the wellhead equipment. There is technology to determine the speed of the plunger arrival so that operational changes can be made, if necessary, to mitigate this problem.

14.5.2   Sucker rod pumping

Many thousands of oil wells and more and more gas wells are produced by sucker rod pumping. This is becoming a common method of artificial lift for gas well deliquification. Automation of sucker rod pumping systems for oil wells started in the 1970s; it is a very well advanced technology. Virtually the same sucker rod automation systems are used for gas wells. Fig. 14.22 shows a schematic of a typical sucker rod pumped well equipped for automation.

Measurements

The primary measurements are the load and the position of the polished rod. The load is usually measured with a polished rod load cell mounted on the top of the polished rod or a strain gauge mounted on the beam of the pumping unit. The polished rod load cell is preferred as it is more accurate. The position is normally measured with an angular position transducer, an accelerometer, or a mercury switch mounted on the beam, or a position switch mounted on the A-frame. The accelerometer or beam mounted switch is preferred; it is more accurate.

Some systems use a dual measurement system that measures the load and position with one device. This is lower cost, but not as accurate as a polished rod load cell and separate position measurement.

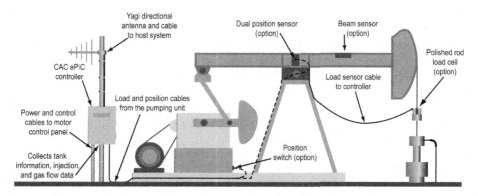

Figure 14.22 Schematic of sucker rod pumping system equipped for automation.

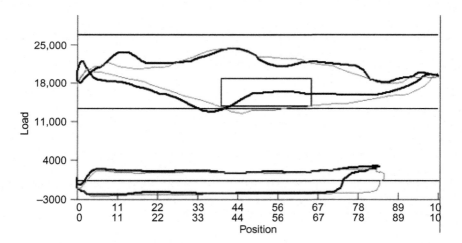

Figure 14.23 Method to detect pump-off.

A number of secondary or optional measurements may be used: a vibration switch to detect unit vibration, a stuffing box leak detector to detect liquid accumulation in the stuffing box, a tubing or flowline pressure transducer, a casing pressure transducer, and other measurements if there are local gas flow meters, etc.

Control

The primary means of automatic control for sucker rod pumping wells is "pump-off" control. The system detects when the pump is no longer filed with liquid on the downstroke, which means that the pump did not completely fill on the upstroke. This is illustrated in Fig. 14.23 where the load on the polished road is carried by the rod part way into the downstroke, and not transferred to the tubing. When the traveling valve strikes the fluid level in the pump barrel part way into the downstroke, some "pounding" may occur; this is referred to as "fluid pound." Compare the green plot (gray in print version) (full card) with the orange plot (black in print version) (pumped off) case. The upper set of plots is measured at the surface; this is the "surface" card. The lower set of plots is calculated at the depth of the downhole pump; this is the "pump" card. The difference in length between the surface and downhole is due to rod stretch.

In most cases the pump-off detection is based on the surface card. However, some systems calculate the downhole card on every pump stroke and make this determination on the downhole or pump card.

This method of control has disadvantages. It depends on some degree of pump-off occurring on every pump cycle. When pump-off occurs, the downhole pump and rods are stressed. Also, if a well is producing gas, which of course is the case for gas wells, the fluid in the pump barrel may be mostly gas and the well may be in a continual state of pump-off. If there is very little liquid compared with the amount of gas, the pump may become "gas locked" and fail to pump liquid at all.

An alternative method of control is with a variable speed drive (VSD). In this case the speed of the pump is adjusted to keep the liquid level just above the pump intake so pump-off does not occur. In a gas well with a small amount of liquid production, even this method may be difficult to use.

Unique hardware

Sucker rod automation uses special RTUs or pump-off controllers. See Fig. 14.24. These units are unique in that they contain only the hardware needed for the required measurements and special logic to monitor and control sucker rod pumping wells. Typically, these units are installed in special NEMA 4 (environmentally sound) cabinets, have battery backup, and are outfitted for radio communications with the host computer system.

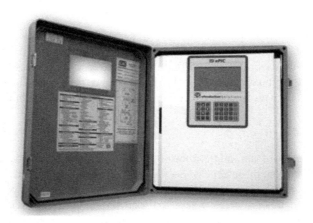

Figure 14.24 Typical rod pump controller.

Unique software

RPCs contain unique software for managing pump operations. To determine if a well is "pumped off," they must collect and process at least 20 pairs of load and position data points per second. Some units process much more data than this. The units check for a number of specific sucker rod pumping alarms (see Section 14.5.2.5) and shutdown conditions. For example, if the rod load is too high, this may indicate a stuck pump.

They check for pump-off on every pump stroke by comparing the load at some preset position with a predefined "pump-off" load limit. If the load is above the limit, implying that the load is still being carried by the traveling valve and rods, the pump is stopped. It will remain off for a predetermined "pump-off idle time" and then automatically restart.

Since most RPCs communicate with the host automation system by radio, pump cards are buffered in the unit's memory, so they can be transmitted to the host at relatively slow radio transmission speed.

Specialized alarms

Sucker rod pumping automation systems perform specialized alarming. A few of the more interesting ones are:

- Run time is too short. This implies that the well is not producing as much as expected. There may be an inflow problem.
- Run time is too long. This implies that the well is producing more than expected, or that the pump is leaking.
- Position signal problem. This implies that the measured position is not sufficiently close to the predicted position.
- Calibration problem. This implies that the design program is not calibrated with the measured data.

Surveillance

The primary purpose of surveillance is to detect problems, so they can be addressed. Some sucker rod pumping automation systems have a very interesting method for problem detection.

Over the years a large library of "problem cases" has been compiled. Each surface or downhole pump card is compared with the cases in the library and the "best" fits are reported to the operator. Thus, for example, the morning report may contain a notation for a well that the pump is sticking, or the pump is leaking, or the pump has gas interference. An example of this comparison is shown in Fig. 14.25 where the actual surface card in red (black in print version) is compared with a library card in blue (gray in print version). The list on the left shows possible library cards and how well each "fits" with the surface card.

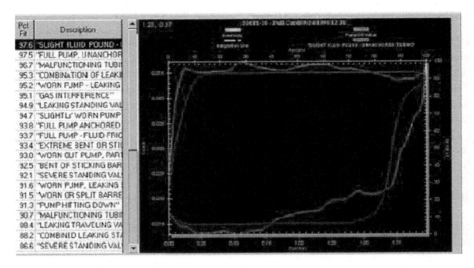

Figure 14.25 Comparison of surface card with library.

Analysis

Sucker rod pumping automation systems focus on analysis of the surface pumping unit, the rod string, and the downhole pump. They determine such things are the torque on the gear box, the stress on each taper of the rod string, the downhole pump fillage, etc. Each calculated value is compared with a target or limit value, so problems can be detected. For example, the peak torque is compared with the torque rating of the gear box.

Another result of the analysis is the degree that the pumping unit is in or out of balance. If a unit is out of balance, this can contribute to overload of the gear box. It can also cause inefficiency and excess power usage.

Design

There are excellent sucker rod pumping design programs. Some production automation systems can calibrate the design program to the analysis program by matching the calculated downhole card with the one produced by the design program. When the program is calibrated, it can produce very accurate designs for determining the effect of pump size, pump speed, stroke length, etc. on beam loads, torque, rod loads, etc.

Some automation systems have the design program as part of the system. In other cases the design program is part of a separate system and information must be shared between the automation system and the system where the design program is run.

Another aspect of design, when sucker rod pumping is used for gas well deliquification, is how best to deal with the gas. In practice the best way is to set the pump intake below the perforations. The gas tends to rise up the annulus and the liquid flows down into the pump intake. This can be augmented by a gas separator. This

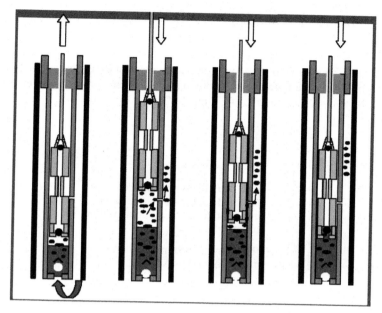

Figure 14.26 Gas vent pump design.

is normally a pipe below the pump intake that further assists with gas/liquid separation.

In addition, there are special pump designs that are better equipped to handle gas. An example is shown in Fig. 14.26. A small vent hole is placed in the pump barrel. The pump has twin plungers. When the pump plunger travels on the downstroke, it pushes some of the gas through the vent hole into the annulus. This can prevent the pump from becoming gas locked.

Optimization

Optimization of a sucker rod pumping systems addresses the following questions:

- What is the optimum pump size?
- What is the optimum stroke length?
- What is the optimum pump speed?
- What is the optimum motor size?
- What is the optimum pumping unit size?

Some automation systems are well equipped to help answer these questions. The first step is to calibrate the design program. Then the design program can be run for many cases of different pump sizes, stroke lengths, pump speeds, motor sizes, unit sizes, etc. The optimum solution is usually the one that produces the desired amount of liquid with the lowest capital and operating costs.

Use of sucker rod pumping on highly deviated or horizontal wells

Some operators are using sucker rod pumping on highly deviated or horizontal wells. While the normal approach is to install the pump in the vertical part of the well, above the "kickoff" point, some operators are installing the pump below the "kick off" point to obtain greater drawdown and increased production. There are at least three issues with this (Figs. 14.27 and 14.28).

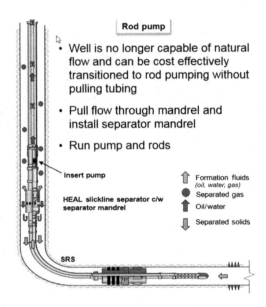

Figure 14.27 Rod pumping a horizontal well.

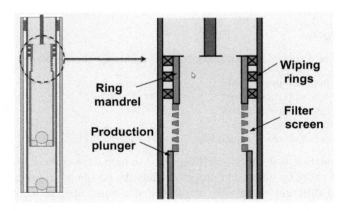

Figure 14.28 Handling sand in a sucker rod pumping well.

- First, the sucker rods must be protected to prevent excessive drag on the tubing. This can be done by using rod guides which help to keep the rods from dragging on the tubing and causing wear on the rods or holes in the tubing. Various types of rod guides are available in the industry.
- Second, there is often significant amounts of gas production associated with horizontal wells, and this gas production often comes in slugs. Sometimes there will be large slugs of gas, followed by slugs of liquid and periods of no production. Shown here is one method to prevent slugging. Other approaches are to install the pump intake on the low side of the tubing so the gas flows above the pump or by using a special gas separator. Some operators have used "sumps" at the bottom of the vertical well section and below the "kickoff" point. The liquid flows down into the sump where it can be pumped, and the gas flows up the tubing to the surface.
- Third, typically large amounts of sand are used to frac (fracture treat) horizontal wells. Special methods must be used to prevent sand from entering the pump and damaging the pump barrel and plunger.
- There are sand exclusion technologies available in the industry. Shown here is one where the large sand particles screened out and prevented from going between the plunger and barrel to prevent damage to the pump plunger and barrel.

14.5.3 Progressive cavity pumping

Progressing or sometimes known as progressive cavity pumping (PCP) is gaining in use for deliquification of gas wells and producing some oil wells. Fig. 14.29 shows a schematic of a rod-driven PCP installation. Some of the data items on the schematic are not activated.

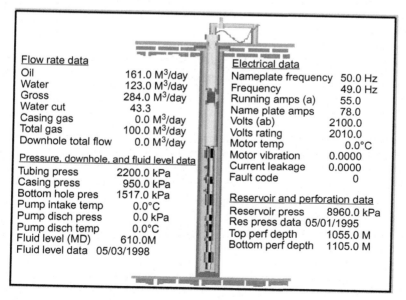

Figure 14.29 Schematic of PCP system.

The primary advantages of PCPs are:

- They can do a better job of handing gas than most other types of pumping systems. Because the flow-through the pump is continuous, there is less likelihood of gas locking.
- They can do a better job of handing solids than most other types of pumping systems. There are no valves and seats that can be eroded by solids.

Their primary disadvantages are:

- They have temperature limitations. The elastomers used to construct the pump stators have a limited temperature range, although this range is continually being extended.
- They have difficulty with highly aromatic liquids, again due to limitations imposed by the elastomers. This can be a problem if a gas well produces condensate.
- They have limitations in the amount of head they can produce, thus the depth of application may be limited.

PCPs are implemented in two different ways.

1. The traditional and most common way is to drive the PCP with a sucker rod from the surface. The rods are rotated at approximately 600 rpm to drive the PCP pump.
2. The other method which is gaining some popularity is the use of a downhole ESP motor. A gear reducer is used to convert the 3600 rpm ESP motor output to 600 rpm for the pump.

These differences have an impact on measurements, control, etc.

Measurements

For surface-driven PCPs, primary measurements are tubing pressure, tubing temperature, casing pressure, flow rate (if possible), torque on the sucker rod, and RPM of the sucker rod.

For downhole ESP motor-driven PCPs, measurements include surface pressures and flow rate, and in most cases downhole instruments are used to measure pump intake pressure (PIP), pump intake temperature, etc.

Control

With PCPs, it is critical that the well not be allowed to "pump off" as this will increase the temperature in the pump and may lead to elastomer swelling. With surface drives, some units use fixed speed drives (FSDs), but the pump must be operated to prevent "pump off." An enhanced method is control of the rod rpm with a VSD. With the VSD the pump discharge can be limited to keep the fluid level above the pump. The challenge is to know where the fluid level is. This is normally addressed by taking periodic fluid level measurements, although there are methods to determine the fluid level on a continuous basis, especially if downhole pressure is measured.

With a downhole drive the pump is controlled by using a VSD to control the rpm of the ESP motor. With downhole electrical motors, it is common to include a downhole measurement system, so the PIP is known, and the fluid level can be determined from this.

Figure 14.30 Surface PCP drive systems.

Temperature is an important limiting variable. Most PCP control systems can stop the pump or change its speed if the temperature is too high. Torque is also a limiting value. If the torque on a rod-driven system is too high, this may indicate a stuck pump.

Unique hardware

For rod-driven PCPs a drivehead is required to rotate the sucker rods. See Figs. 14.30 and 14.31 for examples of drive heads. In addition, a wellhead RTU is used for data acquisition and control, and a VSD may be used to control the rotational speed of the drive rods.

For downhole-driven PCPs a downhole electrical motor is connected to the PCP via a gear reducer and seal section. This is shown in Fig. 14.32. In this case a wellhead RTU is required for data acquisition and control, and a VSD may be used to control the speed of the PCP pump. Also, it is normal in this case to include a downhole measurement system to measure PIP, pump intake temperature, pump discharge pressure (PDP), and other variables.

Unique software

Wellhead RTUs for PCPs contain unique software to control and protect the pump with special shutdown conditions; they also gather and process data for pump alarms, surveillance, and optimization. These specialized RTUs are available from various service companies, so it would be counterproductive to build one from scratch.

There is also special "host" software for PCP analysis and optimization.

Figure 14.31 Surface PCP drive unit.

Specialized alarms

A feature of PCP automation is the detection of special alarm conditions:

- A hole in the tubing or a worn pump may be detected when torque drops, rpm increases (if not controlled), and production declines. The PIP will increase.
- Pump-off may be detected when torque increases gradually, then levels out, and then sharply increases eventually leading to a low-speed shutdown. During this time, rpm remains unchanged and production declines.
- A rod failure may be detected when torque drops, then levels out, and then sharply increases eventually leading to low-speed shutdown. Differential between PIP and PDP declines.
- A plugged flowline or WAX build-up may be detected when torque increases but production drops.
- A gas flow restriction or gas pressure build-up in the annulus may be detected when the gas pressure increases, and PIP stays the same as the pump speed adjusts.
- Periodically, PIP and PDP are collected with the pump stalled. A hole in tubing or fluid slippage can be detected when an increase in PIP and a decrease in PDP are observed. Casing gas pressure may also be measured during the stall test.
- A history of breakout torque at every start-up is recorded to analyze any variations over time; this helps in detecting pump slippage and failure.

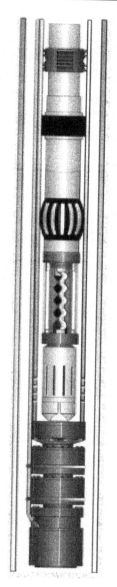

Figure 14.32 Electrical submersible PCP.

Surveillance

PCP surveillance is performed by continuously checking for special alarm conditions and reviewing trend plots. Fig. 14.33 shows a typical surveillance panel from a PCP automation system. It contains status information, trend plots, and operating parameters for a PCP installation.

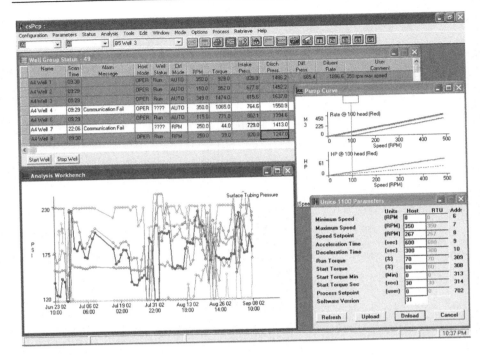

Figure 14.33 PCP surveillance panel.

Analysis

Some PCP automation systems incorporate a set of pump performance curves from the manufacturers and compare current pump operation with theoretical operation. This permits the operator to spot pumps that are worn out or are not operating up to specification for some reason. Fig. 14.34 shows a typical PCP analysis panel.

Fig. 14.35 shows the inputs to this analysis capability; Fig. 14.36 shows the outputs.

Another form of analysis may be performed with the PCP-RIFTS (PCP Reliability Information and Failure Tracking System). This is a system sponsored by an industry consortium for collecting PCP failure data and providing tools for analyzing the causes of failures. Objectives of PCP-RIFTS are to:

- Develop "standards" related to PCP utilization, including:
    - standard practices and guidelines to collect, classify, and analyze run life and failure data, such as the system developed by C-FER for ESPs in the ESP-RIFTS joint industry project (JIP) (http://www.esprifts.com) (see Section 14.5.4);
    - standard ways to evaluate elastomers with respect to the main failure mechanisms;
    - standard ways to name and classify elastomers that would help in their selection for specific applications; and
    - standard pump inspection and failure reporting practices.

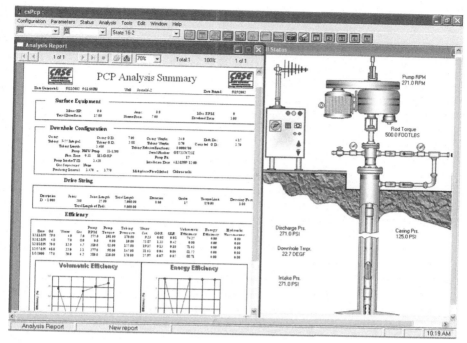

Figure 14.34 PCP analysis pane.

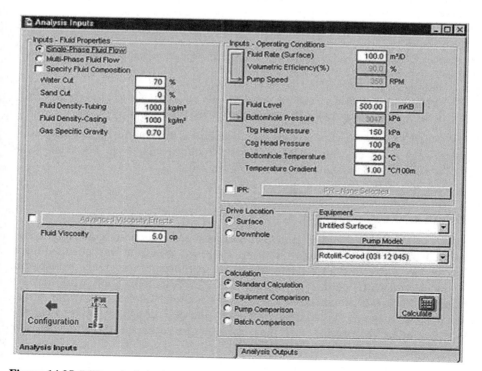

Figure 14.35 PCP analysis inputs.

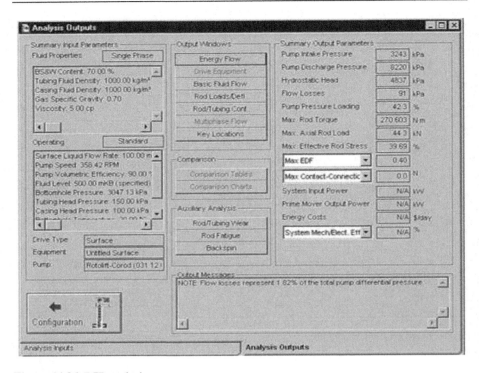

Figure 14.36 PCP analysis outputs.

- Further investigate ways to reduce the failure frequency of the most severe mechanisms. This might include:
  - investigation, collection, and review of past field trials and
  - conducting controlled laboratory and field tests on some of the relatively new PCP technologies, such as pump-off control systems and rotor coating materials.

PCP-RIFTS is not part of a production automation system, but some automation systems can interface with RIFTS and provide information to it. Feedback from RIFTS to automation systems is yet to be developed.

Design and optimization

Unlike sucker rod automation systems, most PCP automation systems do not contain PCP design and optimization software. However, there is at least one (and there may be more) excellent PCP pump design and optimization programs available in the industry. This program also contains a PCP analysis tool. Some work may be required to link PCP automation systems to this analysis and design program. When this is done, it should be possible to calibrate the PCP design program to current PCP operating information and then use this for accurate PCP designs and optimization.

14.5.4 Electrical submersible pumping

ESP is the method of choice for many oil production applications where it is necessary to produce large volume of liquid. It is also used quite extensively for deliquification of gas wells where large volumes of water must be produced. Fig. 14.37 shows a schematic of an automation system for an ESP system. This particular schematic is of a relatively complex system designed for wells that produce gas and some sand.

The primary advantages of ESP systems are:

- Can produce large volumes of liquid.
- Can produce from great depths.

The primary disadvantages are:

- Difficulty handling large volumes of gas.
- Difficulty handing solids.
- Difficulty with high temperatures—the motor must be cooled.

ESP systems use two types of drive systems—FSD and VSD.

FSD systems are used on the majority of ESP systems. However, most operators try not to turn ESPs off and on. Too many restarts may damage the system. So,

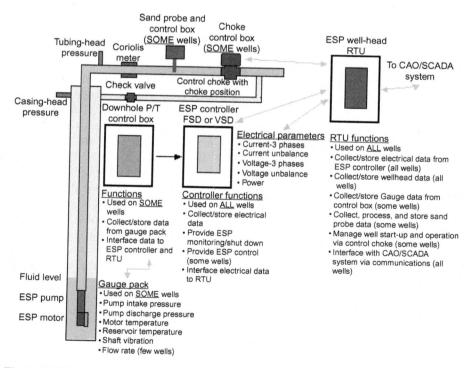

Figure 14.37 Schematic of ESP automation system.

special precautions must be used on wells with FSDs that may cycle due to gas production.

Many newer applications use VSDs to provide added control flexibility, especially if the well produces gas, as of course gas wells do. The choice of control system has an impact on the required instrumentation, hardware, and software.

Measurements

Three types of measurements are input into the automation system on most ESP installations:

1. Surface instruments are used to measure tubing-head pressure, casing pressure, temperature, and some others such as flow rate (on some wells), and sand production rate (on some wells).
2. Electrical parameters are measured by the FSD or VSD controller. Typically, many parameters are available from the controllers; the most commonly recorded are current (sometime all three phases), voltage (sometimes all three phases), current unbalance, voltage unbalance, and power consumption.
3. Downhole instruments are used on almost all ESP wells that have been installed in the last several years. These are used to measure PIP, pump intake temperature, motor winding temperature, PDP, motor shaft vibration, and downhole flow rate (on some wells).

Control

Three primary types of control are used on ESP systems.

Start, stop, and safety shutdown

The FSD or VSD controller provides start, stop, and safety shutdown capabilities. ESPs can be started manually or automatically. They can be started locally or remotely. They can be stopped manually or automatically, locally or remotely. And, ESPs can be shut down if the controller detects an unsafe operation such as low-current load, high-current load, low or high voltage, unbalance, etc. A majority of ESPs are run with only this form of control. If, for example, the well produces a significant amount of gas, the system may employ a form of "pump-off" control by shutting down on low-current load, waiting for a preset idle time, and then starting again.

ESP automation systems are designed to interface with the FSD or VSD controllers, download parameters to them (e.g., for low-current load shutdown limit), transmit commands to them (e.g., for remote start or stop), obtain electrical measurements from them (e.g., current, voltage, etc.), and monitor the status of the ESP system (e.g., numbers of starts and stops).

Control of wells with FSDs

If the production rate from the well must be limited, this can be done by controlling the pumping rate or the fluid level. It may be necessary to do this to limit sand production or gas interference. The pumping rate can be controlled by controlling the back pressure on the wellhead with a choke or back-pressure control valve. This is inefficient but effective. The biggest concern is the production rate must be kept high enough to provide adequate ESP motor cooling. The fluid level can be controlled by adjusting the back pressure on the annulus or by recycling some liquid back into the well's annulus. Again, both of these methods are inefficient. However, they are used relatively frequently because so many wells use FSD control.

Control of wells with VSDs

In recent times (certainly in the 2000s), most wells that require production rate control use VSDs. Here the ESP speed and output can be controlled, over a reasonable range, by adjusting the frequency of the electrical current which controls the speed of the ESP. The most common control variable is the fluid level. It can be determined with reasonable accuracy by knowing the surface casing pressure, the measured downhole PIP, and the composition of the gas and liquid in the annulus. The pump discharge can be controlled to accurately maintain the fluid level at the desired elevation above the pump intake. See Fig. 14.38 for an example of how the fluid level is determined based on knowledge of the surface casing pressure and the measured downhole (pump intake) pressure.

VSDs are also used on wells with gas production. The ESP can be slowed down and speeded up to work a potential gas lock through the pump.

Control of wells on start-up

In some cases, especially with sandy or gassy wells, it is important to start them slowly enough to avoid excessive reservoir pressure drawdown and excessively high inflow rates, but fast enough to avoid motor cooling problems.

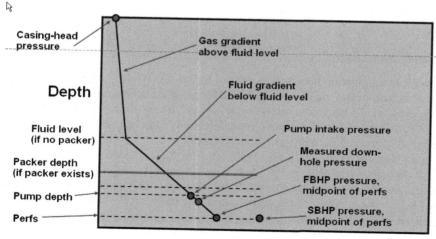

Figure 14.38 Casing pressure gradient.

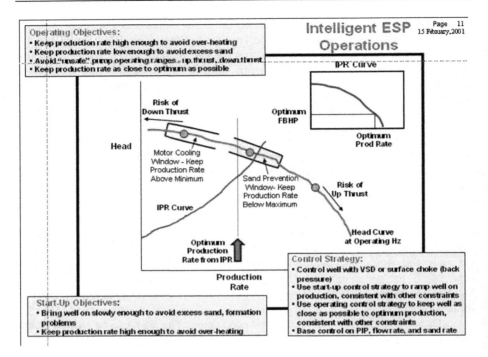

Figure 14.39 ESP start-up guidelines.

Fig. 14.39 shows start-up guidelines for such cases. The production rate must be kept high enough to avoid downthrust and to provide adequate motor cooling. And, it must be kept low enough to prevent upthrust and high reservoir pressure drawdown that can lead to sand influx or excessive gas production.

Some ESP automation systems provide control for this special start-up mode.

Unique hardware

Specialized automation hardware is needed for an ESP system. The "good news" is that some companies make special RTUs for several methods of artificial lift. So, while the specific hardware and software are unique for each type of artificial lift, the RTUs' physical components, communications interface, etc. are the same or similar for several different types of artificial lift.

In most cases, as shown in Fig. 14.40, an ESP RTU is interfaced to an ESP controller and other components. In some cases, companies are working on enhanced ESP controllers that also provide RTU functionality. ESP controllers are normally provided by the supplier of the ESP system, not by the RTU supplier. If use of an enhanced controller to provide both ESP control and RTU functions is desired,

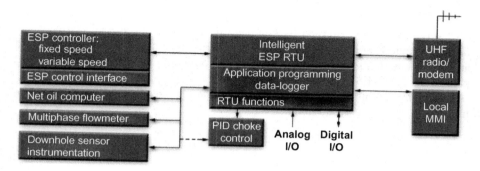

Figure 14.40 Required RTU and associated components.

check with the ESP supplier. However, make certain that the ESP controller can really provide all of the needed RTU functions.

Unique software

Clearly the RTU for ESP automation must contain special software logic. Fortunately, this has been very well developed and proven. It would make no sense to develop this from scratch. This special logic interfaces with the ESP controller (either FSD or VSD), reads the surface instrumentation, interfaces with the downhole instrumentation, provides some forms of control when needed (e.g., control of a surface back-pressure choke), and interfaces with the operator via a local interface and commutation with the host automation system.

As indicated in Section 14.5.4.3, some companies are working on ESP controllers that can provide the functionality of both the controller and the RTU. This has the potential to reduce both cost and equipment "footprint" at the well. However, the precaution mentioned earlier must be used to be certain when the required controller and RTU functionality are provided.

Specialized alarms

A typical ESP automation system may produce dozens of alarms. Some are relatively typical; for example, high current, low current, high pressure, low pressure, etc. However, several are special for the ESP operation:

- Low production rate. Risk of overheating motor due to inadequate cooling.
- Very low production rate. Risk of downthrust.
- High production rate. Risk of excessive drawdown that may increase sand or gas production.
- Very high production rate. Risk of upthrust.
- High motor temperature. Another indication of potential motor overheating.
- High motor shaft vibration. An indication of motor–protector–pump alignment problems. This may lead to premature bearing wear out or failure.

Surveillance

ESP surveillance is performed by continuously checking for special alarm conditions and reviewing trend plots. Fig. 14.41 shows a typical surveillance panel from an ESP automation system. It contains status information, trend plots, and operating parameters for an ESP installation.

In addition to this type of specialized panel, normal alarm reports, status reports, trend plots, etc. are used for ESP surveillance.

Analysis

ESP analysis is performed to answer several questions:

- How well is the ESP well performing?
- How well is the ESP pump system performing?
- How well are the ESP system and the well working together as a system?

Figs. 14.42 and 14.43 are designed to help answer the first question. Fig. 14.42 shows the pressures and pressure gradients in the well, below and above the pump. It shows the pressure (head) increase crated by the pump. It shows the current fluid level. The liquid beneath the fluid level exerts a pressure (back pressure) on the formation which inhibits inflow from the reservoir to the wellbore.

Fig. 14.43 shows the inflow performance relationship (IPR) for the well, based on the Vogel IPR method. It shows the current flowing bottomhole pressure (FBHP), what this would be if the pump were operating perfectly according to the

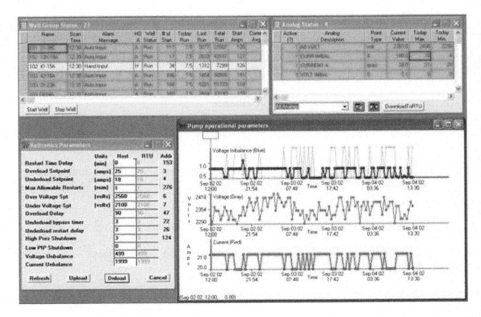

Figure 14.41 ESP surveillance panel.

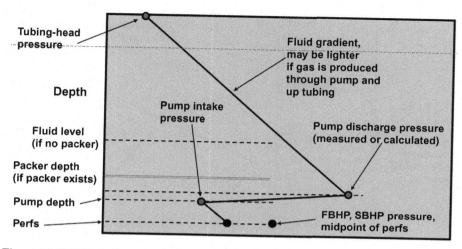

Figure 14.42 ESP wellbore pressures and gradients.

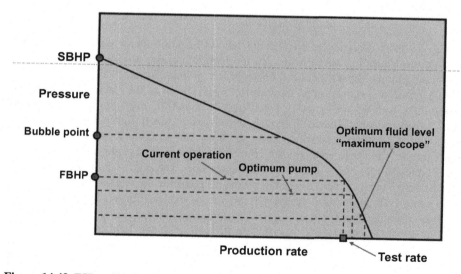

Figure 14.43 ESP well inflow performance relationship.

manufacturer's specifications (this is discussed further in the next paragraph), and what this would be if the well were operating with the "optimum" fluid level. In this example the well has a relatively high bubble point pressure so the IPR curve is significantly curved downward below the bubble point. Therefore the potential increase in production rate by operating at a lower bottomhole pressure is less than it would be if the well had a very low bubble point pressure and the IPR curve were closer to a straight line.

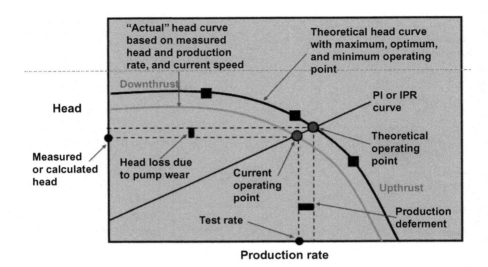

Figure 14.44 ESP head curve.

Fig. 14.44 is designed to answer the second question about pump performance and the third question about how well the well and pump are working together. It shows the pump's theoretical head curve based on manufacturer's data.

The reduced [orange (gray in print version)] curve is the "actual" head curve determined from the measured or calculated head and the measured production rate from a well test or wellhead flow rate. It shows the downthrust point, the upthrust point, and the optimum operating point. It shows the well's IPR or PI (productivity index) curve; the curve appears different on this plot since the plot shows head versus production rate rather than pressure versus rate. The point where the IPR curve and the pump head curve cross is the operating point for the well.

In some ESP automation systems, this plot is expanded further to show the degradation of the head curve due to gas interference or pumping of heavy oil.

Another form of analysis may be performed with the ESP-RIFTS. This is a system sponsored by an industry consortium for collecting ESP failure data and providing tools for analyzing the causes of failures. As of March 2014, the consortium had 24 member companies. The ESP-RIFTS database contained information on over 105,645 ESP installations in 758 oil fields around the world. In addition to conventional ESP systems, the database includes a number of "nonconventional" ESPs, including subsea, coiled tubing deployed, water source well ESPs, downhole oil/water separation ESP systems, and PCP ESP systems.

ESP-RIFTS is used all over the world, as shown on this map.

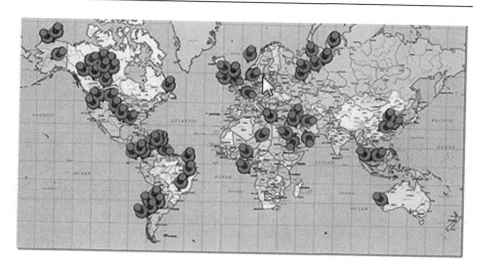

Objectives of ESP-RIFTS are to:

- Facilitate sharing of ESP run life and failure information among operating companies.
- Ensure quality and consistency of data in the system.
- Incorporate useful analysis tools.

Benefits of the system are:

- Benchmark the performance of a company's ESPs against that of other operators.
- Improved decision-making, based on actual reliability data.
- Enhanced capability to predict run life in new applications.
- Ability to predict workover frequency in existing applications.

ESP-RIFTS is not part of a production automation system, but some automation systems can interface with RIFTS and provide information to it. Feedback from RIFTS to automation systems can also be provided.

Design and optimization

Unlike sucker rod automation systems, most ESP automation systems do not contain ESP design and optimization software although at least one automation system does allow use of "what if" questions. For example, what will be the effect on production rate if the type of pump, size of pump, or number of pump stages is changed?

However, there are at least two (and there may be more) excellent ESP pump design and optimization programs available in the industry. These programs also contain an ESP analysis tool. Some work may be required to link ESP automation systems to these analyses and design programs. When this is done, it should be possible to calibrate the ESP design program to current ESP operating information and then use this for accurate ESP designs and optimization.

14.5.5 Hydraulic pumping

Hydraulic lift is an artificial lift method in which the energy to lift fluids to the surface is provided by a power fluid (generally, recirculated pressurized produced fluids). There are two common forms of hydraulic lift—jet pumps and hydraulic piston pumps. Of these, jet pumps are by far the more common and are applicable over a wide range of rates and setting depths. In certain cases the pumps can be retrieved by reverse-circulating to surface or via wireline. Following retrieval the pumps can be repaired at the wellsite. The system is suitable for wells with high deviation, making it a choice for offshore applications. Also, jet pumps can be adapted to existing bottomhole assemblies and sliding sleeves. The systems are excellent for producing viscous crude. Limitations include potentially complex completions, high-pressure surface equipment, and potentially problematic fluid measurement. Depending on the accuracy of flow measurement, it can be difficult to determine what portion of the flow stream is produced by the formation and what portion is provided as power fluid.

While hydraulic lift completion architecture can be complex and varied, two major classes of completion are most common: (1) open and (2) closed completions. In an open completion, power fluids are pumped down a conduit to the downhole pump where they commingle with produced fluids and the combined flow stream is produced to the surface. In generally, in such a completion, the power fluid is pumped down the tubing and produced fluids flow up the tubing/casing annulus. In a closed completion, power fluid is pumped down a conduit to the pump and commingled with the produced fluids prior to being pumped up another closed conduit. Both the produced and power fluids are isolated from the tubing-casing annulus. Because of the need to isolate produced fluids from the produced gas, closed completions are most common in gas well deliquification applications. In a typical deliquification application a closed completion is run without a packer. This enables the power fluid to be pumped down one tubing string, produced fluids to flow up a second parallel tubing string, and gas to flow up the casing-tubing annulus.

The use of hydraulic pumps in deliquification applications is rare. This is due to the following limitations of the system. First, the need for a closed system makes hydraulic lift impractical in many applications where casing sizes are small. Second, the clumping of solids such as coal fines and clogging of the pump intake makes hydraulic lift impractical for coal bed methane applications and other deliquification applications in which a significant number of solids are produced.

Hydraulic lift is the least common artificial lift method, with worldwide well-count estimated at less than 5% of the overall population. As a result, hydraulic lift is one of the most underserved lift methods in terms of automation technology. There are no known real-time host applications or wellsite controllers designed specifically for use in hydraulic lift applications. Instead, where automation exists, operators use existing automation equipment and human/machine interfaces to provide basic surveillance and control capabilities.

Surveillance

Basic parameters for surveillance of hydraulic lift systems include:

* flowing wellhead pressure,
* wellhead injection pressure,
* wellhead temperature,
* production choke position,
* surface pump suction pressure,
* surface PDP, and
* surface pump status.

Operators may install permanent downhole gauges and monitor parameters such as pump intake and discharge pressures. However, such installations are rare, due in large part to the complex completion configurations required and the ability to reverse-circulate pumps to surface (pump reliability is of less concern than in other lift methods).

Control

Basic control parameters for hydraulic lift systems include:

* surface pump status (on/off),
* surface pump speed, and
* production choke position.

14.5.6 Chemical injection

In many gas wells, deliquification can be achieved by injection of surfactants. The basic principle is that foam reduces the surface tension between the liquid and gas phases. Because surface tension is directly proportional to the critical velocity for liquid loading, this will reduce the critical velocity of the well. Provided the critical velocity can be reduced below the in situ gas velocity, this may be sufficient to unload the well and prevent future liquid loading from occurring.

Automation of chemical injection systems is limited. In many cases the surfactant is injected on a periodic basis by means of a spooling unit, where no data are directly uploaded into the automation system. In other cases a chemical injection system may be permanently installed, whereby a control line is run from the surface to a downhole chemical injection valve. At the surface a surfactant drum and small injection pump are installed and plumbed into the wellhead. Such systems operate autonomously and require no external control. No specialized host systems exist for monitoring these systems. Should an operator wish to integrate these devices into their automation systems, basic surveillance and/or control would be provided through existing control systems and human/machine interface.

Surveillance

Basic parameters for monitoring chemical injection systems include:

- injection pump status (on/off),
- chemical injection rate, and
- gas production rate.

Control

Control of these systems is limited to turning on and off the injection pump.

14.5.7   Gas-lift

Gas-lift is the second most used method of artificial lift, after sucker rod pumping. Fig. 14.45 shows a schematic of a typical gas-lift system. This is referred to as a "closed loop" system since the gas is compressed, injected, produced, gathered, recompressed, and recirculated around the system.

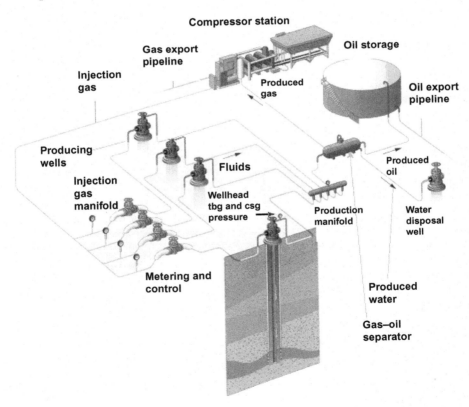

Figure 14.45 Closed loop gas-lift system.

The primary advantages of gas-lift are:

- It "likes" gas. The more gas a well produces, the better.
- It can be installed in almost any well deviation. For wireline installation of gas-lift valves the wellbore deviation must not be greater than about 70 degrees from vertical.
- It can be used in any well depth.
- Downhole gas-lift equipment (valves, orifices, etc.) can be installed by wireline; a workover rig is not required.
- It can handle some degree of sand production. There are no parts in the flow stream so sandy fluid can flow up past the gas-lift valves without damaging them.
- In gas wells, gas can be injected to achieve and maintain critical velocity.
- It is not necessary to cycle wells, as with plunger and pumping systems. Thus the wells stay on production all of the time.

The primary disadvantages or challenges are:

- There must be a source of high-pressure gas; this is usually provided by a compression station or another source of high-pressure gas.
- A good control system is required to maintain optimum gas-lift operation.
- Diligence is required to keep the gas-lift system design in line with the operation of the well. A common problem is a too large injection port or orifice that can lead to unstable (heading) operation.

Measurements

For effective gas-lift operation, there are certain required measurements. There are also optional measurements that can enhance the operation. Some typical measurement devices are shown in Section 14.3.

Required measurements are:

- Gas-lift injection rate. This may be measured at the wellhead or at an injection manifold.
- Gas-lift injection pressure. Normally, this is the casing-head pressure. It should be measured at the wellhead, downstream of any pressure drop devices such as chokes and control valves.
- Production pressure. Normally, this is the tubing-head pressure. It should be measured at the wellhead, upstream of any pressure drop devices such as chokes.

Optional measurements are:

- Gas injection temperature. This may be required to compensate the gas injection rate measurement.
- Production temperature. This is used by some operators to provide surveillance of gas-lift liquid production rates.
- Production rate. In some cases, it is possible to measure (or accurately estimate) the liquid production rate on a continuous basis. This has obvious advantages where it can be provided.
- Downhole variables. In some cases, downhole measurements (and sometimes control) are possible. There are systems that provide downhole pressure, temperature, and gas injection rate measurements. In some of these the rate of gas injection through the operating gas-lift valve or orifice can be controlled.

Control

There are two fundamental methods of gas-lift: continuous and intermittent. There are multiple variations on each of these.

For continuous gas-lift the objective is to control the rate of gas injection on a continuous basis and keep it as stable as possible. The three goals are to inject gas as deep as possible, keep it stable, and keep the injection rate optimized, in that order. To keep the injection rate stable the injection rate at the surface must be in balance with the gas flow capacity of the downhole injection gas-lift port or orifice.

For intermittent gas-lift, there are two primary control methods: choke control and time-cycle control. With choke control the rate of gas injection on the surface is controlled by a choke or control valve. The volume (and pressure) of gas gradually increases in the annulus until the pressure is high enough to open the intermittent gas-lift operating valve. Fig. 14.46 shows a pilot gas-lift valve which is typically used for intermittent gas-lift. When the valve opens a large "slug" of gas is injected into the production stream (tubing) beneath the accumulated slug of

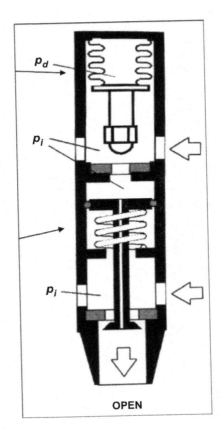

Figure 14.46 Pilot gas-lift valve.

liquid. With time-cycle control, injection is cycled by the use of a time-cycle controller on the surface.

The advantage of choke control is that the gas flow rates and pressures remain relatively stable in the gas-lift distribution system. The disadvantage is that the downhole gas-lift injection valve must be changed to change the injection cycle frequency or volume of gas per cycle.

The advantages of time-cycle control are that the injection frequency and the volume of gas per cycle can be controlled on the surface. The disadvantage is that the rates and pressures in the gas-lift distribution system may vary significantly, potentially causing upsets to the system and other wells served by the system.

Unique hardware

Gas-lift requires use of gas-lift mandrels, valves, and other downhole components.

In addition, special gas-lift automation hardware is required.

Gas-lift mandrels are installed in the tubing string. There are two primary types: conventional and side-pocket. Conventional mandrels require that the gas-lift valves be installed in the mandrels on the surface and run with the tubing. Side-pocket mandrels are designed to permit gas-lift valves and other downhole components to be installed with wireline. A schematic of a side-pocket mandrel is shown in Fig. 14.47.

There are several types of gas-lift valves. The most common are unloading valves, continuous gas-lift operating valves or orifices, and intermittent gas-lift pilot valves.

Unloading valves may be injection pressure (the common) or production pressure operated. Their purpose is to unload liquid from the well's annulus so gas can be injected at desired operating depth, deep in the well. These valves are intended to be used only for unloading and should remain closed at all other times, so gas can be injected below them in the operating valve or orifice. A schematic of an injection pressure operated unloading gas-lift valve is shown in Fig. 14.48.

Typically an orifice valve (actually a gas-lift valve with no stem so it cannot close) is used at the operating depth. Since this valve cannot close, it cannot throttle the gas injection rate into the well; it is always fully open. It is important that the size of the orifice be designed to inject the right amount of gas, so there is a balance between the rate injected at the surface and the rate that can flow through the operating valve or orifice. Otherwise the well will tend to become unstable.

For intermittent gas-lift a pilot operated valve is normally used for the operating valve. This valve can instantaneously go "full open" so the "slug" of gas can be rapidly injected beneath the "slug" of liquid that has accumulated in the wellbore during the "down" portion of the intermittent gas-lift cycle.

Typically, facility and wellhead RTUs are used to provide gas-lift measurement and control. When gas-lift injection is measured and controlled at an injection manifold, this is normally done with a facility RTU or DCS at the manifold station. Typically a small wellhead RTU is used to measure the casing-head injection

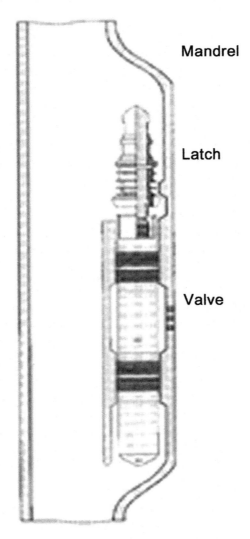

Figure 14.47 Side-pocket mandrel with valve installed.

pressure, the production pressure, and potentially the injection temperature, production temperature, production rate, and in some cases, downhole variables.

Unique software

Specialized software is required for gas-lift automation. This includes software in the RTU(s) to monitor and control, and to provide alarm detection, surveillance, analysis, design, and optimization. This software has been developed by several companies and is available in the industry.

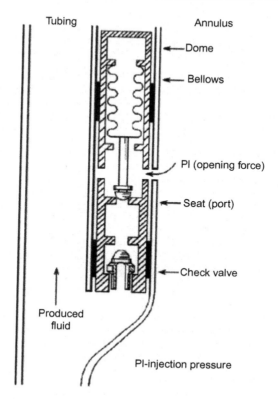

Figure 14.48 Injection pressure operated unloading gas-lift valve.

Specialized alarms

There are several specialized alarms that are important for gas-lift. Some of these include:

- Injection pressure heading. The injection pressure is fluctuating; this will cause inefficiency in the well and may indicate multipoint injection—injection through more than one gas-lift valve all or part of the time.
- Production pressure heading. This almost always occurs in association with injection pressure heading. It may also occur by itself, especially if the tubing size is too large for the current production rate.
- Injection gas freezing. When a pressure drop is taken across a gas control choke or valve, the temperature drops due to the Joule–Thompson effect. If there is water vapor in the gas, it may freeze. The resulting hydrates can block the injection path.
- Gas blowing around. This can occur if gas is being injected through an upper gas-lift valve, or a tubing leak. Gas is being injected but no liquid is being produced.
- Many others. There are many other alarms that are common in gas-lift systems. They are too numerous to list in detail in this section.

Surveillance

Gas-lift surveillance is primarily focused on keeping the wells operating properly and detecting any alarm conditions that need to be addressed. For continuous gas-lift wells, this means keeping the gas-lift injection deep, stable, and optimum. Fig. 14.49 shows a typical gas-lift surveillance plot with the injection pressure fluctuating or heading.

For intermittent gas-lift wells, it means keeping the injection cycles properly timed with the desired volume of gas per cycle.

For gas wells, another objective is to keep the gas injection rate sufficient to maintain critical velocity.

An effective method for gas-lift surveillance is the Well Tracer technique. Here a small amount of CO_2 is injected with the injection gas-lift gas. The time for the CO_2 to return to the surface is determined. This gives an induction of the depth of the gas-lift injection, whether it is through a gas-lift valve or orifice, through multiple gas-lift valves, or through a hole in the tubing. An example is shown in Fig. 14.50, which indicates multipoint injection through the first and fifth valves.

Analysis

The purpose of gas-lift analysis is to determine the root cause(s) of operating problems. A good way to do this is to analyze the performance downhole in the well. Fig. 14.51 shows a "downhole" plot of injection pressure profile(s) and production pressure profile(s) taken at different points on the plot of injection and production pressure versus time.

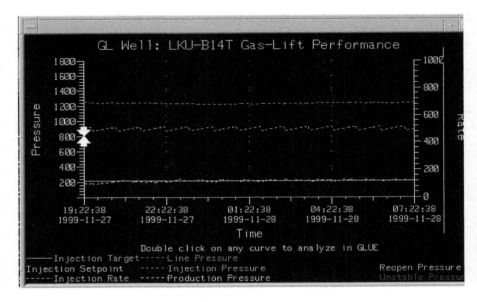

Figure 14.49 Example of injection pressure heading.

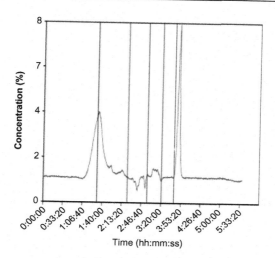

Figure 14.50 Well Tracer plot.

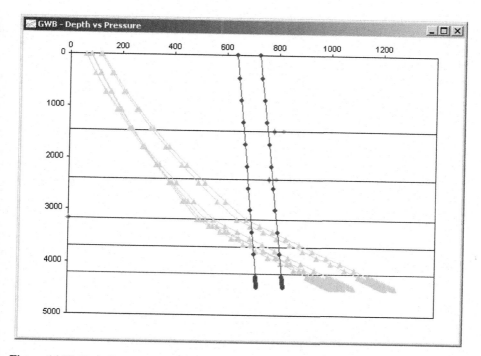

Figure 14.51 Downhole plot of injection and production pressure profiles.

With this the operator can determine where gas is being injected from the annulus into the tubing, and if this may be occurring at different depths when a well is unstable or heading, as shown in this example.

Analysis is also used to help determine or diagnose the causes of various gas-lift alarms. For example, there are many "shapes" of injection and production pressure plots versus time that are indicative of specific types of gas-lift problems.

Design

There are two primary objectives with gas-lift design. The first is to determine the desired spacing of gas-lift mandrels. This is done before the well is completed, or when it is recompleted after a workover. There are several gas-lift design programs available in the industry. There are also several industry courses offered on gas-lift design.

The second objective is to design the unloading gas-lift valves and the operating valve or orifice. Typically, this is done when it is necessary to place a well on gas-lift and it may be redone several times over the life of the well as the well's conditions change. Again, there are several programs available to do this. And there are courses that cover this aspect of design.

An issue is that most gas-lift design programs and courses focus on use of gas-lift for oil wells. Some adjustment, particularly of the selection and design of the operating valve or orifice, may be required for use of gas-lift to deliquefy gas wells.

Use of gas-lift in horizontal wells

Gas-lift can be used in horizontal wells. Often this means installing the gas-lift below the packer as shown in Fig. 14.52. There are several configurations for use of gas-lift in horizontal wells. The advantages are: (1) gas-lift is not affected by the horizontal configuration, as it is with most pumping systems, (2) gas can be injected very deep in the well, sometimes even near the toe of the well, and (3) the injected gas helps to sweep liquid slugs out of the horizontal part of the well, thus preventing blockage to gas and liquid flow.

Dual gas-lift

Gas-lift can be used in dual wells where there are two completions in one wellbore. Gas can be successfully injected in both parts of the dual, especially if the right type of gas-lift valve is used to prevent one some of the dual from "robbing" all or most of the gas and preventing it from entering the other side. A type of valve that is good for this application is the "venture orifice" valve.

Optimization

In theory, optimization is the process of determining and using the right amount of gas to optimize the economic operation of the well. In practice, for oil wells, it is

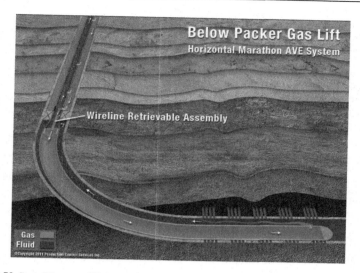

Figure 14.52 Installing gas-lift below the packer in a horizontal well.

better thought of as the optimum allocation of gas, given that the amount of gas available for lift will rarely be exactly equal to the sum of the optimum amounts for each well.

For gas wells, two conditions must be considered. If there is a liquid level in the well, the first objective is to remove it, at least down to the maximum depth of gas-lift injection. During this process, classical optimization is not important. The goal here is to unload the well and remove the liquid column. When the liquid column has been removed, the goal becomes to use the optimum amount of gas to maintain critical velocity. Critical velocity is the rate of gas flow (produced gas plus injected gas) that is required to continuously remove liquid from the wellbore and maintain a minimum FBHP. Critical velocity can be determined by using the Turner or Coleman equations, depending on the well's conditions.

14.5.8 Wellhead compression

Compression is one of the most effective means of deliquifying gas wells. Benefits of compression are twofold. First, by reducing the FBHP, more drawdown is achieved, resulting in a higher production rate. Second, the reduction of in situ pressure throughout the production string reduces the critical velocity required to remove fluids.

No specialized host systems exist for monitoring wellhead compression systems. Should an operator wish to integrate these devices into their automation systems, basic surveillance and/or control would be provided through existing control systems and human/machine interface.

Surveillance

Basic parameters for monitoring wellhead compression systems include:

- suction pressure,
- discharge pressure,
- suction temperature,
- discharge temperature, and
- compressor speed.

Control

Control of these systems is limited to turning on and off the compressor and adjusting the compressor speed.

14.5.9 Heaters

Thermal lift is a recent development in the world of gas well deliquification. The method entails the installation of an ESP power cable in the wellbore, either by strapping the cable to the tubing or running inside coiled tubing. The principle is simple: electricity from the power cable generates heat, which in turn raises the temperature of the produced gas above the dew point. The amount of heat generated is a function of the power supplied to the cable, which in turn is dictated by the voltage (and frequency) of the system. While this method is relatively inefficient, it has proven effective in the field and may be a good choice for certain applications.

No specialized host systems or control systems exist for automating thermal lift systems. Control for such systems can be provided by either adjusting the tap settings in a switchboard or adjusting the operating frequency of a VFD. Surveillance would be by existing host systems.

Surveillance

Basic parameters for monitoring thermal lift systems include:

- voltage,
- current,
- frequency,
- gas flow rate,
- flowing wellhead pressure, and
- flowing wellhead temperature.

Control

Control of these systems is limited to adjusting electrical parameters in the switchboard or VFD which affect output power. Because current is a function of the operating load (which is negligible), voltage is the key parameter affecting the output power and heat generated by the system. If a switchboard is used, voltage is controlled by adjusting the tap settings in the switchboard. If a VFD is used, voltage can be controlled by adjusting either the base frequency of the drive or (more likely) the operating frequency. Either of these parameters can be adjusted either manually in the drive or by means of changing setpoints from the host system.

14.5.10 Cycling

When wells initially load with liquids, it is sometimes possible to unload them by cycling. Cycling, also known as stop-cocking or intermitting, refers to the process of intermittently cycling the well between flowing and shut-in conditions. When the well is shut in, bottomhole pressure increases and pressurized gas accumulates in the annulus. This increased well pressure pushes all or part of the fluids back into the formation, allowing the well to flow again once the well is opened to production. In essence, cycling is akin to using plunger lift without the plunger. The key consideration in optimizing a cycling well is to maximize the amount of time the well is producing while minimizing the amount of time that the well is flowing below the critical velocity for unloading.

No specialized host systems exist for monitoring wells on cycling production. Should an operator wish to automate such a well, basic surveillance and/or control would be provided through existing control systems and human/machine interfaces.

Surveillance

Basic parameters for monitoring wells on cyclic production include:

* flowing wellhead pressure,
* casing-head pressure,
* gas flow rate, and
* production control valve state (open/closed).

Control

Control of these wells is limited to opening and closing an automatic control valve. This can be accomplished by a simple time-cycle controller or through more sophisticated control logic such as that contained in a plunger lift controller.

14.5.11 Production allocation

Production allocation is the process of determining or estimating the production of each well in a system, based on the actual measured production from the system (e.g., production facility) and the measured or estimated production of each well.

Production automation can assist with this process by obtaining the measurements of actual production from the system and the measurements or estimates of production of each well. The allocation is then a simple mathematical process using the following equation:

$$Q_{ai} = Q_{mi} \times \sum (Q_{mi})/Q_s$$

where Q_{ai} is the allocated production to well i; Q_{mi} is the measured or estimated production for well i; and Q_s is the measured production for the production system.

This normally works well for gas since often the volume of gas produced by each well is measured with some relative degree of accuracy. It may be a problem when attempting to determine the allocated production of oil, condensate, and water, since these volumes are typically not measured on a continuous basis, and often only with well tests, which may be taken infrequently.

If production rates are "measured" with well tests, it is also necessary to know or determine the "on production" time of each well to calculate the estimated production volume over the time period of the allocation. Again, the production automation system can assist with this if it has some way to determine the actual "on production" time of each well from its surveillance capabilities.

There are some new techniques being considered in the industry to permit the continuous measurement (or at least accurate estimate) of both liquid and gas production rates from each well. When this information can be gathered by a production automation system, it can permit accurate, continuous, or at least daily allocation to each well. Stay tuned for this to become available in the next few years.

14.5.12   Other unique applications

Production automation can assist with other gas well applications. One example is automatic adjustment of gas production when this is necessary to meet specific demands for gas delivery. The automation system cannot only adjust the production rates to meet the "system" delivery objectives, it can do this in a way to maximize the efficiency of deliquification. For example, if the overall production rate must be temporarily reduced, it may be possible to do this by pinching back on free-flowing wells where liquid loading is not yet an issue, and permitting wells with artificial lift systems for deliquification to continue to work at their optimum level.

14.6   Automation issues

There are a number of issues that must be understood and addressed in considering, defining, justifying, designing, building, installing, using, and maintaining a production automation system. A common fault of many systems is that focus is placed on purchasing and installing equipment without giving adequate consideration to all of the issues. If some are ignored, the system may fail or be underutilized, not due to poor equipment but because of inadequate attention to other details needed for an overall successful system. The purpose of this section is to define and discuss some of the more important issues with the hope that they will be included in the overall production automation project plan.

14.6.1 Typical benefits

Several types of benefits may be realized from a gas production automation system. For management the most important are tangible, quantitative, economic benefits that justify the cost of the system and provide a direct economic payout. For others, there are intangible, qualitative benefits that may be as or more valuable but are difficult to quantify in monetary terms. The production automation system must be designed to provide both types of benefits. The purpose here is to briefly discuss some of the more important tangible and intangible benefits. Actual monetary values can only be placed on these in the context of specific field conditions.

Tangible benefits

Tangible benefits can be measured in economic terms. Some of the more common are:

- Increased production. Production can be increased by:
  - Early detection of downtime and correction of problems so wells are on production a high percentage of time.
  - Keeping artificial lift systems operating at peak efficiency at all times.
  - Keeping wells flowing (producing) at or above critical velocity or keeping wells "pumped off" to avoid accumulation of liquid in the wellbore.
  - Keeping wells on production until their true economic limits are reached, thus increasing ultimate recovery from the reservoirs.
- Reduced operating costs. Operating costs can be reduced by:
  - Needing fewer people to monitor and control the wells.
  - Needing less automotive, boat, or helicopter travel to visit the wells.
  - Optimizing the use of energy (e.g., electricity, gas, etc.) to operate artificial lift systems.
  - Optimizing the use of expendables (e.g., chemicals).
- Reduced maintenance costs. Maintenance costs can be reduced by:
  - Keeping artificial lift systems (especially pumping systems) operating within their safe operating envelope.
  - Detecting problems before they become failures.
- Reduced capital costs. In some cases, capital costs can be reduced by:
  - Deploying more expensive artificial lift systems only when needed.
  - Not overdesigning artificial lift systems.
- Artificial lift specific benefits. There are some benefits that are unique to each type of artificial lift system. Some of the following have been documented by the industry for oil well production:
  - Sucker rod pumping: 7% production increase, 20% energy reduction, 35% maintenance cost reduction.
  - ESP: 3% production increase.
  - Gas-lift: 5% production increase, 10% reduction in gas usage, reduced compressor capital expenditure (CAPEX).

These benefits are not "over and above" those listed earlier. They come from things such as reduced downtime and improved efficiency. However, they confirm that these types of benefits are real.

Intangible benefits

Intangible benefits are more difficult to quantify, but they may be as or more important: Some of the more common are:

- Safety
    - Operators need to visit wells less frequently. This reduces travel hazards.
    - When there is a problem, operators know about the problem in advance and can be prepared with the right equipment to deal with it.
- Environmental protection
    - Production can be remotely (automatically) stopped if a problem (e.g., leak) is detected.
- Personnel
    - Operators and others gain a better understanding of their wells and equipment.
    - This can be an incentive when seeking to hire good people.

14.6.2   Potential problem areas

It is important to be aware of potential problem areas, so they can be avoided and/ or addressed. Some of these are best avoided or addressed by using automation experts—either from within the operating company or from a service company or consultant. Some of the potential problem areas are given in the following sections.

Automation system design

An important consideration in system design is to "keep it simple," but not too simple. The primary objective must be to design a system that can achieve the benefits defined in Section 14.6.1. If the system cannot achieve the required benefits, it will not be fully utilized. If it is not used, it will not be maintained. If it is not maintained, it will fail.

The recommended process is as follows:

- Define the gas production operation to be automated.
- Describe the benefits to be achieved.
- Design the system to achieve these benefits.

Instrumentation selection

As stated in Sections 14.3.1 and 14.3.3, instrumentation is the core of the system. Unless gas and oil production variables can be reliably and accurately measured and controlled, the rest of the system is worthless.

The following process is recommended for instrument selection:

- Define the variables that need to be measured and controlled.
- Select instruments of proven reliability. Reliability is more important than absolute accuracy. And it is more important than initial cost. When purchasing instruments, cheaper is not necessarily better.
- Use an experienced instrument engineer to design and implement instrument installation and commissioning.

Automation hardware and software selection

As discussed in Sections 14.3.4 and 14.3.5, there are many suppliers of RTUs, PLCs, and host production automation computer systems.

In general, RTUs and PLCs should be selected based on the specific application for which they will be used. That is, select an RTU that is specifically designed and programmed to be an RPC for a sucker rod operation. Select an RTU that is specifically designed and programmed to be a plunger lift controller for plunger lift. Select an RTU that is specifically designed and programmed to be a gas-lift controller for gas-lift. This means that if a field uses two or more types of artificial lift, it may have two or more types of RTUs or PLCs. This may cause problems with supply, spare parts, and training of operations and maintenance staff. However, these problems are small compared to the problems that will arise if an attempt is made to "force fit" one type of RTU or PLC to serve various types of operations for which it is not designed. And, if appropriate care is taken in the communications area (see Section 14.6.5), one production automation system can easily support multiple kinds of RTUs/PLCs.

Concerning selection of the "host" production automation computer system, the recommended approach is to select a system that is designed and programmed to support the types of gas and oil production operations in the field. For example, if a field has flowing wells, plunger lift wells, chemical injection wells, pumping wells, and gas-lift wells, a "host" system should be selected that is designed and programmed to serve these operations with one common "look and feel" (e.g., user interface) and one common set of automation software services (e.g., alarming, reporting, plotting, etc.). In addition, the applications for the different types of gas and oil production operations (e.g., artificial lift systems, well test systems, etc.) should communicate with one another when there is a need to share information. This may exclude some "host" systems that are offered by companies that, for example, focus on one form of artificial lift. But this is far preferable to having to support multiple host systems in the same field.

Environmental protection

Typically, automation equipment for gas and oil production operations must function in extreme environmental conditions of temperature, humidity, wind, dust, corrosion, etc. Fortunately, as with instruments and controllers, high-quality RTUs, PLCs, and communications equipment are designed to withstand challenging environments. Here again, quality is the key word. This is not a place to accept low bid unless it is also of high quality.

Communications

As discussed in Section 14.3.6, there are many options for communication between the instruments, the RTUs/PLCs, the host computer systems, and other systems. It is not possible to issue general guidelines. It is recommended that a communications system study be performed by a communications expert to select the best combination of communications methods, communications standards, protocols, and data security. Poor communications can lead to lack of system acceptance. This can be avoided by selection of the appropriate communications components.

Project team

Staffing is discussed in Section 14.6.9. The recommendation here is to select an automation project team that has all of the requisite skills, and to equip the team with strong management backing. It is clear that the team must have strong buy-in from operations, but it must also have strong support from management. Without support from both ends, success is doubtful.

Integration into the organization

An automation system will fail if it is viewed as the property of one engineer, or even one part of the organization. To be successful over the long term, it must be viewed as essential by operations, maintenance, well analysis, production engineering, reservoir engineering, accounting, well services, and management. This is easy to say but challenging to accomplish.

14.6.3 Justification

Each company has its own criteria for justifying and approving projects, so this will not be discussed here. The economic justification is based on the tangible benefits discussed in Section 14.6.1 and the capital, operating, and maintenance costs are discussed in Sections 14.6.4 and 14.6.5. Often the intangible benefits are used to enhance the overall justifications of the project. Therefore a project approved solely based on economics can be easily justified if safety, environmental project, and other qualitative benefits are taken into consideration.

The impact of time

An important factor in project justification is time. This must be considered in several ways:

- First there is the time value of money. Again, each company has its own way to handle this.
- The increased production benefits will be realized over time.
- The additional production due to achieving greater reservoir recovery will be realized over time. An issue here is acceleration versus reserve addition. This is discussed later.
- The reduced operating costs will be realized over time.
- Operating and maintenance costs will be incurred over time.
- And over time it is likely that there will be additional capital expenditures. With the rate of obsolescence of equipment, it is likely that some or all of the capital components will need to be replaced every few years.

Thus to prepare a proper project justification, one must develop a life cycle plan for realization of benefits and incurring of costs.

Acceleration versus increased recovery

Another aspect of project justification comes in the question of production acceleration versus reserve addition. If the production automation system helps to produce
the same amount of gas and oil, but produce it sooner, this is acceleration. To properly account for this, one must have a before and after production forecast and a
forecast of prices and costs over time. None of these are easy to come by.

If the automation system helps to produce more gas and oil by helping to
increase ultimate recovery, this may have a significantly higher economic impact.
If, for example, the system can help keep liquid out of the wellbore by helping to
keep the gas production rate above critical or by helping to keep the well "pumped
off," there may be an excellent opportunity for increased ultimate recovery from
the reservoir.

The role of "pilot" tests

Some companies are reluctant to expend large sums on an automation project without proving that the projected costs and benefits are real. In former times, "pilot"
projects to "test" the cost and benefit assumptions were common. Sometimes they
were worthwhile:

- Sometimes expected costs were confirmed although often costs were higher than
  expected.
- Sometimes expected benefits were confirmed although often benefits were difficult to
  measure because many other things were happening in the field at the same time, for
  example, more drilling, major workover programs, secondary recovery projects, instilla-
  tion of new artificial lift systems, etc.

In general the biggest benefits from most pilot projects have been the experiences gained by the project staff. Having been through a project once, they have
been better equipped to do it again on a larger scale. One of the major shortcomings
has been deferral of benefits. It can take a few years to fully define, install, operate,
and evaluate a pilot project. While this is happening, the benefits that could be realized in the rest of the field or area are not being obtained.

In modern times, many production automation systems have been installed.
Many of them have been successful (see Section 14.7.1). Those that have been less
than fully successful (Section 14.7.2) or have been failures (Section 14.7.3) have
largely been so due to lack of attention to the issues discussed in this section. If the
recommendations and guidelines in this section are followed, it is not necessary to
conduct pilot tests of production automation systems for gas and oil production
operations.

There is one exception. While it is not necessary to pilot test an automation system, it may be appropriate to pilot test special new hardware, software, artificial lift
systems, etc. Often the supplier of these new systems is very anxious to have the
technology tested and proven; so is very willing to participate in field tests; and
may even be willing to share in the cost of the test.

14.6.4   Capital expenditure

A CAPEX is an expenditure that creates future benefits. Companies use CAPEX to acquire or upgrade physical assets such as equipment or property. In accounting a capital expenditure is added to an asset account or is *capitalized*, increasing the asset's basis—the cost or value of an asset as adjusted for tax purposes.

In performing an economic evaluation for a new automation project, there are two distinct forms of CAPEX to consider: (1) those capital expenditures (or savings) impacting the income stream and (2) those capital expenditures impacting the expense stream.

One of the major objectives of automation is to improve the profitability of an asset by reducing CAPEX over time. An example would be installation of pump-off controllers or other types of wellsite intelligence to prevent wear and tear on downhole equipment, thus increasing run life and reducing the number of recompletions required over the life of the well. This incremental savings in CAPEX would be an example of a CAPEX which affects the project's income stream. Such savings in CAPEX are normally realized over an extended period of time.

When deploying an automation project, there are a variety of capital expenditures that impact the expense stream of the project. In generally, these expenses are incurred early in the project, at or prior to the time of deployment. Typical items which affect the project's CAPEX include:

- instrumentation,
- controls,
- wellsite intelligence (RTUs, PLCs, VFDs, and other controllers),
- wiring,
- communications infrastructure (radios, towers, fiber optics, modems, etc.),
- servers,
- desktop computers and/or hardware upgrades,
- software license fees (host systems, historians, and others), and
- related services which are not treated as operational expense (OPEX) (dependent on corporate accounting practices).

14.6.5   Operational expense

An OPEX is the ongoing cost of running a business, product, or system. Similar to the discussion of CAPEX, there are operational expenses (or savings) which impact the income stream of a project and expenses which impact the expense stream of a project.

In addition to reducing long-term capital expenditures, automation projects also strive to improve an asset's profitability over time by reducing the ongoing OPEX. Such savings in OPEX would impact the project's income stream. Examples of savings include reduction in energy consumption through more efficient operation of artificial lift systems, reduction in man-power requirements, reduced equipment maintenance, and others.

As with CAPEX, there are a variety of operating expenses that are incurred in conjunction with deployment of a new project. These forms of OPEX impact the project's expense stream. Examples include:

* IT support,
* project management,
* systems integration,
* telecommunications fees,
* web-hosting services, and
* software leasing fees.

14.6.6 Design

Design of an automation system is a critical step that can have long-term consequences on the asset and the organization. Yet, surprisingly this step is often overlooked, allowing automation systems to be deployed which are too costly, are incompatible with existing infrastructure or overall IT needs, or fail to deliver the benefits that were originally envisioned. For this reason, it is important that careful attention be given to the design of the automation system and that the system is designed with the overall needs of the organization in mind.

There are three major factors that influence the success or failure of a technology project: (1) people, (2) process, and (3) technology. Effective design of automation systems considers the needs of each.

People

For an automation project to be successful, it is important that all of the key stakeholders be engaged in the design process at the earliest possible stage. This includes end users such as engineering and operations staff, information technology staff, management, and the various providers of system components. For each of these stakeholders, it is important to determine how the system will be used and what these individuals hope to accomplish with the system. While this may seem basic, more often than not projects are deployed without truly considering the needs of the people using the systems or the overall goals of the project. This very often leads to automation systems which are a technical success but fail to meet the business needs of the enterprise.

Perhaps the most important group of stakeholders is the actual end users of the technology. Often, new technology can have a disruptive and sometimes threatening impact on individuals in the organization. This can lead to lack of use or misuse of the technology once it is deployed. For end users to adopt new technology, they need to feel they have an emotional stake in the success of the project and that it is in their best interests to see the project succeed. Engaging end users in the functional specification and design of the system yields long-term rewards in the form of a system that is fit-for-purpose and a workforce that is eager to see the project succeed.

For a project to be successful, it must have management buy-in. For this to happen the project needs to be designed so it supports the organization's business objectives. If, for instance, the organization's goal is to reduce expenses, the system needs to control costs while achieving the technical aims of the project. Also, most successful projects have an executive sponsor. By engaging individuals in the management team in the design of the project, it is far more likely that these key influencers will do what they can to ensure the project's success.

Another key group of stakeholders is the company's information technology team. This group manages the deployment of information technology throughout the organization as well as setting the overall IT strategy for the enterprise. They are uniquely positioned to guide the development and deployment of technology projects. This ensures that existing technology is properly leveraged, reducing the overall cost of the system and minimizing any disruption to IT services while also ensuring that the system will continue to work as the underlying infrastructure evolves.

A final group of stakeholders are the actual vendors of the various pieces of automation technology. These organizations typically have significant experience and expertise in the automation domain and first-hand knowledge of what works and what does not. By engaging these organizations early in the design process, operating companies can gain insight into what types of solutions are best suited to their particular needs and avoid making costly mistakes.

Process

In any given production operation, certain processes have been established over time to govern how work gets done. Often, these processes are undocumented and may not be recognized by the individuals performing them. Nonetheless, these processes exist and govern all of the basic functions of that operation. When new technology is deployed in an asset, it disrupts these processes and challenges people to do their work in new and different ways. If the project is implemented without paying consideration to these underlying processes, it is likely that individuals will go back to doing their work the way they have always done it, causing the technology to be unused. Successful technology projects recognize the need to integrate the technology with company work processes, and include these workflows in the design of the system.

In one such project a major operator sought to build a real-time system for managing their subsurface maintenance program for a large onshore operation. Prior to deploying the project, this organization spent nearly 1 year examining the various processes of the organization and mapping out the various roles and responsibilities of those individuals responsible for these activities, particularly as they related to this new system. These processes were ultimately incorporated in the functional requirement specifications for the real-time system. In addition, new roles were created to address organizational gaps, extensive training was provided to users and standard operating procedures were developed which described exactly how key tasks would be performed with the new system and by whom. While this may seem like a significant amount of work, the result was a technology project that was quickly adopted throughout the organization and easily integrated into their day-to-day business.

Technology

Automation technology is evolving at a startling rate. The result is a myriad of options for virtually every component in any system. This gives operating companies flexibility in designing a system to suit their needs, but also makes the task enormously complex. The challenge is to design a system selecting technology that is fit-for-purpose, rather than choosing technology for technology's sake.

To aid in the evaluation of new technology, the Society of Petroleum Engineers Real-Time Optimization Technical Interest Group (RTO TIG) has proposed a methodology for classifying and assessing various system components. This allows organizations to understand the entire scope of a technology project and identify opportunities for improvement. This methodology divides the automation system into seven major components:

1. measurement (sensors),
2. telemetry,
3. data handling and access,
4. analysis,
5. visualization,
6. automatic control, and
7. integration and automation.

For each of these system components the technology is assessed and assigned a level for comparison and tracking progress. These levels are defined as follows:

- *Level 1*—ad hoc (manual or disjointed system)
- *Level 2*—multifunctional
- *Level 3*—integrated

Once these major system components are evaluated, a spider diagram can be constructed, such as the one shown in Fig. 14.53.

Each of the system components moves from an initial level (dashed line) to a higher level of maturity, following the implementation of a technology project. The axes represent technology level (0, 1, 2, 3). Movements along any given axis represent some improvement in that technology which may be economic or may be related to some other noneconomic project objective. By using tools such as this one, operators can easily identify which key areas require investment in technology and budget accordingly. Also, once the project is complete, this provides a benchmark for evaluating the success of the project from a technology standpoint.

14.6.7 Installation

Often the installation process begins many weeks or months prior to deployment of equipment and software in the field. The various automation vendors painstakingly

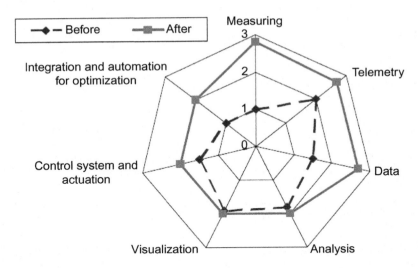

Figure 14.53 Spider diagram illustrating technology status.

test the components in their own labs to ensure they will function as designed and work together as a system. Once they are certain that the various systems can function as envisioned, they will work hand-in-hand with the operator's IT department and other service providers to commission the systems and get everything up and running.

Throughout this process a variety of tasks are performed by numerous individuals. Among these are the following:

- Obtain historical data.
- Set up test servers.
- Load historical data into host system.
- Site acceptance testing—(host system, RTUs, and other end devices).
- Installation of instrumentation.
- Installation and configuration of controllers.
- Installation and configuration of communication devices.
- Commissioning host system.
- Customization and configuration of host.
- Installation of client software.
- User training.

14.6.8 Security

Because of the sensitive nature of automation equipment and technology, security is always a concern of operators. This relates to both security of the physical assets and security of proprietary data. To address these concerns a variety of steps are taken in the course of deploying an automation project.

Field devices

The hardware that is deployed in the field—particularly in remote locations—can often pose a tempting target for thieves. Items such as solar panels, pressure transducers, RTUs, and copper wiring can all be easily stolen, and often are. To safeguard against theft, operators may take a variety of steps. Devices can be installed inside steel cages, bunkers, or other enclosures where they are out of site and not easily reached. In other cases, operators select hardware that is unlikely to attract the attention of thieves. In some cases, regular patrols are made by security personnel or other company employees to deter theft.

Because of the mission critical nature that is performed by end devices in the field, operators also want to ensure that only authorized individuals are able to adjust settings within those devices. For this reason, virtually all RTUs, PLCs, and other controllers provide security features that require user login, govern who can access what features, and document any changes that were performed to the device settings, who made the changes, and when they were made.

Host systems

Host systems provide multiple layers of security. The following summarizes the most common security features pertaining to host systems:

- Configuration of the system is based on access privilege governed by the host system, itself.
- Users are able to access only those features and perform those functions which are governed by their host system login credentials.
- Any changes made to system settings or data are logged in the system, identifying when the change was made and by whom.
- Systems are deployed within the corporate firewall, often within a corporate intranet. These systems are not accessible over the Internet or from individuals outside the company.
- Prior to logging-in to the host system, users must first log in to their corporate network with their normal user credentials.
- In certain cases the host system can only be accessed via Citrix connection or Remote Desktop session, requiring yet another level of user log-in.
- In many cases, operators employ technologies such as smart cards, tokens with rotating passwords, and even biometrics to further restrict access to the system.

14.6.9 Staffing

At least three teams are recommended for a production automation project. Table 14.6 provides the recommended members of each team and whether or not the members should be "in-house" staff or may be others, for example, from service companies or consultants.

The three teams are the steering team or steering committee, the automation team, and the surveillance team. These teams share some of the same members, have different responsibilities and operate over different time periods.

Table 14.6 Production automation teams

Champion	If possible	Facilitator	Advisor	
Management	Definitely	Chair		
Project engineer	If possible	Yes	Chair	
Engineering	If possible	Yes	Yes	Yes
Automation specialists		Yes	Yes	
Technicians	If possible		Yes	
Automation support			Yes	
Operations	If possible	Yes	Yes	Yes
Maintenance	If possible	Yes	Yes	Yes
Well analysis	If possible	Yes	Yes	Chair
Well servicing				
Accounting/finance		Yes		
Other/service company			Sometimes	Sometimes

Steering committee

The purpose of the steering committee is to provide overall priority, justification, direction, and focus for an automation project. It must have representation from a broad spectrum of stakeholders, managers or leaders, automation providers, and users. It must be chaired by a member of the operating company's management team to assure strong management support. The steering committee may exist over the life of an automation system, but its primary function will be during the early months/years of the project when it is being defined, justified, staffed, etc.

Companies with successful production automation projects have an automation champion. This is a person who is strongly committed to the overall automation effort. This person may be in management, engineering, or operations. In most cases, this person will be a member of the operating company staff, but in some cases, he/she may be from a third party. The representative of management must chair the steering committee, but the champion will be the facilitator to call meetings, set agendas, "drive" the project schedule, etc.

Automation team

The automation team is responsible for execution. They define, design, build, test, implement, commission, and maintain the system. Typically the champion will be an advisor to this team and a project engineer will be its chair. This team must exist from the very start of an automation project and must continue, in one form or another, for the entire life of the system, since the jobs of making enhancements, providing maintenance, and training staff are never finished.

This team must have members with special skills from each of the engineering disciplines involved, including applications, instrumentation, communications, hardware, software, and training. Some of these functions may be provided by third-party staff (e.g., experts in instrumentation, communications, hardware, software, and training). However, someone on the project team must oversee each of these, assure that they are performed and meet the project objectives.

The team must have members from operations, maintenance, and well analysis to provide input and feedback as the project is implemented. The system must meet their needs and they must "buy in" to the system.

Surveillance team

The surveillance team (it may have other names) uses the system. They use it every day to monitor, control, and optimize the gas and oil production operations. This team may have many members; many of them may not see themselves as part of a formal team and may not attend team meetings. But, they are the ones who use the system to gain the benefits for the company.

This team, like the others, must have a chair or focal point. In some companies, this person is called a well analyst. He/she may also be a production engineer, production technologist, well surveillance specialist, automation specialist, or lead operator. The name does not matter, but the function is vital. This person must assure that people are continuously assigned and motivated to use the automation system for routine daily monitoring, control, and optimization; they have the training they need; and they have the support they need from other functions in the company or from third parties for troubleshooting, maintenance, well servicing, system enhancements, training, etc.

14.6.10   Training

Each person on the three teams defined in Section 14.6.9 must receive training. Three levels of training are defined: aware, knowledgeable, and skilled. Those that must receive each type of training are given in Table 14.7.

Aware

Awareness training consists of the following:

- Attend "high-level" production automation course or seminar.
- Maintain awareness of important production automation issues.
- Have good understanding of:
  - Relative merits of each form of production equipment, artificial lift, and production automation economics.
  - Why production automation has been chosen for this field.
  - Interdependencies between production automation system and other production systems in the field.
  - Skills and personal characteristics needed by *knowledgeable* and *skilled* production automation staff.
  - Value of proper production automation system deployment, including production automation selection, design, installation, operation, optimization, troubleshooting, and surveillance.

Table 14.7 Automation training requirements

Staff position	Training level required		
	Aware	Knowledgeable	Skilled
Champion			■ Know entire system
Management	■		
Project engineer		■	
Engineering		■	
Automation specialists			■ Know entire system
Technicians			■ Know components
Automation support			■ Know support
Operations		■	
Maintenance			■ Know components
Well analysis			■ Know applications
Well servicing	■		
Accounting/finance	■		
Other/service company	■	■	Depends on jobs

Knowledgeable

To be knowledgeable, the following training is needed:

- Attend "high-level" production automation course or seminar.
- Attend "intermediate-level" production automation course that provides thorough understanding of production automation selection, design, installation, operation, optimization, troubleshooting, and surveillance.
- Maintain awareness of key production automation technologies and practices.
  - Spend time actually working in one or more facets of production automation.
  - Obtain full set of awareness that is required for the "aware" level.
  - Have detailed knowledge of both technical and business issues involved in production automation.
  - Have ability to advise people who are directly involved in production automation engineering and/or operations, by assisting them in:
    - obtaining needed resources,
    - prioritizing their work, and
    - evaluating the economics of their projects, etc.

Skilled

To be skilled, the following training is needed:

- Attend "high-level" production automation course or seminar.
- Attend "intermediate-level" production automation that provides thorough understanding of production automation selection, design, installation, operation, optimization, troubleshooting, and surveillance.
- Attend "comprehensive" production automation courses that provide thorough and detailed understanding of all of the facets of production automation.

- These courses should provide significant "hands-on" training in performing the various aspects of production automation.
- Obtain the full set of "awareness" and "knowledge" that are required for the "aware" and "knowledgeable" levels of competency.
- Maintain awareness of key production automation technologies and practices by *continuing education.*
  - Attend company and/or industry production automation workshops and/or seminars and sessions for sharing best practices.
  - Be fully conversant with key recommended practices produced and maintained by various sources in industry.
  - Spend time working under direct tutelage of an expert production automation engineer, well analyst, technician, or operator.
  - Obtain practical, hands-on experience with each aspect of production automation with which the person is involved.
  - Receive "feedback" on activities performed, in terms of evaluations of actual production automation installations.
  - Develop ability to train "new" staff in effective production automation engineering and/or operations.

There are a limited number of formal training programs in production automation. However, there are courses available from several service companies on specific subsystems or applications, and there are consultants who offer comprehensive automation system training.

14.6.11 Commercial versus "in-house"

In former times, say in the 1960s, 1970s, and early 1980s, there was not much choice between commercial versus "in-house" for supply of automation systems. There were very few commercial systems. If an operating company wanted a system, it had to develop it, and many operating companies did just that.

Beginning in the middle to late 1980s and 1990s, and certainly in the 21st century, the reverse is true. There are several commercial systems and very few operating companies develop their own systems. This is certainly true for automation systems for oil production and is becoming increasingly so for gas.

In current times, many operating companies have no choice. They do not have the staff or internal expertise to design and develop their own systems, so they are dependent on the suppliers. The "good news" is that operating companies who do not have the staff or expertise can acquire and use good automation systems. The "bad news" is that often they are constrained to use the systems offered by the suppliers, with little chance to optimize the systems to best meet their requirements.

The "best" approach, which is happening in a few cases, is a partnership between the operating company and the service company. They work together to "fine-tune" the system to meet the requirements of the operating company in terms of its specific technical requirements, geographical constraints, staffing needs, etc. This "teamwork" approach is recommended whenever a significant gas production automation project is undertaken.

14.7 Case histories

There have been some notable production automation success stories, some failures, and some systems that work but have not reached their potential. The purpose of this section is to summarize some of these experiences without providing specific references to companies or locations. The hope is that we can learn from our successes and failures.

14.7.1 Success stories

Automation has been deployed in upstream oil and gas applications for several decades, with an acceleration in activity over the last 20 years. Many operators have achieved significant enhancements in operating efficiency. The following are some examples.

Rod pump controllers

An operator in the East Linden (Cotton Valley) field in East Texas purchased and installed RPCs, replacing time clocks, on 15 wells. These wells are 10,000 ft deep, producing 42- to 46-degree API gravity oil, with a 5%–50% water cut. With contract pumpers and using timers set by trial and error, this operator saw the upside potential of using RPCs. Before installing RPCs, well problems were identified during daily visits. Timers were used to operate the wells—set based on a single visit to each well during a 24-hour period. The trial-and-error process to set these timers would often result in some wells pounding fluid for long periods of time. Another problem was "under-pumping" some wells where pumping time was as little as 3 h/day. Gas "breakout" during this downtime resulted in the unrecognized need to pump the wells longer to move the gas as well as all available fluids. Factor in any pump wear—also requiring additional run time—and the possibility of "under-pumping" were very real.

The installation of RPCs eliminated the guesswork and trail and error involved in setting time clocks for correct well control. The RPCs also automatically adjusted idle time based on buffered data in the controller from historical cycle times. Observed benefits from the use of RPCs by this operator include reducing rod and tubing failures by 31% and electrical cost by 40% ($50,000 per year). For wells that were maintaining a high fluid level due to incorrect timer cycles, the RPCs increased fluid production.

Plunger lift automation

An operator in the San Juan Basin in Northwestern New Mexico implemented a major automation project for a field producing coal bed methane via plunger lift. This project incorporated wellsite control using plunger lift controllers and EFM, host software, and telemetry between the host and the wells.

Plunger lift controllers contained self-adjusting algorithms for plunger arrival time, after flow, and shut-in time based on preconfigured parameters to maximize gas production. This resulted in a sustained production uplift of 4 MMSCFD for 40 wells. In 30 tubing flow control installations, an average uplift of 130 MSCFD per well was achieved by automatically controlling the casing valve based on well conditions.

Host system/workflow management

A mature asset in the San Joaquin Valley, California, was producing over 1000 wells via reciprocating sucker rod pump. Most of these wells were utilizing RPCs which were communicating with a host system. In addition, a variety of tools were used in the asset to manage operational issues such as rig management. This project sought to consolidate various tools that were used throughout the organization into a single web-based platform which would standardize the overall workflow used for surveillance, optimization, and well services management. In addition, the organization sought to create a single "system of record" that would provide an electronic repository for all operational data created during the day-to-day operations of the field. One of the primary goals was to enable the ability to perform score-carding against this data so that the operator could uncover hidden performance trends and enable continuous improvement within the asset.

From a business standpoint the operator hoped to reduce the field's failure rate, minimize downtime and production deferrals, and minimize time to return wells to production.

Prior to designing this system the operator conducted a comprehensive examination of the organization's business practices and workflows. Based on this study a system was specified which incorporated the business practices in place. Rather than adapting the organization to the software, the operator sought to implement a software system that reflected the organization's dynamics.

The tangible evidence of success in this project was the reduction in well failure rate which followed deployment of the software system. The improvement in failure rate was from 0.15 to 0.1 failures per well per year in the field. This corresponds to a cost saving for repairing failures of approximately $0.5 million per year in this field. As a result of this initial success, the system was deployed across the entire business unit. (Scaling this performance improvement up to the whole unit represents an annual saving of $6 million.) Since the software system enables online surveillance as well as automated standardized well service management planning and execution, production deferment on failures should substantially reduce. The time taken to diagnose a problem and schedule the right job to address, it is much faster with the new system. An additional means of speeding the repair process is that contractors have access to the system and can see the appropriate scheduled jobs as soon as they are approved for action. An estimate of the annual saving in deferred production for the unit is $3 million.

14.7.2   Failures

While people love to talk about success stories, they rarely discuss failures. This is unfortunate because we learn more from the projects that fail than from the ones that succeed. The following are three cases where the projects' failures yield clear lessons.

Beam pump optimization

A real-time optimization project was implemented in an onshore field in Western Venezuela. This system incorporated RPCs which communicated to a dedicated host system in the field office via radio. Operators were able to use a variety of sophisticated features such as management by exception and downhole pump card diagnostics.

Initially, extensive training was conducted along with the regular support of end users in the field. During the early years of the project, the system was an integral part of the production management process, with numerous documented benefits.

However, over time, personnel transferred out of the asset and much of the local knowledge of the system were lost. As new personnel were brought into the asset, they did not receive sufficient training. In addition, the automation provider interfaced primarily with the company IT department, who were principally concerned with the technical performance of the software and servers. As a result, the automation company remained insulated from end users' concerns. As one might predict, this system fell into disuse and, over time, reached a state where well test data was no longer being imported into the system and none of the wells were configured.

The lesson learned is that once a system is deployed, the work is not finished. For an automation system to yield value over the life of the asset, there needs to be an evergreen process for training and support.

Progressing cavity pump optimization

A real-time PCP optimization system was deployed in a 300-well field in Latin America. This system incorporated electronic downhole gauges, VFDs, automatic well testing using multiphase flow meters, and a sophisticated host system. Although a data historian had been installed in the field, the system was designed such that most of the analog data were fed directly into the host system, bypassing the historian. Data were collected in 1-minute intervals from each of the 10 analog instruments per well. As a result the host system was only able to store approximately 2.5 days' worth of data before overwriting its circular buffer. This meant that to preserve these data for future analysis, a technician had to travel to the field location and manually download data to a DVD several times per week. While it was then possible to find the data from a specific point in time and recover this for analysis (i.e., pressure transient analysis), it was not possible to perform long-term surveillance within the host system. This rendered many of the features of the host system useless and prevented engineers and operators from evaluating any long-term performance trends.

The lesson learned is that it is important to include all key stakeholders in the design of an automation system and ensure that the system architecture supports the needs of the users as well as the project's business objectives.

Gas-lift automation

In one project an operator sought to deploy a real-time optimization project for a large population of wells. The intent was to build an integrated model which included well performance, network performance, and facilities performance while regularly updating the model with real-time data from the automation system and well test database.

After an extensive review of various proposed solutions the operator implemented a "home-grown" solution in which a team of consultants built a custom human/machine interface that interfaces with the facilities software, well analysis software, and network optimization software. The result was a system that took significantly longer to build and cost significantly more than anticipated, while requiring a staff of dedicated full-time consultants to keep the system running. Further, once the system was deployed, many of the intended capabilities were not achieved and use of the system resulted in minimal improvement in field-wide performance.

The main lesson learned from this case speaks to the issue of "build versus buy." There are pros and cons to building a system from scratch just as there are for buying one that is available "off-the-shelf." However, if a suitable system exists in the market that meets the majority of needs, it is generally better to buy the off-the-shelf solution. This provides at least an "80% solution" at a fraction of the cost and time to deployment. Also the burden of supporting this product is born by the vendor rather than by the organization, while costs of ongoing improvements to the product are shared by all of its customers.

14.7.3 Systems that have not reached their potential

Some automation systems are technically qualified as a success, yet fail to achieve their full potential. Often this is due to lack of use resulting from inattention to soft rather than hard issues. This is often because the individuals expected to use the system are never properly trained in its use, are not engaged in the specification or design process, are fearful that the technology may displace them, or all of the above.

In one project an operator had implemented a successful pilot project in a large onshore asset. This project utilized wellsite intelligence and sophisticated host systems. As a result, they were able to realize substantial reductions in downtime and increases in incremental production. Based on the initial success, they opted to implement the same tools across each of the remaining assets in the business unit. Much to their surprise, none of the other fields achieved measurable improvement in the key performance metrics.

After investigating the usage of the system across the business unit, it became clear that the problem was one of managing soft issues. While the same technology was deployed across all the assets, the approach to change management was not the same. In the pilot project a team concept was encouraged where operators and engineers worked as equal stakeholders with common goals which were communicated in daily operational review meetings. All of the individuals in the team were fully engaged in the design and implementation of the system and felt that they had a personal stake in its success. However, when the system was implemented in other assets within the business unit, the technology was deployed but the organizational approach was not. The lesson learned was that in technology projects, soft issues can be the difference between success and failure.

14.8 Summary

Considering the value of gas and oil, the costs of staff and services, the negative impact of liquid loading on gas production and ultimate recovery, the requirement to remove liquid loading over the long term, the importance of gas and oil well monitoring, control, and optimization, and the difficulty of performing these manually, the business case for production automation is compelling.

Production automation equipment and applications exist to assist with effective management of gas and oil well operations. This equipment and these applications must be carefully selected, configured, installed, operated, and maintained. This requires attention to many details, including project staffing and training.

The cost of automation systems is low enough that every gas and oil well, and especially every well that has an artificial lift system, should be automated.

Further reading

The following references have been used in compiling this chapter. All are acknowledged with gratitude.

Section 14.2
Dunham CL, Anderson SR. The generalized computer-assisted (CAO) system: a comprehensive computer system for day-to-day oilfield operations. In: Paper SPE 20366 presented at the 1990 SPE Petroleum Computer Conference, Denver, 1990 25−28 June.
Dunham CL. Supervisory control of beam pumping wells. In: Paper SPE 16216 presented at the 1987 SPE Production Operations Symposium, Oklahoma City, Oklahoma, 1987 8−10 March.

Section 14.3.1
Examples of transmitters, <http://www.foxboro.com>.
Information on transducers, <http://sensors-transducers.globalspec.com>.

Section 14.3.2

Requirements for EFM: American Petroleum Institute Manual of Petroleum Measurement Standards, Chapter 21—flow measurement using electronic metering systems, Section 1—Electronic gas measurement.

Section 14.3.3

Burrell GR. *Chapter 16: Automation of lease equipment. Petroleum engineering handbook.* 3rd ed. Society of Petroleum Engineers; 1992.

Examples of control valves, <http://www.emersonprocess.com/fisher/oilandgas.html>.

Examples of motor controller, switchboard and VFD, <http://www.weatherford.com>.

Section 14.3.6

Information on communications, <http://www.freewave.com/>.

Section 14.3.7

Buchmann A. Real time database systems. In: Rivero LC, Doorn JH, Ferraggine VE, editors. *Encyclopedia of database technologies and applications.* Idea Group; 2005.

Carpron HL, Johnson JA. *Computers: tools for the information age.* 5th ed. Prentice Hall; 1998.

Stankovic JA, Son SH, Hansson J. *Misconceptions about realtime databases, cybersquare.* University of Virginia; 1999.

Section 14.3.8

Information on data exchange, <http://posc.org/>, <http://www.energistics.org/posc/PRODML_'07_Work_Group.asp?SnID = 6378497>.

Section 14.4.1

Example of user interface: Microsoft Windows user interface standard.

Section 14.4.3

Example alarm display: shell global solutions. Fieldware, <http://www.shell.com/home/Framework?siteId = globalsolutions-en>.

Section 14.4.5

Information on trend plots: SPE 49463, real-time artificial lift optimisation, W.J.G.J. der Kinderen, H. Poulisse, C.L. Dunham, Member SPE, SIEP-RTS, The Netherlands.

Section 14.4.6

Example display: shell global solutions. Fieldware, <http://www.shell.com/home/Framework?siteId = globalsolutions-en>.

Section 14.4.7

Example data historian access, <http://www.osisoft.com/Products/PI + Process + Book.htm>.

Section 14.5.1

Information on plunger lift: weatherford artificial lift training course.

Plunger lift images: eProduction solutions, <http://www.ep-solutions.com/>.

Section 14.5.2

Schematic of sucker rod pumping well, method for detecting pump-off, example of rod pump
controller: eProduction Solutions, <http://www.ep-solutions.com/>
Example of sucker rod surveillance: shell global solutions. Fieldware, <http://www.shell.
com/home/Framework?siteId = globalsolutions-en>.
Example of gas vent pump to overcome gas locking: Harbison−Fischer, <http://www.
hfpumps.com/>.

Section 14.5.3

Schematic of progressing cavity pumping well: shell global solutions. Fieldware, <http://
www.shell.com/home/Framework?siteId = globalsolutions-en>.
Examples of PCP surface drive systems: Baker Hughes centrilift, <http://www.bakerhughes.
com/centrilift/PCPS/driveheads.htm>.
Thru-tubing conveyed progressive cavity ESP − operational issues, a short story, J. Ryan
Dunn, et al., 2002, ESP workshop, Houston, Texas.
Special PCP alarms: eProduction solutions, <http://www.ep-solutions.com/>.
Information on PCP-RIFTS and PCP pump design: C-FER technologies, <http://www.cfer-
tech.com/> and <http://www.pc-pump.com/>.
Information on controls: Burrell, G.R., Chapter 16: Automation of lease equipment. In:
Petroleum engineering handbook, 3rd ed. Society of Petroleum Engineers, 1992.

Section 14.5.4

Schematic of electrical submersible pumping well: eProduction solutions, <http://www.ep-
solutions.com/>.
Casing pressure gradient. Automation training program. Oilfield automation consulting,
<www.oilfieldautomation.com>.
RTU components: eProduction solutions, <http://www.ep-solutions.com/>.
ESP surveillance panel: eProduction solutions, <http://www.ep-solutions.com/>.
ESP surveillance plots: automation training program. Oilfield automation consulting, <www.
oilfieldautomation.com>.
Information on ESP-RIFTS: C-FER technologies, <http://www.cfertech.com/>.

Section 14.5.5

Simpson DA, Lea JF, Cox JC. Coal bed methane production. In: Paper SPE 80900, presented
at production and operations symposium, Oklahoma City, March 23−25, 2003.
Information on hydraulic lift: weatherford hydraulic lift course.

Section 14.5.7

Schematic of gas-lift system, schematic of gas-lift mandrel, schematic of gas-lift valve,
example of injection pressure heading, downhole plot of injection and production pressure profiles. Gas-lift course materials. OGCI Petroskills, <http://petroskills.com/>.
Turner RG, Hubbard MG, Dukler AE. Analysis and prediction of minimum flow rate for the
continuous removal of liquids from gas wells. *J Petrol Technol* 1969;**21**:1475−82.
Coleman SB, Clay HB, McCurdy DG, Norris III HL. A new look at predicting gas-well load
up. *J Petrol Technol* 1991;**43**:329−33.

Section 14.5.9

Pigott MJ, Parker MH, Vincente D, Dalrymple LV, Cox DC, Coyle RA. Wellbore heating to prevent liquid loading. In: SPE 77649, presented at the SPE annual technical conference and exhibition, San Antonio, TX, September 2, 2002.

Section 14.6.6

Mochizuki et al. Real time optimization: classification and assessment. In: SPE 90213, presented at the SPE annual technical conference and exhibition, Houston, TX, September 28–29, 2004.

Ormerod et al. Real-time field surveillance and well services management in a large mature onshore field: case study. In: SPE 99949, presented at the 2006 SPE intelligent energy conference held in Amsterdam, The Netherlands, April 11–13, 2006.

Section 14.6.9

Table of production automation teams: automation training program. Oilfield automation consulting, <www.oilfieldautomation.com>

Section 14.6.10

Table of production automation training: automation training program. Oilfield automation consulting, <www.oilfieldautomation.com>

Appendix A: Development of critical velocity equations

A.1 Introduction

This appendix summarizes the development of the Turner[1] equations to calculate the minimum gas velocity to remove liquid droplets from a vertical wellbore.

A.1.1 Physical model

Consider gas flowing in a vertical wellbore and a liquid droplet transported at a uniform velocity in the gas stream, as illustrated in Fig. A.1.

The forces acting on the droplet are gravity, pulling the droplet downward, and the upward drag of the gas as it flows around the droplet.

The gravity force is

$$F_G = \frac{g}{g_C}\left(\rho_L - \rho_G\right) \times \frac{\pi d^3}{6}$$

and the drag force is given by

$$F_D = \frac{1}{2g_C}\rho_G C_D A_d (V_G - V_d)^2$$

where g = gravitational constant = 32.17 ft./s^2; g_C = 32.17 lbm-ft./lbf-s^2; d = droplet diameter; ρ_L = liquid density; ρ_G = gas density; C_D = drag coefficient; A_d = droplet projected cross-sectional area; V_G = gas velocity; V_d = droplet velocity.

The critical gas velocity to remove the liquid droplet from the wellbore is defined as the velocity at which the droplet would be suspended in the gas stream. A lower gas velocity would allow the droplet to fall, resulting in liquid accumulation in the wellbore. A higher gas velocity would carry the droplet upward to the surface and remove the droplet from the wellbore.

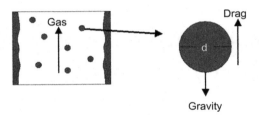

Figure A.1 Liquid droplet transported in a vertical gas stream.

Thus the critical gas velocity V_C is the gas velocity at which $V_d = 0$. In addition, since the droplet velocity is zero, the net force on the droplet is also zero. The defining equation for the critical gas velocity is then

$$F_G = F_D$$

or

$$\frac{g}{g_C}(\rho_L - \rho_G)\frac{\pi d^3}{6} = \frac{1}{2g_C}\rho_G C_D A_d V_C^2$$

Substituting $A_d = \pi d^2/4$ and solving for V_C gives,

$$V_C = \sqrt{\frac{4g}{3}\frac{(\rho_L - \rho_G)}{\rho_G}\frac{d}{C_D}} \qquad (A.1)$$

This equation assumes a known droplet diameter. In reality the droplet diameter is dependent upon the gas velocity. For liquid droplets entrained in a gas stream, Ref. [2] shows that this dependence can be expressed in terms of the dimensionless Weber number

$$N_{WE} = \frac{V_C^2 \rho_G d}{\sigma g_C} = 30$$

Solving for the droplet diameter gives

$$d = 30\frac{\sigma g_C}{\rho_G V_C^2}$$

and substituting into Eq. (A.1) gives

$$V_C = \sqrt{\frac{4}{3}\frac{(\rho_L - \rho_G)}{\rho_G}\frac{g}{C_D}30\frac{\sigma g_C}{\rho_G V_C^2}}$$

or

$$V_C = \left(\frac{40gg_C}{C_D}\right)^{1/4} \left(\frac{\rho_L - \rho_G}{\rho_G^2}\sigma\right)^{1/4}$$

Turner assumed a drag coefficient of $C_D = 0.44$ that is valid for fully turbulent conditions. Substituting the turbulent drag coefficient and values for g and g_C gives

$$V_C = 17.514\left(\frac{\rho_L - \rho_G}{\rho_G^2}\sigma\right)^{1/4} \text{ ft./s} \qquad (A.2)$$

where ρ_L = liquid density (lbm/ft.³); ρ_G = gas density (lbm/ft.³); σ = surface tension (lbf/ft.).

Eq. (A.2) can be written for surface tension in dyne/cm units using the conversion lbf/ft. = .00006852 dyne/cm to give

$$V_C = 1.593\left(\frac{\rho_L - \rho_G}{\rho_G^2}\sigma\right)^{1/4} \text{ ft./s} \qquad (A.3)$$

where ρ_L = liquid density (lbm/ft.³); ρ_G = gas density (lbm/ft.³); σ = surface tension (dyne/cm).

A.2 Equation simplification

Eq. (A.3) can be simplified by applying "typical" values for the gas and liquid properties.

From the real gas law the gas density is given by

$$\rho_G = 2.715\,\gamma_G\frac{P}{(460 + T)Z} \text{ lbm/ft.}^3 \qquad (A.4)$$

Evaluating Eq. (A.4) for typical values of

Gas gravity, γ_G	0.6
Temperature, T	120 F
Gas deviation factor, Z	0.9

gives

$$\rho_G = 2.715 \times .6\frac{P}{(460 + 120) \times .9} = .0031P \text{ lbm/ft.}^3$$

Typical values for density and surface tension are

Water density	67 lbm/ft.^{3}
Condensate density	45 lbm/ft.^{3}
Water surface tension	60 dyne/cm
Condensate surface tension	20 dyne/cm

Introducing these typical values and the simplified gas density Eq. (A.4) into Eq. (A.3) yields

$$V_{C,\text{water}} = 1.593 \left(\frac{67 - .0031P}{(.0031P)^2} 60 \right)^{1/4} = 4.434 \frac{(67 - .0031P)^{1/4}}{(.0031P)^{1/2}} \text{ ft./s}$$

$$V_{C,\text{cond.}} = 1.593 \left(\frac{45 - .0031P}{(.0031P)^2} 20 \right)^{1/4} = 3.369 \frac{(45 - .0031P)^{1/4}}{(.0031P)^{1/2}} \text{ ft./s}$$

A.3   Turner equations

Turner et al.[1] found that for his field data, where wellhead pressures were typically ≥ 1000 psi, a 20% upward adjustment to the theoretical values was required to match the field observations. Applying the 20% adjustment then yields

$$V_{C,\text{water}} = 5.321 \frac{(67 - .0031P)^{1/4}}{(.0031P)^{1/2}} \text{ ft./s}$$

$$V_{C,\text{cond.}} = 4.043 \frac{(45 - .0031P)^{1/4}}{(.0031P)^{1/2}} \text{ ft./s}$$

However in the original paper by Turner,[1] the coefficients were found to be 5.62 for the critical water velocity equation aforementioned and 4.02 for the condensate critical velocity aforementioned, but these values are slightly in error as developed earlier.

A.4   Coleman et al. equations

Coleman et al.[3] found that Eq. (A.3) would fit their data. This was without the 20% adjustment that Turner made to fit his data at higher average wellhead pressures. Therefore if the "corrected" Turner equations are written without the 20% adjustment, then the Coleman equations can be written as later if the same simplifications and typical data are used as earlier.

$$V_{C,\text{water}} = 4.434 \frac{(67 - .0031P)^{1/4}}{(.0031P)^{1/2}} \text{ ft./s}$$

$$V_{C,\text{cond.}} = 3.369 \frac{(45 - .0031P)^{1/4}}{(.0031P)^{1/2}} \text{ ft./s}$$

References

1. Turner RG, Hubbard MG, Dukler AE. Analysis and prediction of minimum flow rate for the continuous removal of liquids from gas wells. *J Petrol Technol* 1969;1475−82.
2. Hinze JO. Fundamentals of the hydrodynamic mechanism of splitting in dispersion processes. *AICHE J* 1955;1(3):289.
3. Coleman SB, Clay HB, McCurdy DG, Norris III HL. A new look at predicting gas-well load up. *J Petrol Technol* 1991;329−33.

Appendix B: Nodal concepts and stability concerns*

B.1 Introduction

Gas in a gas well must flow through the reservoir rock matrix, then through the perforations and gravel pack, possibly through a bottomhole standing valve, then through the tubing, possibly a subsurface safety valve, and through the surface flowline and flowline choke to the separator. It is possible for nodal to model all of these parts of gas well system.

One useful tool for the analysis of well performance is system nodal analysis. Nodal analysis divides the total well system into two subsystems at a specific location called the "nodal point." One subsystem considers the inflow from the reservoir, through possible pressure drop components and to the nodal point. The other subsystem considers the outflow system from some pressure on the surface down to the nodal point. For each subsystem the pressure at the nodal point is calculated and plotted as two separate, independent pressure—rate curves.

The curve from the reservoir to the nodal point is called the "inflow curve," and the curve from the separator to the nodal point is called the "outflow curve." The intersection of the inflow and outflow curves is the predicted operating point where the flow rate and pressure from the two independent curves are equal. The inflow and outflow curves are illustrated in Fig. B.1.

Although the nodal point may be located at any point in the system, the preferred position is at the top of the perforations or possibly at the tubing or pump intake. With this nodal point the inflow curve represents the flow from the reservoir through the completions into the tubing and the outflow curve represents the flow from the node to a surface pressure reference point (e.g., separator), summing pressure drops from the surface to the node at the mid-perforations depth.

The nodal analysis method employs single or multiphase flow pressure drop correlations for tubing flow, as well as correlations developed for the various components of reservoir, well completion, and surface equipment systems to calculate the pressure loss associated with each component in the system. This information is then used to evaluate well performance under a wide variety of conditions that will lead to a more optimum single well completion and production practices. It follows that nodal analysis is useful for the analysis of the effects of liquid loading on gas wells.

* ™—Macco Schlumberger.

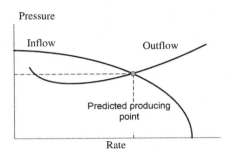

Figure B.1 System nodal analysis.

B.2   Tubing performance curve

The outflow or tubing performance curve (TPC) shows the relationship between the total tubing pressure drop and the surface pressure value, with the total liquid flow rate. The tubing pressure drop is essentially the sum of the surface pressure, the hydrostatic pressure of the fluid column (composed of the liquid "holdup" or liquid accumulated in the tubing and the weight of the gas), and the frictional pressure loss resulting from the flow of the fluid out of the well. For very high flow rates, there can be an additional "acceleration term" to add to the pressure drop but the acceleration term is usually negligible compared to the friction and hydrostatic components. The frictional and hydrostatic components are shown by the dotted lines in Fig. B.2 for a gas well producing liquids. Duns and Ros[1] and Gray[2] are examples of correlations used for gas well pressure drops that include liquid effects.

Notice that the TPC passes through a minimum. To the right of the minimum the total tubing pressure loss increases due to increased friction losses at the higher flow rates. The flow to the right of the minimum is usually in the mist flow regime that effectively transports small droplets of liquids to the surface.

At the far left of the TPC the flow rate is low and the total pressure loss is dominated by the hydrostatic pressure of the fluid column brought about by the liquid holdup, or that percent of the fluid column occupied by liquid. The flow regime exhibited in the left-most portion of the curve is typically bubble flow or some high holdup flow regime, that carries more liquids in the tubing than, for instance, mist flow.

It is common practice to use the TPC alone, in the absence of up-to-date reservoir performance data, to predict gas well liquid loading problems. Points on the curve to the left of the minimum in the curve are unstable and prone to liquid loading problems. Conversely, flow rates to the right of the minimum of the TPC are considered to be stable in the friction dominated portion of the tubing curve.

Understandably, this method is inexact but is useful to predict liquid loading problems in the absence of better reservoir performance data. Therefore you can just

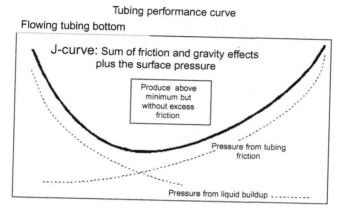

Tubing performance curve

Flowing tubing bottom

J-curve: Sum of friction and gravity effects plus the surface pressure

Produce above minimum but without excess friction

Pressure from tubing friction

Pressure from liquid buildup

Figure B.2 Tubing performance curve.

select the flow rate range you are measuring currently and see if it is in a favorably predicted portion of the TPC or not regardless of having the reservoir inflow curve.

With reservoir performance data, however, intersections of the tubing outflow curve and the reservoir inflow curve allow a prediction of where the well is flowing now and into the future if reservoir future inflow performance relationship (IPR) curves can be generated.

B.3 Reservoir inflow performance relationship

In order for a well to flow, there must be a pressure differential from the reservoir to the wellbore at the reservoir depth. If the wellbore pressure is equal to the reservoir pressure, there can be no inflow. If the wellbore pressure is zero, the inflow would be the maximum possible—the absolute open flow or AOF. For intermediate wellbore pressures the inflow will vary. For each reservoir, there will be a unique relationship between the inflow rate and the wellbore pressure.

Fig. B.3 shows the form of a typical gas well IPR curve. The IPR curve is often called the deliverability curve.

B.3.1 Gas well inflow performance relationship equations

The equation for radial flow of gas in a well perfectly centered within the well drainage area with no rate dependent skin is

$$q_{sc} = \frac{0.000703 \, k_g h (P_r^2 - P_{wf}^2)}{\mu Z T \ln\left(0.472 \frac{r_e}{r_e} + S\right)} \tag{B.1}$$

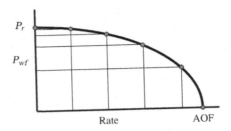

Figure B.3 Typical reservoir IPR curve.

where q_{sc} = gas flow rate (Mscf/D); k_g = effective permeability to gas (md); h = stratigraphic reservoir thickness (perpendicular to the reservoir layer) (ft); P_r = average reservoir pressure (psia); P_{wf} = flowing wellbore pressure at the mid-perforation depth (psia); μ_g = gas viscosity (cp); Z = gas compressibility factor at reservoir temperature and pressure; T = reservoir temperature (ǫR); r_e = reservoir drainage radius (ft); r_w = wellbore radius (ft); S = total skin.

When considering the lateral of a horizontal well with multiple fractures, the aforementioned equation does not have the inputs to model such a situation.

Eq. (B.1) can be used to generate an inflow curve of gas rate versus P_{wf} for a gas well if all the aforementioned data are known. However, often the date required to use this equation are not well known, and a simplified equation is used to generate an inflow equation for gas flow that utilizes well test data to solve for the indicated constants.

$$q_{SC} = C(P_r^2 - P_{wf}^2)^n \tag{B.2}$$

where q_{SC} = gas flow rate, in consistent units with the constant C; n = a value that varies between about 0.5 and 1.0. For a value of 0.5, high turbulence is indicated and for a value of 1.0, no turbulence losses are indicated.

Lacking data, some authors have used this simplified equation to model inflow from a horizontal well.

This equation is often called the "backpressure" equation with the radial flow details of Eq. (B.1) absorbed into the constant C. The exponent n must be determined empirically. The values of C and n are determined from well tests. At least two test rates are required since there are two unknowns, C and n, in the equation, but four test rates are recommended to minimize the effects of measurement error.

If more than two test points are available, the data can be plotted on log–log paper and a least squares line fit to the data, to determine n and C.

Taking the log of Eq. (B.2) gives

$$\log(q_{SC}) = \log(C) + n\log(P_r^2 - P_{wf}^2) \tag{B.3}$$

On a log–log plot of rate versus $(P_r^2 - P_{wf}^2)$, n is the slope of the plotted line and $\ln(C)$ is the Y-intercept, the value of q when $(P_r^2 - P_{wf}^2)$ is equal to 1.

For two test points the n value can be determined as follows:

$$n = \frac{\log(q_2) - \log(q_1)}{\log(P_r^2 - P_{wf}^2)_2 - \log(P_r^2 - P_{wf}^2)_1} \tag{B.4}$$

This equation may also be used for more than two test points by plotting the log–log data as described and picking two points from the best-fit line drawn through the plotted points. Values of the gas rate, q, and the corresponding values of $P_r^2 - P_{wf}^2$ can be read from the plotted line at the two points corresponding to the points 1 and 2 to allow solving for n.

Once n has been determined the value of the performance coefficient C may be determined by the substitution of a corresponding set of values for q and $P_r^2 - P_{wf}^2$ into the backpressure equation. See more detail in Appendix C, Plunger troubleshooting procedures.

If pseudo-stabilized data can be determined in a convenient time, then this equation can be developed from test data easily. Pseudo steady state indicates that any changes have reached the boundary of the reservoir, but practically it means for wells with moderate to high permeability, that pressures and rates recorded appear to become constant with time. If the well has very low permeability, then pseudo-stabilized data may be near impossible to attain, and then other means are required to estimate the inflow of the gas well. Rawlins and Schellhardt[3] provide more information on using the backpressure equation. For more details on the backpressure equation, see Appendix C, Plunger troubleshooting procedures.

In truth, many operators do not find the time or know the expense involved with testing low-pressure gas wells. Instead of loading analysis, they use the critical rate correlations and examine the decline curves. However, for sizing compression and tubing size, it is advantageous to have an IPR for the well. If one knows the approximate shut-in pressure of a well, then a flowing bottom-hole pressure can be calculated as a point on the IPR and if using the backpressure equation, with an assumed value(s) of the "n," then an IPR can be constructed with calculations and without testing. Better success is obtained with this approach if done before the well is loaded however.

Some operators have noted the similarity of the Vogel IPR curve for liquid and have applied it to gas production.

Many times the shut-in pressure is not known, so one may resort to a two test set of data to solve simultaneous equations for the C and P_r value in the backpressure equation given the n value as input. This shows the value of the P_r at zero producing rate. You can do the same with the Vogel where two sets of test data (q, P_{wf}) can be used to solve for the Q_{max} and the P_r in the Vogel equation.

The earlier discussion if for pseudo steady-state conditions and is not for non-steady-state conditions.

B.3.2 Future inflow performance relationship curves

For predicting backpressure curves at different shut-in pressures (at different times), the following approximation from Fetkovitch[4] can be used for "future" inflow curves.

$$q = C\left(\frac{P_r}{P_{ri}}\right)(P_r^2 - P_{wf}^2)^n \tag{B.5}$$

where q = current gas rate; C = coefficient consistent with the gas rate and pressure units; P_r = average reservoir pressure, at current time (psia); P_{wf} = flowing current wellbore pressure (psia); P_{ri} = initial average reservoir pressure use to determine C and n (psia).

However, it is not known if this equation finds use for horizontal or shale wells. Ref. [5] indicates a method for developing transient IPRs for transient unconventional reservoirs.

Ref. [6] shows how to generate transient IPRs for a unconventional well if a "typical" decline curve is available for the field. A plot from this paper is shown in Fig. B.4.

In summary, however, it is difficult to generate transient IPRs for an unconventional declining reservoir. Still tubing and AL performance curves can be very useful to compare one lift method against another. AL performance tubing curves are flowing tubing curves but with the pressure boosted at intake of the tubing with pumps or the gradient adjusted by gas injected using gas lift. More detail of AL performance curves is found in the chapters on pumps, gas lift, and plunger in the book.

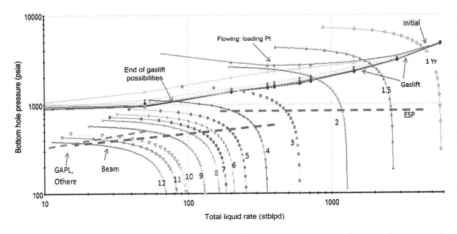

Figure B.4 Transient IPRs with artificial lift performance curves superimposed.

B.4 Intersections of the tubing curve and the deliverability curve

Fig. B.4 shows a TPC intersecting a well deliverability inflow curve (IPR). The figure shows the tubing curve intersecting the inflow curve in two places. Stability analysis shows that the intersection between points C and D is stable while the intersection between A and B is unstable and, in fact, will not occur.

For example, if the flow rate strays to point D, then the pressure from the reservoir is at D but the pressure required to maintain the tubing flow is above D. The added backpressure against the sand face of the reservoir then decreases the flow back to the point of stability where the two curves intersect. Similarly, if the flow temporarily decreases to point C, the pressure drop in the tubing is decreased, decreasing the pressure at the sand face prompting an increase in flow rate back to the equilibrium point. Note also that the stable intersection between C and D is to the right of the minimum in the tubing curve. When the intersection of the TPC and the IPR curve occurs to the right of the minimum in the "J" curve, the flow tends to be more "stable" and stable versus erratic flow most always means more production (Figs. B.6 and B.7).

If, on the other hand, the flow happens to decrease to point A, the pressure on the reservoir is increased due to an excess of fluids accumulating in the tubing. The increase in reservoir pressure decreases the flow further, thus increasing the pressure on the reservoir further until the well dies. Similarly, if the well flows at point B, the increased pressure against the reservoir reduces the flow which again increases the pressure drop in the tubing until ultimately the well again dies. The crossing point on the left side, therefore, uniformly tends toward zero flow, consistent with the minimum point of the TPC (Fig. B.5).

Thus the crossing point of the IPR and the TPC, to the right of the minimum of the "J" curve, represents a stable flow condition where liquids are effectively

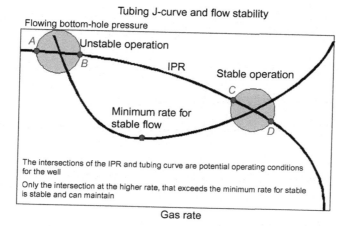

Figure B.5 Tubing performance curve in relation to deliverability curve.

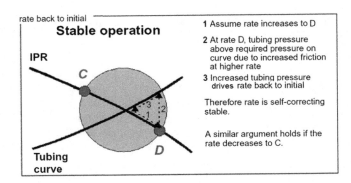

Figure B.6 Stable flow.

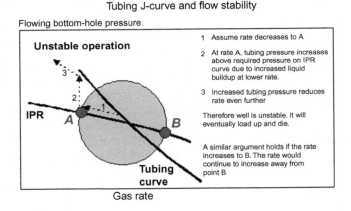

Figure B.7 Unstable flow.

transported to the surface and that to the left of the minimum represents unstable conditions where the well loads up with liquids and dies.

B.5   Tubing stability and flowpoint

Another way to summarize unstable flow is presented in Fig. B.8. Here the difference between the two curves is the difference between the flowing bottom-hole pressure and the flowing tubing surface pressure. The apex of the bottom curve is called the "flowpoint." Greene[7] provides additional information on the "flowpoint" and also for more information on gas well performance, as well as fluid property effects on the AOF of the well.

The reasons for the flow rates below the "flowpoint" not being sustainable are explained at each rate by the slopes of the inflow and outflow curves, as shown in

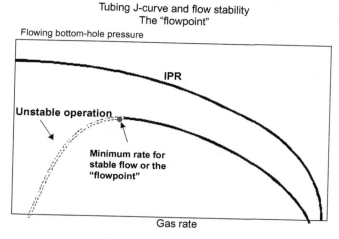

Figure B.8 Flowpoint or minimum stable flow rate for gas well with liquids production.

Section B.4. From Greene,[7] "a change in the surface pressure is transmitted down-hole as a similar pressure change, but a compatible inflow rate in the same direction as the pressure change does not exist." The result is an unstable flow condition that will either kill the well or, under certain conditions, move the flow rate to a compatible position above the "flowpoint" rate.

This discussion on stable rates is not the same as flowing below the "critical rate," as discussed in Chapter 3, Critical velocity. There the mechanism is flowing below a certain velocity in the tubing that permits droplets of liquid to fall and accumulate in the wellbore instead of rising with the flow. Discussed here is the interaction of the tubing performance with the inflow curve and reaching a point where the well will no longer flow in a stable condition.

However, the instability is brought on by regions of tubing flow where liquid is accumulating in the tubing due to insufficient gas velocity, so although the arguments are dissimilar, the root causes of each phenomenon are similar. Many nodal programs will plot the "critical" point on the TPC. Often this point is on the minimum of the tubing curve or to the right of the minimum in the tubing curve.

When analyzing a gas well, check for nodal intersections that are "stable" and check for critical velocity at the top and bottom of the well. For instance, if selecting tubing size, choose one that will allow the well to flow above the "critical flow rate" and also to be "stable" from nodal analysis.

B.6 Tight gas reservoirs

A possible exception to the above stability analysis is the tight gas reservoir. A tight gas reservoir is generally defined as one where the reservoir permeability is less than

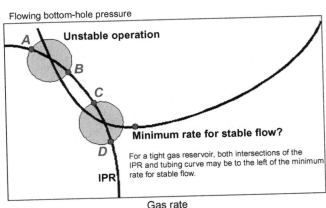

Figure B.9 Tight gas well modeled with nodal intersections.

0.01 md. Tight gas reservoirs having low permeability have steep IPR relationships and react to changes in pressure very slowly. A possible tight gas inflow curve is shown in Fig. B.9. This figure shows that the right-most crossing of the TPC and the IPR might be to the left of the minimum of the "J" curve. The above "slope" arguments would lead the conclusion that the right-most intersection is unstable but the well is flowing to the left of the minimum in the TPC. For tight gas wells, pseudo steady-state data are usually impossible to obtain to get a good inflow curve and often just using the "critical velocity" concept is the best tool to analyze liquid loading.

B.7 Nodal example—tubing size

From the earlier analysis, it is clear that size (diameter) of the production tubing can play an important role in the effectiveness with which the well can produce liquids. Larger tubing sizes tend to have lower frictional pressure drops due to lower gas velocities that in turn lower the liquid carrying capacity. Smaller tubing sizes on the other hand have higher frictional losses but also higher gas velocities and provide better transport for the produced liquids. Chapter 5, Compression, provides additional information on sizing the tubing following this introductory example.

In designing the tubing string, it then becomes important to balance these effects over the life of the field. To optimize production, it may be necessary to reduce the tubing size later in the life of the well.

Fig. B.10 shows TPCs superimposed over two IPR curves. For the higher pressure IPR curve, C, D, and E tubing curves would perform acceptably, but D and E would have more friction and less rate than would the TPC C. Curves A and B may

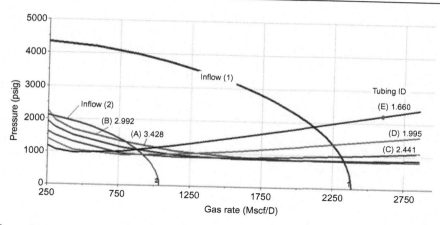

Figure B.10 Effect of tubing size on future well performance.

be intersecting to the left of the minimum in the TPC and this is thought to generate unstable flow.

For the low-pressure or "future," IPR curve, curves A, B, C, and D are all showing intersection below the minimum for the TPCs and as such would not be good choices. TPC E, the smallest tubing, performs acceptably for the low-pressure IPR curve. Curve E could intersect a little lower on the low-pressure IPR curve, and does not in this case because of a fairly high (400 psi) surface tubing pressure.

B.8 Summary

Systems nodal analysis is for pseudo steady-state or steady state conditions. It can be a model of one well. Tubing correlations and the TPC can indicate if one is in a stable flow or producing regime or not. IPRs allow an approximate indication of what rate the well should be producing but the development of IPRs for unconventional wells is more difficult than for conventional wells.

References

1. Duns Jr H, Ros NCJ. Vertical flow of gas and liquid mixtures in wells. In: *Proc. sixth world pet. congress*; 1963. p. 451.
2. Gray HE. Vertical flow correlations in gas wells. API user's manual for API 14, Subsurface controlled subsurface safety valve sizing computer program, Appendix B; 1974.
3. Rawlins EL, Schellhardt MA. *Back pressure data on natural gas wells and their application to production practices*, 7. Bureau of Mines Monograph; 1935.
4. Fetkovitch MJ. The isochronal testing of oil wells. In: *SPE paper No. 4529, 48th annual fall meeting of SPE of AIME*. Las Vegas, NV; September 30–October 3 1973.

5. Shahamot S, Tabatabale SH, Matter L, Motomed E. Inflow performance relationship for unconventional reservoirs (Transient IPR). In: *SPE-175975-MS, presented at the SPE/ CUR unconventional resources conference*, Calgary, Alberta; 20–22 October, 2015.
6. Lea J, Brock M, Kannan S. *Artificial lift with declining production*. Lubbock, TX: SWPSC; April 2016.
7. Greene WR. Analyzing the performance of gas wells. In: Presented the annual SWPSC, Lubbock, TX, April 21–21, 1978.

Appendix C: Plunger troubleshooting procedures*

C.1 Motor valve

C.1.1 Valve leaks

When motor valves leak, there are two possible sources. Under normal conditions a valve will have from 20 to 30 psi on the diaphragm section of the valve, and much higher pressures on the body of the valve. External leaks are most commonly found at the packing section located between the diaphragm and the body of the valve. This occurs when the packing around the stem wears and leaks due to the high pressure from the body of the valve. All valves have some type of packing around the stem of the valve. In some cases, it is possible to tighten a packing nut and stop the leak, but generally, it is necessary to replace the packing to eliminate the leak. Contact the valve manufacturer or the plunger lift company to help with the repair and/or parts.

The diaphragm portion of the valve can leak at one of two places. Either the valve will leak around the flange where the two portions of the diaphragm assembly are connected, or at the breather hole (located on the opposite side from where the supply gas enters the diaphragm). In the latter case the leak indicates a ruptured diaphragm. It is possible that a leak occurring at the flange can be the result of loose bolts, so it may be corrected by simply tightening the bolts and nuts, eliminating the need for replacing the diaphragm.

C.1.2 Internal leaks

The most common leaks encountered in motor valves are internal leaks. Often ball and seat configurations are normally used as the sealing element. Because of the extreme pressure differential and high flow rates, the seat area is subject to fluid cut or erosion, which can be aggravated by abrasive materials. If the valve has an insert seat it will have an O-ring seal, which is also susceptible to cutting or deterioration due to gas composition.

If a valve is suspected to be leaking, the leak can be isolated by simply putting pressure on the upstream side with the valve closed and checking to see if there is

---

* Revised from Ref. [1].

any flow through the valve. If flow is identified, the leak is likely across the seat and can be corrected in the following ways.

Check the valve adjustment. Depending on the size of the seat, the size of the diaphragm, and the flow path, there is a maximum pressure that a particular valve will hold. Manufacturers have charts for determining this differential pressure.

If the valve seat is subject to a higher pressure difference, it is possible that the diaphragm and spring cannot contain the pressure. If the valve is equipped with an adjustment bolt on top of the diaphragm, tightening down on this bolt will put more pressure on the ball and seat to seal against the higher pressure. Be careful not to screw the bolt all the way in as it will restrict the valve from fully stroking open.

Also consider using a smaller seat. It is the differential pressure across the area of the seat that prevents the seat from holding. A smaller seat can dramatically reduce the force against the diaphragm spring. If a smaller seat is objectionable, consider larger diaphragm housing. The larger housing will have a larger spring and can hold a higher differential pressure.

The valve may be turned around in the flow. This will put higher pressure on top of the seat and that pressure will act to help hold the valve closed. Caution should be exercised, however, because if the pressure is in fact too high for the particular seat, then it will prevent the valve from opening. This is a last resort before new equipment is installed as this idea will make the valve chatter.

Another cause of a leaking ball and seat can be the formation of hydrates (an "ice" formed of hydrocarbons and water) in the seat area. An extreme pressure drop, across the ball and seat, in some service will prompt the formation of hydrates. Correcting the leak under these conditions is a matter of dissolving the ice or hydrate at the valve. With the hydrates removed the valve should hold.

- The prevention of ice or hydrate formation presents a somewhat more complex problem. The formation of hydrates might be prevented by either reducing the pressure differential across the valve or by increasing the temperature. Simply using a larger trim in the valve will not reduce the pressure drop. The best solution is to lower the operating pressure of the entire system. This is not always possible, however, since operating pressure directly affects plunger system efficiency.
- A common, but expensive, method to solve hydrate problems is to inject methanol just upstream of the freezing point. Alternately a choke (larger than the valve seat) can be placed downstream of the valve. This will reduce the pressure drop across the valve seat and can reduce or eliminate the formation of hydrates by spreading out the pressure drop.

C.1.3   Valve will not open

Generally, there are four factors that play a part in the opening or closing of a motor valve.

1. The size of the diaphragm
2. The amount of pressure applied to the diaphragm
3. The compression of the diaphragm spring
4. The line pressure acting with or against the valve trim

A malfunction of any one or combination of these components can prevent the valve from properly opening.

As pointed out earlier, too much line pressure acting on top of the trim of the valve could hold the valve closed. In this situation, it is possible to increase the supply gas pressure to the diaphragm to assist in opening the valve. Do not to exceed 30 psi diaphragm pressure when attempting this procedure. If the valve still does not open and the adjusting screw has been backed out, then change to the next smaller seat or use a larger diaphragm. Exceeding the 30 psi limit placed on the diaphragm gas pressure can cause the valve to bang open, which can cause damage, or rupture the diaphragm.

Another reason for a motor valve not opening is the adjustment of the compression bolt. The compression bolt puts tension on a closing spring that is connected to the trim by a short stem. If the compression bolt has been over-tightened, the valve will not open fully. When flowing over the seat, the tension should be at a minimum.

Finally, if the above items have been checked and the valve still does not open, then the valve may have severe mechanical problems, such as a bent stem or a clogged valve. A bent stem or a frozen or clogged valve, although not common, is not out of the question.

C.1.4 Valve will not close

Many of the steps mentioned earlier are appropriate for troubleshooting a valve that will not close. In addition, line pressure that is out of the operating range for the diaphragm size can prevent closure.

The top adjusting bolt unscrewed too far could also prevent the valve from closing. Under certain conditions, it is not uncommon for ice to form in the trim preventing the ball and seat from making a complete seal, thus keeping the valve open. Sand, paraffin, welding slag, or other foreign objects can get lodged between the ball and seat preventing valve closure. If the controller is not allowing the supply gas to bleed, the valve will not close. If this problem is suspected, the compression nut on the copper tubing link to the motor valve should be loosened while operating the controller open and closed. This should free the controller to bleed the supply gas.

C.2 Controller

The most complex part of the plunger lift system is the controller. There are many commercial controllers, and description and analysis of each is beyond the scope of this book. The following discussion covers only basic components that may apply to the majority of controllers.

Basically, all controllers have similar operational characteristics. Generally, most controllers use a 20–30 psi pneumatic source, usually gas, which is utilized to open and close a motor valve. The motor valve is opened by directing supply gas through the controller to the valve diaphragm to force the valve open. The motor valve is closed when the controller blocks the supply gas and bleeds the gas from the diaphragm that opened the valve, thus allowing the valve to close. The discussion of controller troubleshooting will be in two sections: the electronic and pneumatic sections.

C.2.1   Electronics

When the controller does not appear to be working properly and faulty electronic equipment is suspected, the first thing to check is the LCD (or LED) display. In addition to showing the time, most controllers are designed so that the display will indicate the mode of the controller (whether it is on or off), if it has power, or whether there are any outside switch contacts. No display may simply mean no power, so check batteries for charge and proper contact.

In general, there are so many controllers available nowadays, one must consult operating procedures or the manufacturer for troubleshooting.

C.2.2   Pneumatics

All controllers use some type of interface valve to control the pneumatic signal. The two most common are the latching valve and the slide valve. Each operates differently, but essentially performs the same function. The latching valve is made up of an electromagnet and a small poppet valve. The valve operates when an electric "on" pulse from the electronics module activates the magnet and pulls the poppet off its seat then latches it back, directing supply gas to the motor valve. The off pulse from the electronics reverses the polarity of the electromagnet releasing the poppet and a spring moves it to the closed position. In the closed position the poppet valve blocks the supply gas to the diaphragm and vents the gas, closing the motor valve.

The slide valve consists of housing and a small piston that slides through a cylinder in the housing. The movement of piston is limited by end plates. The piston is fitted with three O-rings, one at each end for power and one in the middle. The position of the piston, either at one end of the cylinder or the other, directs the gas or determines whether the valve is in the open or closed position. A solenoid is fixed to each end plate of the housing. When either solenoid receives an electronic signal from the controller, it directs a shot of gas to the power end of the shift piston, pushing it to the opposite end of the cylinder, thus opening or closing the valve. When the valve slides to the open position, supply gas is directed to the diaphragm of the motor valve, when it slides to the closed position, the supply gas is blocked and the diaphragm is bled.

Troubleshooting and maintenance of these valves is performed in the same manner. If a pneumatic problem is suspected, the gauges on the bottom of the controller should first be analyzed. With supply gas being fed to the controller, when the controller is pulsed to "on" both gauges should read the same pressure. Then if the controller is pulsed to "off" the pressure on the right hand gauge should drop to zero. If this is not the case then the likely problem is a faulty valve (shifter) in the controller.

The fact that the shifter (latching valve or slide valve) is not working does not necessarily mean that it is damaged. The shifter does require voltage. Once you have determined the shifter is not operating, the next step is to check its supply voltage. Check the wiring to ensure there are no loose connections or broken wires. Next, with a voltmeter, check to see that there is power being supplied to the shifter from the electronics module. There should be no power supplied to the shifter until the controller is pulsed on or off when only a brief pulse is issued. If no pulse is evident then the electronic module must be replaced. If the pulse is being fed to the shifter and it is not operating then the problem is with the shifter.

The most common problem encountered with the shifters is fouling from contaminated supply gas. Fortunately, shifters are easily disassembled and cleaned. After a thorough cleaning the slide valve must be lubricated with a thin coat of lightweight grease (such as Parker O-ring Lube). It is not recommended to disassemble the solenoid valves, located at both end plates of the sliding valve, for cleaning. The solenoid valves rarely malfunction, but when they do, they must be replaced.

To ensure smooth operation of either type shifter, it is recommended that a filter be installed in the supply gas line to keep impurities from entering the shifter mechanism. The supply gas should also be maintained as dry as possible. If casing head gas is to be used, it is good practice to install a drip pot upstream of the controller and keep it blown dry.

New controllers may contain features for which the aforementioned discussion does not apply.

C.3   Arrival transducer

The arrival transducer is a device that plays a very important role in most plunger lift installations. The function of the switch is to detect the arrival of the plunger in the lubricator. This then typically signals the controller either to shut-in the well (oil well), or to switch valves or just to register the cycle in a plunger counter (gas well). Most commercially available switches use a magnet to close a set of contacts on an electric switch. This switch closure completes a circuit that sends a signal to the controller. These switches are normally trouble-free, but mechanical malfunctions are possible. There are some switches based on vibration.

To isolate a malfunction, the first step should be to determine if the switch is even capable of operation. This is done by removing the switch from the housing on the catcher nipple, and touching it to the lubricator. If making contact causes, the controller to turn off or record arrival, then it can be concluded that the wiring is functional, and that the switch is at least capable of operation. Some styles cannot be unplugged. This type must be shorted manually by placing a small piece of metal across the switch (inside the catcher nipple). Again, if shorting the switch causes the controller to turn off or signal arrival, then the switch is capable of operation.

A closed magnetic switch and the entire off-time may be displayed if this is a controller function. If this occurs first, determine whether the plunger is up in the lubricator, which would indicate normal operation. If the plunger is not in the lubricator, then remove the switch from the housing and see if normal operation resumes. If the controller does not immediately start counting down, then disconnect the wiring from the controller. The countdown should resume unless the problem is within the controller itself. If the controller restarts the countdown and the plunger is not in the lubricator, then the problem is either in the switch or the wiring.

If the aforementioned procedures have been followed and all components appear to be functional, further investigation is required to isolate the apparent malfunction with respect to the operation of the entire system. It might be possible for a plunger to travel at speeds too fast for the switch to detect. Most switches are sensitive enough to detect a plunger traveling at speeds in excess of 1000 ft/min. There are controllers on the market, however, that are not capable of detecting these high reaction rates. In such cases, it is possible for the arrival of the plunger to go undetected. Slowing down the plunger travel speed is the best way to determine if this is the cause of the apparent magnetic arrival transducer malfunction.

Another possible system malfunction that could mistakenly be attributed to a magnetic sensor problem is when the plunger does not surface or does not travel high enough in the lubricator for detection. In the former case the best method to determine, whether the plunger is truly arriving at the surface, is by physical inspection. Although it is possible to receive an indication of the plunger surfacing on a chart or recorder, these indications are not totally reliable. The well response can indicate the plunger surfacing on the chart without the plunger actually making it to the surface.

When the plunger is surfacing, but not going far enough into the lubricator to trip the switch, adjustments must be made to the system to allow the plunger to pass further into the lubricator. To ensure the plunger travels far enough into the lubricator to make contact with the magnetic sensor switch, it is recommended that the upper flow outlet be open to allow flow to go past the sensor, carrying the plunger past the magnetic switch and allowing the sensor to signal arrival of the plunger. Some plunger wellheads try to use only one outlet, but the dual outlet used as described is a better setup for the arrival transducer.

C.4 Wellhead leaks

Wellhead leaks must be repaired to maintain a safe and clean environment at the well site. On most wellhead hookups, leaks are generally due to faulty threads. Leaking around the wellhead bolts is typically caused by improper torque on the bolts, improperly repaired wellhead, or damaged bolts.

Other than the bolt connections the most common place for wellhead leaks is at the catcher assembly or where the lubricator screws into the flow collar. The catchers usually are attached through some type of packing gland (not unlike those found on many valves). Leaks that occur at the catcher can normally be fixed by tightening the packing nut. If not, it may be necessary to replace the catcher assembly.

The lubricator upper section has a quick connect, which has an O-ring seal. These can leak and need to be replaced periodically. In most cases, tightening will not stop a leaking O-ring.

Wellhead connections may be screwed on or flanged. Flanged wellheads are thought to be safer if the plunger arrives dry.

C.5 Catcher not functioning

For plunger inspection, but not general operation, the catcher should be able to hold the tool in the lubricator to accommodate its removal. Plunger catchers catch or trap the tool and hold it in one of two ways.

Some catchers use a spring-loaded cam-type device. To activate the catcher and catch the plunger, either a thumbscrew is unscrewed (which activates the catcher) or a catcher handle is released. In both cases a cam is extended into the path of travel of the plunger. As the plunger moves past the catcher, the cam is pushed back allowing the plunger to move past it. Once the plunger has moved past the catcher the spring-loaded cam flips out beneath the plunger, preventing it from falling back downhole.

The other type catcher, commonly found on older installations, uses a friction catch to hold the plunger at the surface. The friction-type catcher consists of a ball extending into the sidewall of the catcher, pushed by a coil spring. As the plunger moves past the ball, the compression of the spring on the ball causes friction against the side of the tool, preventing it from falling.

Before troubleshooting catcher problems, first it is important to verify that the plunger is arriving at the surface and then to make sure that upon arrival, the plunger is traveling far enough into the lubricator for the catcher to engage. One way to assist the plunger to go further up into the lubricator is to open the flow outlet above the catcher. Also, closing the lower outlet will direct all the flow through the upper outlet, driving the tool higher into the lubricator. Under these conditions, if the catcher still fails to capture the plunger, further inspection of the catcher itself is required.

The first step in troubleshooting the catcher is through visual inspection to determine if ice or paraffin or other produced solids have clogged the catcher. The removal of foreign material should restore catcher operation.

Next inspect the catcher nipple while manually engaging and disengaging the catcher. The nipple should move all the way out of sight and stay there. In the run position when the catcher is activated, the cam (or ball) should extend into the path of the plunger. If it is not extending or retracting back into the housing, then it requires repair or replacement.

Never operate with the plunger surfacing and the wellhead open at the surface.

C.6   Pressure sensor not functioning

A common method for starting many plunger cycles is with a casing pressure activated switch-gauge. The switch-gauge is a pressure gauge with two adjustable contacts on the face and a pressure indicator (needle), all of which are connected to electric wires. Changes in casing pressure, up or down, cause the needle to move toward one contact or the other. When the needle touches either the high or low contact, it completes a circuit which signals the controller to open or close the motor valve. Switch-gauges seldom have malfunctions, but problems do occur.

As a first step, before examining the switch-gauge itself, check the controller as outlined earlier. Often a properly functioning sensor is blamed for a controller malfunction. If the controller is functioning properly, check the contacts on the gauge. The contacts on either side of the gauge can become fouled and unable to complete the circuit to the controller. To check the contacts, try to operate the controller manually by moving first the high and then the low set points (on the face of the gauge), so that they make contact with the needle. If the circuit is intact, the controller should function. If there is no response, clean the contacts and try again. If there is no response, check the condition of the wires between the switch and the controller. Shorted or crushed wires may break the circuit.

Examine the gauge to ensure that all pressure lines are properly attached. Determine that all pressure valves leading to the gauge are open. Using the switch-gauge as a flow line sensor, check and see if all valves downstream of the gauge are open. Bleeding the line connecting to the gauge should cause an appropriate response by the needle. If no response is registered, make certain that the casing pressure is changing. This can require the temporary installation of a second pressure gauge. If the casing pressure is changing normally but not registering on the gauge, the gauge must be replaced. If only small casing pressure changes (or no changes at all) are being recorded, then the plunger system is not operating normally and must be reoptimized.

The aforementioned discussion may not apply to some controllers.

C.7   Control gas to stay on measurement chart

Controllers may be used to throttle motor valves open or closed while maintaining a set-sensed pressure. When used in conjunction with a plunger lift system, its purpose is to restrict the initial surge of head gas within the pressure limits of the system, in order to prevent the produced gas from going off the sales chart. These controllers are most commonly used on compressors and production units but are finding application with plunger systems. However, it would be better to have an electronic sensor record the bursts of gas, because throttling back the surge of head gas can only serve to have some effect in *reducing the production.*

The unit works by sensing a pressure then converting that signal to a proportional pneumatic pressure to the diaphragm of a motor valve causing the motor valve to throttle. The sensed pressure pushes on a high-pressure flexible element, which in turn operates a pilot valve. This throttling of the pilot valve varies the pressure supplied to the motor valve causing the motor valve to respond in a manner directly proportional to the sensed pressure signal. By throttling the motor valve the unit attempts to maintain a constant-sensed pressure. If the system (well) cannot supply enough pressure to meet the throttling range preset, however, the motor valve will remain wide open. On the other hand, if the sensed pressure exceeds the preset design pressure maximum, the motor valve will close completely.

This system is known to have two disadvantages. First, the supply gas entering the controller is metered through a small choke or orifice. When the gas supply leading to the output signal is slow to respond (build), this orifice should be examined. The choke is very small and is prone to get clogged with debris from dirty supply gas. It can usually be cleared with a small wire. It is good practice to place a filter in the supply gas line upstream of the choke to help prevent this clogging.

If the controller does not respond to sensed pressure, the sensing element should be inspected.

C.8 Plunger operations

C.8.1 Plunger will not fall

Plungers are free-traveling pistons that depend solely on gravity to get back to the bottom of the well. If the plunger remains in the wellhead after the shut-in period or if it is back at the surface very quickly after opening the well, there is likely an obstruction either in the lubricator or downhole keeping the plunger from falling to bottom.

In the event that the plunger returns to surface too quickly, first make sure that the plunger has been given ample time to reach bottom. Ideally a plunger should travel up the hole between 500 and 1000 ft/min. However, plunger fall rates can be considerably slower. Plungers without a bypass, to allow gas to easily flow through the plunger on the down cycle, can fall at rates of only 150–500 ft/min or greater. Plungers equipped with a bypass or collapsible seal may fall at rates between 500 and 2000 ft/min. Fast fall is recommended to optimize a system for high production. If liquids have accumulated in the well during the last bit of after-flow, then for maximum production the well should be opened as soon as the plunger lands on the bumper spring.

Echometer Inc. has devised a system[1] that tracks the plunger during both the rise and fall portion of the cycle. The measurements have been made by both acoustically recording the plunger depth using acoustic pulses generated with a gas gun and the pressure change that occurs as the plunger travels past the tubing joints. Fig. 7.13 is a schematic of the Echometer setup to record plunger travel with time. Fig. 7.14 is an example of the pressure and acoustic trace of a plunger cycle.

C.8.2   Plunger will not surface

Plunger lift operations require the tool to travel the full distance between the bottom-hole spring and the lubricator for each cycle. If the tool is not getting to the surface, some or all of the liquid loads will remain in the well. Determining the source of the problem preventing the plunger from surfacing can be difficult. There are both mechanical and operational considerations.

The ideal travel time for a surfacing plunger is in the 500−1000 ft/min range. This, however, is the ideal rate and many installations operate at much slower speeds. It is important, therefore, that ample time be given for the plunger to travel to the surface. If the plunger has been given sufficient time to surface (corresponding perhaps to an equivalent 100−200 ft/min rise time), then other problems must be investigated.

First inspect the system for mechanical malfunctions. Most of the mechanical problems that would prevent a plunger from falling to bottom would also prevent it from rising to the surface. Debris in the tubing, tubing quality, and plunger damage can all prevent the plunger from reaching the surface. In addition to restrictions, however, conditions that prevent the plunger from sealing in the tubing can prevent its reaching the surface. These would include ballooned tubing, mixed tubing strings with changes in the ID, tubing leaks, gas lift mandrels installed in the tubing, etc. Typically, when the plunger encounters enlargements and looses its seal, it will stop traveling at that point. It is vital that well completion records be checked closely before installing a plunger lift system.

Finally the plunger itself may have been damaged, preventing it from surfacing. Plungers equipped with bypasses may develop leaks, preventing an adequate pressure seal across the plunger. The plunger should be checked regularly for wear and loose parts. Although uncommon, plungers can come apart in the hole. In some cases where the plunger will not surface under normal conditions, it may be possible to bring the plunger up by venting the head gas to a low-pressure separator. This provides extra pressure differential across the plunger that may be sufficient to bring the plunger to the surface. If this fails, the plunger must be wirelined out of the well.

Operational problems that would prevent the plunger from surfacing have all been discussed in the previous sections. It may be necessary to go through the initial kickoff procedure again to ensure that the well is ready to begin normal plunger operations. It is important to make sure that the casing is allowed to reach the required operating pressure. It might be necessary to allow the casing to come to equilibrium before attempting another plunger cycle. If the plunger has been idle for a time, it may be necessary to swab the well and produce most of the liquids or shut the well in for a period to drive liquids in the formation, before attempting to start the plunger cycle.

C.8.3   Plunger travels too slow

The speed with the plunger travels to the surface can greatly affect the performance of the plunger lift system. Plunger travel speeds that fall below the suggested

750 ft/min can significantly reduce the efficiency of transporting the liquids. For high-rate gas wells, this may not be a critical problem since these generally have ample gas production to replace that lost in inefficiencies. On weak or marginal gas wells where gas production is low and all the available casing gas is needed to surface the plunger, this can be a very important issue.

Bear in mind the plunger and liquid slug rise with the aid of the gas stored up in the casing annulus with some help from formation production as well. If there is not a large volume of casing gas available and/or if it takes long shut-in times to rebuild casing pressure, the maximum possible number of plunger cycles per day is less. Experience has shown that the slower the plunger travels, the less efficient it becomes and the more gas it takes to move it to surface as gas leaks past the plunger. The seal between the plunger and the tubing is such that some gas always slips past the plunger reducing its effectiveness since the pressure below the plunger is larger than above the plunger. When the plunger is traveling within the optimal speed range (750–1000 ft/min), this gas slippage is presumed minimal. As the travel speed falls below the optimum, however, the amount of gas slippage is dramatically increased. This means that more of the casing gas is used each cycle, so the shut-in (or build-up) time is longer. Ultimately, this results in fewer cycles per day, which generally amounts to less liquid production per day. It is important to maintain plunger speeds near the optimum so as not to waste valuable casing gas, particularly on low rate wells. *Note that many of these guidelines are from experience and may not have been tested extensively, so questioning and testing standard procedures for your particular wells is not a bad idea.*

There are a number of ways to increase the plunger rise speed while conserving casing gas and maintaining adequate liquid production. The plunger travel speed is a function of the size of the liquid load and the amount of net casing pressure (less sales line pressure), plus the rate that the head gas is removed at the surface. As mentioned earlier, also the plunger speed has a major effect on the efficiency of the plunger seals. First it is important to analyze the well conditions and determine whether smaller slug size or a higher casing operating pressure is warranted.

By raising the casing operating pressure, more effective pressure is exerted against the formation, thus lowering the liquid influx and reducing the slug size.

Reducing the size of the liquid load then allows higher plunger rise speeds. Similarly, lengthening the shut-in time again raises the net pressure on the formation while increasing the amount of gas in storage in the casing annulus. With roughly the same liquid load (or possibly less) but more compressed gas in the annulus at a higher pressure (more energy), the speed of the plunger is again increased.

On the other hand, reducing the sales line pressure has the same effect as increasing the casing pressure on the plunger travel time (more differential pressure across the plunger) without the adverse effects of longer shut-in periods and more pressure against the formation. Plungers with more efficient seals can also operate with reduced plunger travel times, by reducing the amount of gas slippage. Often just replacing plungers with worn seals will have a dramatic impact on performance.

Finally, plunger performance can be improved by rapid evacuation of the head gas (the reverse of trying to choke back gas surges to keep them on a recording chart) above the liquid slug. This could require replacing an existing orifice plate with a larger size, or opening up a choke or enlarging the dump-valve trim to allow greater use of the gas that is available.

Velocity controllers control the flow time and the build-up time to maintain the correct rise average velocity but do not necessarily trend to low-operating pressures and shorter cycles needed for production optimization.

C.8.4 Plunger travels too fast

A plunger traveling up the well too rapidly could have bad consequences. While the efficiency of the plunger sealing mechanism is not dramatically affected by higher speeds, well safety and equipment longevity dictate that the plunger rise speed be maintained below the 1000 ft/min maximum. The plunger and lubricator undergo fairly severe punishment under normal operating conditions. As the plunger speed increases the impact force imposed on the lubricator by the plunger increases roughly by the square of the speed. Although the plunger and lubricator are designed to withstand plunger impacts under normal speeds, higher speeds can quickly wear out and destroy both. In general, the economic benefits brought about through longer equipment life far outweigh those of shorter plunger travel times. A large plunger coming up dry such as a plunger for 2⅞ or 3½ in tubing can cause the most damage.

From an operational standpoint, either decreasing the casing build-up pressure or increasing the size of the liquid slug can reduce plunger travel speed. This can be accomplished by allowing the well to flow for longer periods of time after plunger arrival at the surface. Another way to accomplish this is to reduce the shut-in period. Choking the well, however, to slow the plunger is not recommended, although will sometimes accomplish to objective. Choking or operating with too large of a liquid slug reduces production.

One reason that a plunger is coming up too fast is that even though there was liquid in the tubing when the plunger fell, the liquid can be displaced from over the plunger to the casing during the shut-in period. This could be due to bubbles entering the tubing during shut-in, or perhaps the casing liquid is dropped below the tubing end that might accelerate the loss of liquid from over to under the plunger.

One method to control this is to run a standing valve below the bumper spring. However, a standing valve would trap any random slug that might be too large to lift and you could not then raise the tubing pressure to push the slug below the plunger to start the cycle again. The standing valve would hold the large slug over the plunger regardless of pressure changes.

One common method used to attack this problem is to use a standing valve, but notch the seat of the valve so it will leak. It will then give some resistance to liquids leaking back to below the plunger during the build-up cycle, but liquids can still be forced below the plunger through the leak if the slug should be too large to continue the cycles.

In general, all low-pressure, low liquid rate wells should probably be equipped with a standing valve unless sand or scale, etc. dictate otherwise.

Another method is to use a device: a new spring-loaded seat on the standing valve. The standing valve holds liquids over the plunger during the off cycle, but if the need arises to add tubing pressure to pressure liquids back below the plunger, then enough pressure can be applied to force the seat down and allow the liquid to leak back from over to under the plunger.

C.8.5   Head gas bleeding off too slowly

Bleeding the head gas off too slowly can reduce the differential pressure needed to surface the plunger. The slower the bleed, the less the differential pressure across the tool and the less the chance of the plunger surfacing. The faster the head gas is allowed to bleed, the better the plunger performance.

Small chokes and high flow line pressures act as large barriers for the head gas to overcome, keeping the system from performing at optimum efficiency. Getting rid of the head gas as quickly as possible is critical. If it is necessary to choke the well, the choke should be as large as possible. To accomplish this, it may be necessary to modify the surface facilities, but the benefit of doing so far outweighs the cost. If the sales line pressure is too high, then efforts should be directed toward reducing that pressure although this is a potentially expensive process that may require compression.

C.8.6   Head gas creating surface equipment problems

A common complaint about intermittent operations is that they create problems with the surface equipment and gas measurement. Plunger lift falls within this category.

When a well fitted with plunger lift is first opened, generally a surge of high-pressure gas is forced at high rates through the surface equipment. Often the surface equipment was designed for an average flow rate and cannot handle the short duration surge that ends up going off the charts. One common, but not recommended way, of handling this problem is to install a positive choke in the flow line. Although the choke will restrict the initial gas surge to manageable levels, it will also restrict the flow of the remainder of the gas and the liquid slug. In particular, when a liquid slug passes through a gas choke, the flow is drastically reduced, presenting a wall of liquid to the plunger that has a similar effect as closing a valve. The consequence of this is almost always a loss of production.

Fortunately the problem is often negated once the well has been optimized. If this is not the case, however, other methods can be employed. One of the most effective ways to correct the problem is to install a valve with a throttling controller (discussed earlier) to limit downstream pressure while allowing the motor valve to be opened slowly to minimize production loss. This type of controller can be optimized to adjust to the pressure capabilities of the surface system and can therefore eliminate problems like selling off the chart, overpressuring separators, and surging compressors. Finally, in installations where several wells are on plunger lift, the surge effects can be negated by producing a number of wells through a manifold. In this manner the surges produced by various wells can be timed to occur at different intervals and any single surge will make up a smaller percentage of the total flow and therefore be less likely to peg the sales meter.

C.8.7 Low production

Optimizing or fine-tuning a plunger lift well can make a difference in the production. Consider testing short flow times to bring in small slugs of liquid. Then short build-up times required to build smaller casing pressures are required to lift smaller liquid slugs. The result is a lower average flowing bottom-hole pressure and more production. Limits are that a too short flow period could result in no liquid slug and a too short shut-in period would not allow the plunger to reach bottomhole.

In general, whatever can be done to lower the average casing pressure per cycle will add to gas production.

C.8.8 Well loads up frequently

Many wells are found to be very temperamental, where any small change in the operation can greatly affect their performance. Marginal wells tend to be particularly sensitive and are often easily loaded up. Liquid loading on a plunger lift well is usually a result of too long of a flow time or too little casing pressure during the shut-in period. Also trying to run plunger lift in small tubing can aggravate this problem.

In general, imposing a more conservative plunger cycle can alleviate liquid loading of a plunger lift well. This means, as stated earlier, higher casing operating pressures and longer shut-in periods. Once the cycle has been changed, the well should be allowed to stabilize, which might take several days. Then continue with the optimization procedures outlined earlier making only small incremental changes to the system times and pressures then allowing the well to achieve stability between each change. It is possible to eventually adjust the well back to the original cycle settings once the well has had a chance to clean itself up.

If a well is completely loaded with liquids then it must be brought through the kickoff procedures from the beginning. First, shut the well in and allow it to build pressure. With the well loaded, it may be necessary to swab the well to clean it up before starting the kickoff. Remember to work slowly making small incremental changes to the system then allowing the system to become stable before continuing to the next step. Many new controllers now adjust cycle times and pressures to follow optimization algorithms.

Reference

1. *Ferguson Beauregard plunger operation handbook.*

Appendix D: Gas lift terminology

$P_{pd} \cdot P_{rs}$ — Flowing production pressure at the depth of a valve. If flowing up tubing this would be the flowing tubing pressure at valve depth.

P_{wf} — Flowing bottom-hole pressure.

Q_{liq} — Liquid production rate. More specifically, the liquid rate produced when lifting off of a particular valve. Lifted gradient above and unlifted or formation gradient below.

TGLR — Total gas-to-liquid ratio.

Q_{gi} **(req)** — The required quantity of lift gas required to meet the conditions on that line.

Q_{gi} (VPC, ThH, Wink) — The quantity of lift gas the valve will flow if set at the described conditions.

P_{vo60} — The test rack opening pressure at 60°F for this particular set of design conditions.

$P_{vo\ instld}$ — The actual P_{tro} values that are installed in a well that you are troubleshooting. So, for a new design, you should make $P_{vo60} = P_{vo}$ Instld by pushing the "Update P_{vo}" button on the gas lift design/valve design tab.

P_{gd} — Gas injection pressures at each valve depth starting at the design "GL design pres" listed on the "gas lift design" input dialog.

GLD P_{iod} — The gas injection pressure, at the depth of the valve, required to open the valve under design conditions.

GLD P_{vcd} — The bellows pressure at the depth of the valve under design conditions.

GLD P_{pTran} — The transfer pressure used in the gas lift design. The required flowing tubing pressure to be able to unload to the next valve down the hole.

$F_{lwg}\ C_t$ — The temperature correction factor at the flowing temperature at the valve depth.

$F_{lwg}\ P_{iod}$ — The gas injection pressure, at the depth of the valve, required to open the valve under flowing conditions. Temperature may be different between design and flowing conditions.

$F_{lwg}\ P_{vc}$ — The surface closing pressure (bellows pressure), at flowing temperature conditions.

GLD C_t — The temperature correction factor for the design temperatures at valve depth (could be different than flowing temperature).

GLD P_{io} — The surface gas injection pressure required to open the valve under this set of conditions.

GLD P_{vc} — The design surface closing pressure (bellows pressure), called P_{sc} by most gas-lift vendors. If the surface closing pressures do not drop, the design will not unload. $P_{vc} = P_{vcd}$-(injection gas gradient pressure from surface to valve depth, we often call this the "gas weight").

P_{vc} **DP**	The pressure drop used to set the P_{vc} of the next deeper valve. Some programs use the "Surface Closing Pressure" design method. The surface closing pressure of the top valve is calculated using the P_{iod} and P_{pTran} pressures from the gas lift graph. Then once that top valve's surface closing pressure is determined, all the rest of the surface closing pressures are set by taking a predetermined pressure drop at each deeper valve. Then the required P_{iod} values are back calculated (solving the force balance equation for P_{iod} instead of Bellows Pressure). This guarantees that the surface closing pressure are in the correct order, and as long as none of the P_{iod} pressures are greater than the maximum attainable gas injection pressure at depth, the design will always work.
$P_{pd} - P_{min}$	The objective liquid gradient pressure minus the transfer pressure at each valve depth. If the objective gradient is to the right of the transfer pressure this will be a negative number, which indicates it will be difficult or perhaps not possible to transfer to the next deeper valve.

From Tom Nations, Nations Consulting, common industry practice, and API 11V6.

Index

Printed in the United States
By Bookmasters